Radon Series on Computational and Applied Mathematics 3

Gröbner Bases in Control Theory and Signal Processing

Edited by

Hyungju Park
Georg Regensburger

Walter de Gruyter · Berlin · New York

Editors

Hyungju Alan Park
School of Computational Sciences
Korea Institute for Advanced Study (KIAS)
207-43 Cheongnyangni
2-dong Dongdaemun-gu
Seoul 130-722
Korea
e-mail: alanpark@kias.re.kr

Georg Regensburger
Radon Institute for Computational and Applied Mathematics
Austrian Academy of Sciences
Altenberger Straße 69
A-4040 Linz
Austria
e-mail: georg.regensburger@oeaw.ac.at

Key words:
Multidimensional systems, Gröbner bases, symbolic computation, polynomial ideals, multivariate filter banks, multidimensional linear systems, multivariate polynomial matrices, multidimensional signal processing and control, Quillen-Suslin theorem, polynomial rings, flat systems, stability, randomized algorithms, input/output systems, orthonormal wavelets, linear systems, time-varying systems.

Mathematics Subject Classification 2000:
00B25, 11E25, 13C10, 13F20, 13Pxx, 16Sxx, 35B35, 42C40, 47A65, 68Wxx, 93Bxx, 93Cxx, 93Dxx, 94Axx.

⊗ Printed on acid-free paper which falls within the guidelines of the ANSI to ensure permanence and durability.

Library of Congress Cataloging-in-Publication Data

A CIP catalogue record for this book is available from the Library of Congress.

ISBN 978-3-11-019333-6

Bibliographic information published by the Deutsche Nationalbibliothek

The Deutsche Nationalbibliothek lists this publication in the Deutsche Nationalbibliografie; detailed bibliographic data are available in the Internet at http://dnb.d-nb.de.

Printed in Germany
Cover design: Martin Zech, Bremen.
Typeset using the authors' TeX-files: Kay Dimler, Müncheberg.
Printing and binding: Hubert & Co. GmbH & Co. KG, Göttingen.

Preface

Bruno Buchberger's now well-known article on Gröbner bases in the context of control theory appeared in 1985 in *Multidimensional Systems Theory*. Motivated by this article, many researchers have worked on applications of Gröbner bases and related methods in multidimensional control theory and signal processing. While early work was focused on linear systems theory and partial differential control theory, many interesting new applications have been found more recently in multidimensional signal processing. In particular, Gröbner bases proved to be a powerful tool in designing multidimensional filter banks and wavelets and have also shown potential in solving PDEs.

This volume grew out of the D3 Workshop on *Gröbner Bases in Control Theory and Signal Processing* (May 18–19, 2006) held in Linz, Austria, in the frame of the *Special Semester on Gröbner Bases and Related Methods*. The special semester was organized in the spring and summer of 2006 by the Johann Radon Institute for Computational and Applied Mathematics (RICAM), in close cooperation with the Research Institute for Symbolic Computation (RISC). Directed by Bruno Buchberger (RISC) and Heinz W. Engl (RICAM), it created a unique atmosphere for interdisciplinary cooperation on all aspects of Gröbner bases.

This volume collects survey articles and original research papers by some of the leading researchers in the area and offers a glimpse of the state of the art on the subject. The topics covered include

- Gröbner bases in multidimensional systems,
- the Quillen-Suslin theorem and systems theory,
- statistical signal processing,
- parametric problems in control theory,
- stability of multidimensional input/output systems,
- wavelets and filter design,
- synthesis of multidimensional control systems,
- time-varying linear systems.

For the full workshop program we refer to the RICAM special semester website at:

`http://www.ricam.oeaw.ac.at/specsem/`

It is our pleasure to express our gratitude to all authors for their enthusiasm that undoubtedly will make this volume a valuable contribution to the area. We extend our sincere thanks to Bruno Buchberger, Heinz W. Engl, and the staff at RICAM and RISC for their warm hospitality during the workshop. We cordially thank Robert Plato and Kay Dimler from the publishing house Walter de Gruyter for their professional help from early brainstorming to the final editing stage.

Hyungju Park and Georg Regensburger Seoul and Linz, October 2007

Table of Contents

Radon Series Comp. Appl. Math 3, 1–22

Two decades (1985–2005) of Gröbner bases in multidimensional systems

N. K. Bose

Key words. Multidimensional systems, Gröbner bases, polynomial ideals, multivariate filter banks, multidimensional systems stabilization, multivariate polynomial matrices, multidimensional signal processing and control.

AMS classification. 13P10, 93D15, 13F20, 11E25

1 Introduction

The subject of linear shift-invariant multidimensional systems, originally conceptualized (now expanded along other directions, not considered here), is concerned with a mathematical framework for tackling a broad range of paradigms whose analysis or synthesis require the use of polynomials, rational functions and matrices with either polynomial or rational function entries in several complex variables. Its applications, which may range from the processing of spatial and temporal signals of diverse physical origin to the design of linear discrete multidimensional control systems, are plentiful. The areas of image processing, linear multipass processes, iterative learning control systems, lumped-distributed network synthesis, nonlinear system analysis via multidimensional transforms and geophysical signal processing have benefited from the tools available in the theory of multidimensional systems. The implementation of various tests of multidimensional systems properties, necessitate the development and use of tests for global or local positivity of polynomials of several real variables and, in a broader sense, algebraic as well as semialgebraic sets.

Supported by RICAM (Radon Institute for Computational and Applied Mathematics) and RISC (Research Institute for Symbolic Computation).

Some of the fundamental difficulties which either complicate the development of resources for meeting the needs of diverse applications or are responsible for limiting the scopes of attempts at naive generalizations of one-dimensional (1-D) results to the corresponding n-dimensional (n-D) situations have been recognized during the past three decades. Considerable analytical resources are required to go from 1-D to 2-D results and even after the associated bottlenecks are circumvented, the subsequent generalization from 2-D to 3-D may not be routine. This fact is dramatically illustrated by the coprime factorability of univariate polynomial matrices whose elements are in an Euclidean domain [30], the nontrivial generalization of such results to the bivariate case [19] and the impossibility, in general, of further generalization to the trivariate case [24]. Some of seminal results in [24] are illustrated with examples in [37].

While discussing the n-D real Euclidean space $\mathbb{R}^n$, Daniel Asimov [3] states that "strictly on their own merits, higher-dimensional spaces tend to blur together into multidimensional sameness." He goes on to remark that "it is often among low-dimensional spaces that the most dramatic transitions take place: as the number of dimensions rises, fundamental properties suddenly flash into existence or vanish forever, never to change again." This remark is dramatically illustrated by the celebrated conjecture made in 1904 by the distinguished French mathematician, Henri Poincaré, on a possible simple test to classify the 3-sphere among all 3-dimensional manifolds without boundary. Its generalization was first solved in the case of n-dimensional manifolds in 1960–1961 for $n \geq 5$ (Stephen Smale) and in 1982 for $n = 4$ (Michael Freedman) by ingenious innovative methods. The original conjecture was settled by Grisha Perelman in 2003. The undeniable challenge and the promised reward have propelled the subject-matters of multidimensional systems theory and algorithmic computer algebra to a state of maturity that guarantees a continuing proliferation of increasingly complex and diversified nature of activities.

While involved with the development of multidimensional systems theory, awareness of the potentials of Gröbner bases in computational aspects of the subject was natural. Such bases for polynomial ideals provide an algorithmic method for solving a number of computability and decidability problems concerning the ideal; for example, given a multivariate polynomial $f(\mathbf{z})$ and an ideal $\mathbf{I}$ specified by a finite number of generators, one can perform computations in the original polynomial coefficient field to decide constructively whether or not $f(\mathbf{z})$ belongs to $\mathbf{I}$.

2 Buchberger's algorithm for construction of Gröbner bases

This author's interest in Gröbner bases (Hironaka's standard bases for polynomial ideals) originated during the early nineteen eighties when Professor Buchberger, a former student of Gröbner, visited the University of Pittsburgh as an invitee of the author. On realizing the relevance of Gröbner bases to multidimensional systems theory, this author's invitation to contribute a survey chapter on the subject in a book [6] was gladly accepted by Bruno Buchberger. The highly readable chapter 6 in [5] and its updated version (chapter 4 in [6]) are strongly recommended to the readers of this paper. Other worthwhile readings are the texts [1] and [11], where computational algorithms for the fast construction of such bases are extensively documented. It is worth mentioning as a historical fact that the algorithm developed by Buchberger [10] for the solution of a problem on the multivariate analog of the Euclidean algorithm, posed by Gröbner, led to the naming of Gröbner bases by the humble student in deference to his respected teacher. Justifiably, subsequent researchers have, sometimes, referred to Gröbner bases as Gröbner-Buchberger bases. Many computational problems that are extremely difficult for polynomial ideals generated by arbitrary bases are very easy for polynomial ideals generated by Gröbner bases. One of the early results was a formula expressing the greatest common divisor of a set of polynomials of several variables in terms of a Gröbner basis of the ideal generated by them [16]. More recently, it has been shown how Gröbner bases can be used to investigate zero coprimeness and minor coprimeness of multivariate polynomial matrices [7, 6] leading up to the difficult problem of factor coprimeness (for an up-to-date discussion see the recent text [21]).

Let a polynomial ring involving indeterminates $z_1, z_2 \ldots, z_n$ over an arbitrary but fixed base field K of coefficients be denoted by $R = K[z_1, z_2, \ldots, z_n]$. A monomial $z_1^{i_1} z_2^{i_2} \ldots z_n^{i_n}$ will be called a *term* (sometimes this is called a power product and then the term refers to a coefficient times a power product) whose degree is $i_1 + i_2 + \ldots + i_n$. Define an order $<_T$ on the terms as follows: $z_1^{i_1} \ldots z_n^{i_n} <_T z_1^{j_1} \ldots z_n^{j_n}$ if and only if

$$\deg(z_1^{i_1} \ldots z_n^{i_n}) < \deg(z_1^{j_1} \ldots z_n^{j_n}) \quad \text{or} \quad \deg(z_1^{i_1} \ldots z_n^{i_n}) = \deg(z_1^{j_1} \ldots z_n^{j_n})$$

and $i_1 = j_1, \ldots, i_k = j_k,\ i_{k+1} > j_{k+1}$ for some k with $1 \le k+1 \le n$. The next example illustrates an order $<_T$ defined on the terms.

Example 2.1 $1 < z_1 < z_2 < z_3 < z_1^2 < z_1 z_2 < z_1 z_3 < z_2^2 < z_2 z_3 < z_3^2$
$< z_1^3 < z_1^2 z_2 < z_1^2 z_3 < z_1 z_2^2 < z_1 z_2 z_3 < z_1 z_3^2 < z_2^3$
$< z_2^2 z_3 < z_2 z_3^2 < z_3^3 < z_1^4 < \ldots.$

The above shows an admissible linear ordering of monomials – the total degree ordering, with ties broken by a lexicographical (lex) order on the variables. Other ordering include degree reverse lexicographical [1, pp. 18–25] (degrevlex), antigraded revlex [14] etc.

2.1 Notations

In the following notations t denotes a term, taken to be a monomial $z_1^{i_1} z_2^{i_2} \ldots z_n^{i_n}$ of degree $i_1 + i_2 + \ldots + i_n$ and f, g are polynomials in $\mathbb{R}[\mathbf{z}]$.
$\mathrm{Coef}(t, f) \triangleq$ coefficient of the term t in polynomial f (possibly 0).
$\mathrm{Occ}(t, f) \longleftrightarrow \mathrm{Coef}(t, f) \neq 0$.
$\mathrm{Hterm}(f) \triangleq$ maximal term in f with respect to ordering $<_T$, if $f \neq 0$; 1, otherwise.
$\mathrm{Hcoef}(f) \triangleq \mathrm{Coef}(\mathrm{Hterm}(f), f)$.
$\mathrm{Head}(f) \triangleq \mathrm{Hcoef}(f)\mathrm{Hterm}(f)$.
$\mathrm{Rest}(f) \triangleq f - \mathrm{Head}(f)$.
$\mathrm{Mult}(z_1^{i_1} z_2^{i_2} \ldots z_n^{i_n}, z_1^{j_1} z_2^{j_2} \ldots z_n^{j_n}) \longleftrightarrow i_1 \ge j_1, i_2 \ge j_2, \ldots, i_n \ge j_n$.
$\mathrm{LCM}(z_1^{i_1} z_2^{i_2} \ldots z_n^{i_n}, z_1^{j_1} z_2^{j_2} \ldots z_n^{j_n}) \triangleq z_1^{\max(i_1,j_1)} z_2^{\max(i_2,j_2)} \ldots z_n^{\max(i_n,j_n)}$.

In the following notations F denotes a finite sequence $\{f_1, f_2, \dots, f_r\}$ of polynomials.
$L(F) \triangleq$ length of the sequence $\{f_1, f_2, \dots, f_r\} = r$.
$\text{Mterm}(t, F) \longleftrightarrow \exists f_i \in F, f_i \neq 0$, such that $\text{Mult}(t, \text{Hterm}(f_i))$.
$\text{Normal } f(t, F) \longleftrightarrow \forall t$ with $\text{Occ}(t, f), \neg\text{Mterm}(t, F)$.

The last two items are expanded below for clarity.
$\text{Mterm}(t, F)$ occurs if and only if $\exists f_i \in F, f_i \neq 0$, such that $\text{Mult}(t, \text{Hterm}(f_i))$, i. e. t is a multiple of $\text{Hterm}(f_i)$.
$\text{Normal } f(g, F)$ occurs if and only if $\forall t$ with $\text{Occ}(t, g)$, there is no $\text{Mterm}(t, F)$.

Example 2.2 Let $F = \{f_1, f_2\} = \{z_1^3 z_2 z_3 - z_1 z_2, z_1 z_2 z_3^2 - 2z_3^2\}$. Then, we get

$$\text{Head}(f_1) = z_1^3 z_2 z_3, \quad \text{Head}(f_2) = z_1 z_2 z_3^2, \quad \text{Rest}(f_1) = -z_1 z_2, \quad \text{Rest}(f_2) = -2z_3^2$$

$$\text{and} \quad \text{LCM}(\text{Head}(f_1), \text{Head}(f_2)) = z_1^3 z_2 z_3^2.$$

Next, a reduction procedure is defined as follows:
$f \bullet >^{(1)}_{F,t,i} g \longleftrightarrow 1 \leq i \leq L(F), f_i \neq 0, \text{Occ}(t, f), \text{Mult}(t, \text{Hterm}(f_i)), g = f - a \cdot s \cdot f_i$, where $a = \frac{\text{Coef}(t,f)}{\text{Hcoef}(f_i)}$, $s = \frac{t}{\text{Hterm}(f_i)}$.
$f \bullet >^{(1)}_{F} g \longleftrightarrow \exists\, t, i$ such that $f \bullet >^{(1)}_{F,t,i} g$.
$f \bullet >_F g \longleftrightarrow \exists\, f_0, \dots, f_k, k \geq 0$, such that $f_0 = f, f_k = g, f_i \bullet >^{(1)}_F f_{i+1}$ for $i = 0, \dots, k-1$.
(Case $f = g$ included; if $f \bullet >_F g$, then f M-reduces to g).
$f \bullet >_F \tilde{g} \longleftrightarrow f \bullet >_F \tilde{g}$ and $\text{Normal } f(\tilde{g}, F)$.
(That is, $\tilde{g}$ is a normal form of f with respect to F).
$f\ \nabla^{\text{succ}}_F\ g \longleftrightarrow \exists\, h$ such that $f \bullet >_F h, g \bullet >_F h$.

$$\begin{aligned}\text{Spol}(f, g) \triangleq \text{Hcoef}(g) & \frac{\text{LCM}(\text{Hterm}(f), \text{Hterm}(g))}{\text{Hterm}(f)} f \\ - \text{Hcoef}(f) & \frac{\text{LCM}(\text{Hterm}(f), \text{Hterm}(g))}{\text{Hterm}(g)} g.\end{aligned}$$

If $F = f_1, f_2, \dots, f_r$ is a finite sequence of polynomials, $j_1, j_2, \dots, j_k$ are integers with $1 \leq j_i \leq L(F)$, then $H_F(j_1, j_2, \dots, j_k)$ denotes the LCM of the head terms of $f_{j_1}, f_{j_2}, \dots, f_{j_k}$.

Example 2.3 Let $R = Q[z_1, z_2, z_3]$, where Q denotes the field of rationals. Let $F = f_1, f_2$ denote a sequence of two polynomials,

$$f_1 = z_2^2 + z_2 z_1 + z_1^2, \; f_2 = z_2^2 z_1 + 1.$$

Then, the polynomial $f = z_1 f_1 - f_2 \in \mathbf{I}(F)$, the polynomial ideal set up by the two polynomials in the sequence F of polynomials. However, $f = z_2 z_1^2 + z_1^3 - 1$ is in normal form modulo $\{f_1, f_2\}$ with respect to every term order since the head terms, z_2^2 and $z_2^2 z_1$ have been lifted to their least common multiple $z_2^2 z_1$ and subtracted so that the head monomials formed in f cancel out.

Consider the polynomial ring $R = K[z_1, z_2, \ldots, z_n]$. A system of generators of an ideal $\mathbf{I} \subset \mathbf{R}$ is called a Gröbner basis if the leading monomials of its elements generate the same ideal in R as the leading monomials of all the polynomials in $\mathbf{I}$. The notion of a Gröbner basis admits a number of equivalent reformulations and serve as the foundation of the whole of constructive commutative algebra.

Gröbner basis algorithms are based on the significant fact that if the finitely many differences of the kind illustrated in the above example all reduce to zero then every polynomial in $\mathbf{I}(F)$ reduces to zero to make F a Gröbner basis, as defined next. The Gröbner basis might not be unique but by putting certain conditions on the polynomials in the basis, uniqueness can be guaranteed. In that case the basis becomes a reduced Gröbner basis, which exists for a fixed term order [1, p. 42].

If $f_1, \ldots, f_r$ is a Gröbner basis for the ideal $\mathbf{I} = < f_1, \ldots, f_r >$, then to determine whether or not a given polynomial f is in $\mathbf{I}$ one M-reduces f to normal form which is unique; if this normal form is zero then f is clearly in $\mathbf{I}$ by the definition of M-reduction; if it is not zero then f is not in $\mathbf{I}$. To M-reduce a polynomial to normal form successively eliminate M-terms t from f by executing a step of the form $f \bullet >_{F,t,i} g$ until no M-terms are left. To do this in the most efficient way, eliminate at each step the highest M-term t with respect to the ordering $<_T$; it is clear that in a finite number of steps a normal form will be reached. The theorem below summarizes several equivalent definitions for Gröbner bases.

Theorem 2.4 *Let $F = f_1, f_2, \ldots, f_r$ be a sequence of polynomials. The following conditions are equivalent and each provides a definition for Gröbner bases.*

1. $g \in \mathbf{I}(F) \longleftrightarrow g \bullet >_F 0$.
2. *For* $1 \leq i < j \leq L(F)$, $\mathrm{Spol}(f_i, f_j) \bullet >_F 0$.
3. $h \bullet >_F g, h \bullet >_F f \;\Rightarrow\; g = f$.
4. $h \bullet >_F h_1 \,, h \bullet >_F h_2 \Rightarrow h_1 \nabla_F^{\mathrm{succ}} h_2$.
5. $\forall 1 \leq i < j \leq L(F) \; \exists$ *sequence* $i = j_1, j_2, \ldots, j_k = j$, *such that*

$$H_F(j_1, j_2, \ldots, j_k) = H_F(i, j)$$

and

$$\mathrm{Spol}(f_{j_1} \,, f_{j_{(l+1)}}) \bullet >_F 0, \quad l = 1, 2, \ldots, k-1.$$

The S-polynomials in the reduction procedure are keys to the Buchberger algorithm for basis construction. After the basis is constructed, solutions may be sought regarding the location or clustering of the common zero-set of the original system of polynomial equations over the ground field of interest in a particular application. In systems theory applications considered here, the ground field is either the field of complex numbers (complex algebraic geometry domain) or the field of real numbers (real algebraic geometry domain).

Buchberger's algorithm for the scalar case goes through mutatis mutandis for the vector case. Also one can construct an algorithm which will terminate after a finite number of steps and yield a vector Gröbner basis.

2.2 Why Gröbner bases in multidimensional systems?

Linear multidimensional systems for signal processing and control have considerable use of multivariate polynomials, rational functions, and matrices of multivariate polynomials and rational functions. Basic concepts in such usage include greatest common divisor extraction, factorization, polynomial ideals, algebraic varieties and semi-algebraic sets. In the past, tools such as resultants and subresultants [4] as well as Bezoutiants have been extensively used for the purpose [6]. However Gröbner basis could offer several advantages that will be illustrated through selected examples.The first example is concerned with the finding of common zeros, which are always finite in number for any two (or more) relatively prime bivariate polynomials. Those common zeros can be found by resultant theory. But, reduced Gröbner bases yield the common zeros, usually, with lesser computational effort.

Example 2.5 Consider two bivariate polynomials (in a filter bank design problem), that are known to be relatively prime:

$$a(z_1,z_2) = -0.075(z_1^2z_2 + z_1z_2^2 + z_1 + z_2) - 0.0375(z_1^2 + z_2^2 + 1) + 0.85z_1z_2,$$
$$b(z_1,z_2) = 0.125(z_1z_2 + z_1 + z_2 + 1).$$

Resultant based calculations could be messy! The reduced Gröbner basis for $a(z_1,z_2)$ and $b(z_1,z_2)$, using SINGULAR (computed with respect to degree reverse lexicographical ordering), is

$$g_0(z_1,z_2) = 3z_2^3 - 77z_2^2 - 80z_2,$$
$$g_1(z_1,z_2) = z_1z_2 + z_1 + z_2 + 1,$$
$$g_2(z_1,z_2) = 3z_1^2 + 3z_2^2 - 80z_1 - 80z_2 - 83.$$

The common zeros of $g_0(z_1,z_2)$ and $g_1(z_1,z_2)$ are easy to find at $(-1,0)$, $(0,-1)$, $(-1,80/3)$ and $(80/3,-1)$, and these are also the common zeros of $a(z_1,z_2)$ and $b(z_1,z_2)$.

Similarly, an alternative to finding the common zeros of a set of relatively prime multivariate polynomials through the use of multipolynomial resultants [14] is the Gröbner basis approach of successive elimination of more and more variables, followed by extension.

For a recent survey of the design of multidimensional wavelets and filter banks, see [22]. For N-channel analysis $(\{h_i(\mathbf{z})\}_{i=1}^N)$ and $(\{f_i(\mathbf{z})\}_{i=1}^N)$ synthesis filters in a PR (perfect reconstruction) filter bank,

$$\sum_{i=1}^{N} h_i(\mathbf{z})f_i(\mathbf{z}) = \mathbf{z}^{\mathbf{m}}, \quad \mathbf{m} \in \mathbb{Z}_+^n, \tag{2.1}$$

$\mathbf{z} = (z_1,\ldots,z_n)$ and $\mathbf{z}^{\mathbf{m}}$ is the notation for a n-variate delay $z_1^{m_1}\ldots z_n^{m_n}$. Note that $\mathbf{m}$ is not known a priori and given $\{h_i(\mathbf{z})\}$ the set $\{f_i(\mathbf{z})\}$ plus $\mathbf{m}$ has to be found satisfying the PR constraint. Without loss of generality, consider the $N = 2$, $n = 2$

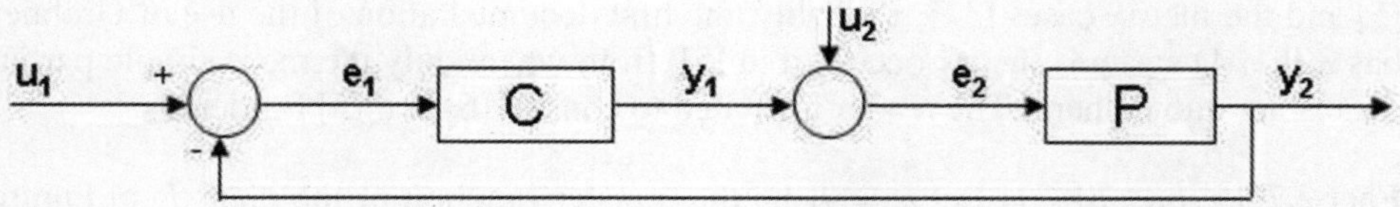

Figure 2.1 Scalar feedback system

(2-channel bivariate case). In this case, $h_1(z_1, z_2)$ and $h_2(z_1, z_2)$ could have common zeros at $(0, 0), (0, \beta_i)$ with $\beta_i \neq 0$ or $(\alpha_i, 0)$ with $\alpha_i \neq 0$. The "Rabinowitsch trick" [28] can be used to solve the problem:

$$h_1(z_1, z_2)f_1(z_1, z_2) + h_2(z_1, z_2)f_2(z_1, z_2) = s^r(z_1, z_2)$$

for $f_i(z_1, z_2)$ and positive integer r, given that a polynomial $s(z_1, z_2)$ vanishes on the variety of the ideal generated by $h_i(z_1, z_2)$, as was done using Gröbner basis in [34] (this paper also contained matrix equation results). Note that no previous knowledge of the positive integer r is needed. Let z_3 be a new indeterminate. Then $(1 - z_3 s(z_1, z_2))$ and $h_i(z_1, z_2), i = 1, 2$, are zero coprime. As a consequence of Hilbert's (Weak) Nullstellensatz, the ideal generated by these three zero-coprime polynomials must be the unit ideal, i.e there exist polynomials $\hat{g}_i(z_1, z_2, z_3)$ for $i = 1, 2, 3$ such that

$$\sum_{i=1}^{2} h_i(z_1, z_2)\hat{g}_i(z_1, z_2, z_3) + (1 - z_3 s(z_1, z_2))\hat{g}_3(z_1, z_2, z_3) = 1.$$

Then, substitute $1/s(z_1, z_2)$ for z_3 and clear out denominator to obtain an equation of the form desired in equation (2.1). This problem is a case of Hilbert's Nullstellensatz.

Theorem 2.6 (Hilbert's Nullstellensatz) *Let K be a field, K_1 an algebraically closed extension of K and $s, h_1, h_2, \ldots, h_N \in K[z_1, z_2, \ldots, z_n]$. For all $\mathbf{z} \in K_1^n$, $h_1(\mathbf{z}) = h_2(\mathbf{z}) = \ldots = h_N(\mathbf{z}) = 0$ implies $s(\mathbf{z}) = 0$, if and only if there exists a positive integer r such s^r belongs to the ideal generated by the n-variate polynomials h_k for $k = 1, 2, \ldots, N$.*

2.3 Multidimensional linear control systems and Gröbner bases

Consider the generic plant/compensator feedback system in Figure 2.1. The plant and compensator could be scalar as in the figure or matrix-valued. In the figure, the compensator could be more general than a quotient of polynomials; in fact it could be a quotient of elements from a different ring. The stabilization of scalar as well as matrix feedback systems for the bivariate case has been documented in detail and the class of all stabilizing compensators have been identified in [6]. The compensator can be constructed by using Gröbner bases and Hilbert's Nullstellensatz for both, the scalar

[5] and the matrix cases [34]. Probably, the first documentation of the use of Gröbner bases in n-D systems theory occurred in [5], from where only the more simple partial results are quoted here. The reader is invited to consult [6] and [5] for details.

Fact 2.7 [5, pp. 70–71] Let $h = \frac{n}{d}$ be the transfer function of the plant P in Figure 2.1, such that n and d are relatively prime polynomials with no common zeros in the bidisc $\bar{U}^2$. One can constructively find polynomials $x, y \in \mathbb{R}[z_1, z_2]$ such that $yd + xn \in \mathbb{R}_s[z_1, z_2]$, the multiplicatively closed set of bivariate polynomials which have no zeros in $\bar{U}^2$. Then, the stabilizing compensators are characterized by the expression $C = \frac{-s_1 d + s_2 x}{s_1 n + s_2 y}$, where $s_1 \in \mathbb{R}[z_1, z_2]$ and $s_2 \in \mathbb{R}_s[z_1, z_2]$ are arbitrary.

Example 2.8 Let

$$\begin{aligned} n(z_1, z_2) &= z_1 z_2 - z_1 - 2z_2 \triangleq G_1(z_1, z_2), \\ d(z_1, z_2) &= z_1^2 z_2^2 - 2z_1 z_2^2 - 2z_1 z_2 + 4z_1 + 4 \triangleq G_2(z_1, z_2), \\ \mathrm{Spol}(G_1, G_2) &= G_2 - z_1 z_2 G_1 = z_1^2 z_2 - 2z_1 z_2 + 4z_1 + 4 \\ &\overset{-z_1 G_1}{\cdot>} z_1^2 + 4z_1 + 4 \triangleq G_3(z_1, z_2). \end{aligned}$$

$G_3(z_1, z_2)$ is in normal form, i. e. no term of G_3 is a multiple of the head terms of $G_i, i < 3$. Backtracking,

$$d(z_1, z_2) - z_1(z_2 + 1)n(z_1, z_2) = z_1^2 + 4z_1 + 4 \in \mathbb{R}_s[z_1, z_2].$$

Example 2.9 Consider the plant transfer function whose numerator and denominator polynomials are

$$n(z_1, z_2) = z_1 + z_2, \quad d(z_1, z_2) = 1 - z_1 + z_2.$$

Common zeros occur at $(1/2, -1/2) \in \bar{U}^2$. Therefore, the plant is not stabilizable by causal compensators (but is stabilizable by weakly causal systems (Guiver-Bose, 1985) [5]).

Fact 2.10 [5, pp. 72–78] In the MIMO bivariate causal case, if $D_L^{-1} N_L$ is an irreducible left matrix-fraction description (LMFD) (maximal order minors of $[D_L \ N_L]$ have no nontrivial common factor and hence have only a finite number of common zeros) of an unstable bivariate plant P, and $X_R Y_R^{-1}$ denotes an irreducible right MFD (maximal order minors of $[X_R^t \ Y_R^t]^t$ have no nontrivial common factor) of a bivariate compensator C, then C stabilizes P if and only if $\det(D_L Y_R + N_L X_R)$ is zero free in $\bar{U}^2$.

For stabilization of MIMO Weakly Causal systems, and updates on other results, including computational methods for determining strong stabilizability of n-D systems, see [6] and works of Ying, Xu and Lin [36] and Ying [35]. Research counterparts in the n-D case ($n > 2$) for constructing a stabilizing compensator for a MIMO plantusing Gröbner bases (following initial work of Xu, Saito and Abe in 1994 [34]) is worth pursuing.

n-D System Stabilizability [36]: In the generic feedback system configuration, plant $p(\mathbf{z}) = f(\mathbf{z})/g(\mathbf{z})$, compensator $c(\mathbf{z}) = h(\mathbf{z})/k(\mathbf{z})$. Do there exist polynomials $h(\mathbf{z})$ and $k(\mathbf{z})$ such that (strong stabilizability), $k(\mathbf{z}) \neq 0$ and $f(\mathbf{z})h(\mathbf{z})+g(\mathbf{z})k(\mathbf{z}) \neq 0$ for all $\mathbf{z} \in \bar{U}^n$? Procedure is based on cylindrical algebraic decomposition of semialgebraic sets.

Example 2.11 The algebraic variety of common zeros of the plant defined by polynomials f and g with zero sets $V(f), V(g)$ is specified to be

$$V(f) \cap V(g) = \left\{ \left(-\frac{1+\sqrt{2}}{2}, \frac{1-\sqrt{2}}{2} \right), \left(\frac{-1+\sqrt{2}}{2}, \frac{1+\sqrt{2}}{2} \right) \right\}.$$

f and g do not have common zeros in $\bar{U}^2$. Let, as in Guiver-Bose [6, p. 47],

$$s(z_1, z_2) = \left(z_1 + \frac{1+\sqrt{2}}{2} \right) \left(z_2 - \frac{1+\sqrt{2}}{2} \right).$$

Then, by Rabinowitsch trick [28, 34], since $f(z_1, z_2), g(z_1, z_2)$ and $1 - z_3 s(z_1, z_2)$ do not have any common zeros in $\mathbb{R}[z_1, z_2, z_3]$, using Gröbner bases (by Hilbert's (Weak) Nullstellensatz)

$$-\frac{1}{4} z_3 f(z_1, z_2) - \frac{1+\sqrt{2}}{2} z_3 g(z_1, z_2) + (1 - z_3 s(z_1, z_2)) = 1.$$

Setting $z_3 = 1/s(z_1, z_2)$,

$$f + 2(1+\sqrt{2})g = -4s.$$

As $V(s) \cap \bar{U}^2 = \Phi$, therefore

$$h = \frac{1}{2(1+\sqrt{2})}$$

is a stable stabilizer.

3 Multivariate polynomial equations and inequalities

Construction of solutions of sets of polynomial equations and inequalities by elementary decision algebra, with reference to the output feedback stabilization and related problems, was considered in [2]. There has been a surge of interest during the last decade in the topics of semidefinite programming, semialgebraic sets (defined by multivariate polynomial equations, inequations and inequalities), robust optimization and sum-of-squares representation of classes of nonnegative definite multivariate polynomials and forms for applications in analysis and synthesis of control systems [25, 20, 26].

Definition 3.1 A basic semialgebraic set is a subset of $\mathbb{R}^n$ defined by a finite number of polynomial equations and inequalities.

Example 3.2 Two basic semialgebraic sets are given by

1. $\left\{(x_1, x_2) \in \mathbb{R}^2 \middle| \frac{x_1^2}{3^2} + \frac{x_2^2}{2^2} \leq 1, x_1^2 - x_2 \leq 0\right\}$ and,
2. given $f(\mathbf{x}) \in \mathbb{R}[\mathbf{x}]$, the set $\left\{\mathbf{x} \triangleq (x_1, \ldots, x_n) \in \mathbb{R}^n \middle| f(\mathbf{x}) > 0\right\}$.

Approaches for (2) include

1. Sum of squares (SOS) representation, when possible to do (SOSTOOLS, SeDuMi), because positive multivariate polynomials or forms may not always be representable as a sum of squares of forms.
2. Semidefinite programming to test feasibility of algebraic sets (C. N. Delzell [15], P. A. Parrilo and B. Sturmfels in [20]).
3. Elementary decision algebra methods (A. Tarski, A. Seidenberg, G. E. Collins, N. K. Bose, B. D. O. Anderson, E. I. Jury) [4, chapter 2], [26].
4. Global lower bound approach (N. Z. Shor [31]).
5. Gram matrix method (N. K. Bose, C. C. Li, M. D. Choi, T. Y. Lam, B. Reznik in [25]).

3.1 SOS decomposition by Gram matrix method

[Bose-Li (1968), Choi-Lam-Reznick (1995), Parrilo (2000)] (see [25] for references).

Fact 3.3 A multivariate real coefficient polynomial $p(\mathbf{x})$ in n real variables

$$\mathbf{x} \triangleq (x_1, \ldots, x_n)$$

and of total degree $2d$ is a SOS if and only if it is representable as $p(\mathbf{x}) = \mathbf{v}^T Q \mathbf{v}$ with the $\binom{n+d}{d}$ vector of monomials

$$\mathbf{v}^T = (1 \;\; x_1 \;\; x_2 \;\; \ldots \;\; x_n \;\; x_1 x_2 \;\; \ldots \ldots \;\; x_n^d)$$

and a symmetric PSD (positive semidefinite) matrix Q.

3.1.1 Nonnegativity of a polynomial $f(\mathbf{x})$ on an algebraic variety

Let $h_i(\mathbf{x}) = 0$ for $i = 1, \ldots, l$ be constraints and let $\mathbf{I}$ denote the polynomial ideal $\mathbf{I} = \langle h_1(\mathbf{x}), \ldots, h_l(\mathbf{x})\rangle$. Then, there exist polynomials $\lambda_i(\mathbf{x}) \in \mathbb{R}[\mathbf{x}]$ such that $f(\mathbf{x}) + \sum_i \lambda_i(\mathbf{x}) h_i(\mathbf{x})$ is a SOS in n-variate polynomial ring $\mathbb{R}[\mathbf{x}]$ if and only if $f(\mathbf{x}) + \mathbf{I}$ is a SOS in the quotient ring $\mathbb{R}[\mathbf{x}]/\mathbf{I}$. Under these equivalent conditions, $f(\mathbf{x})$ is nonnegative on the real variety $\{\mathbf{x} \in \mathbb{R}^n | h_i(\mathbf{x}) = 0, \forall i\}$. [26, pp. 187–188].

Example 3.4 This example is considered in [26, pp. 187–188], where an error in the construction of matrix L below is corrected.

Is $f(\mathbf{x}) = 10 - x_1^2 - x_2$ nonnegative on $x_1^2 + x_2^2 - 1 = 0$? $\mathbf{I} = \langle x_1^2 + x_2^2 - 1\rangle$ in this case of one constraint equation, $h(\mathbf{x}) = x_1^2 + x_2^2 - 1$ is the Gröbner basis of the corresponding ideal

$$\begin{aligned} 10 - x_1^2 - x_2 &= \begin{pmatrix} 1 & x_1 & x_2 \end{pmatrix} \begin{pmatrix} q_{11} & q_{12} & q_{13} \\ q_{12} & q_{22} & q_{23} \\ q_{13} & q_{23} & q_{33} \end{pmatrix} \begin{pmatrix} 1 \\ x_1 \\ x_2 \end{pmatrix} \\ &= q_{11} + q_{22}x_1^2 + q_{33}x_2^2 + 2q_{12}x_1 + 2q_{13}x_2 + 2q_{23}x_1x_2. \end{aligned}$$

In the quotient ring $\mathbb{R}[\mathbf{x}]/\mathbf{I}$,

$$\begin{aligned} f(\mathbf{x}) \,(\mathrm{mod}\, \mathbf{I}) &= (q_{11} + q_{22}) + (q_{33} - q_{22})x_2^2 + 2q_{12}x_1 + 2q_{13}x_2 + 2q_{23}x_1x_2 \\ &= 9 + x_2^2 - x_2 \triangleq \mathbf{v}^T Q_1 \mathbf{v}, \end{aligned}$$

where $\mathbf{v}$ denotes the vector in Fact 3.3 above corresponding to this example and

$$Q_1 = \begin{pmatrix} 9 & 0 & -1/2 \\ 0 & 0 & 0 \\ -1/2 & 0 & 1 \end{pmatrix} = L^T L, \quad L = \begin{pmatrix} 3 & 0 & -1/6 \\ 0 & 0 & \sqrt{35}/6 \end{pmatrix}$$

$$\Rightarrow 10 - x_1^2 - x_2 \equiv \left(3 - \frac{x_2}{6}\right)^2 + \frac{35}{36}x_2^2 \,(\mathrm{mod}\, \mathbf{I})$$

$$\Rightarrow f(x_1, x_2) \text{ is a SOS on } \mathbb{R}[x_1, x_2]/\mathbf{I}.$$

Therefore, SOS on quotient ring $\mathbb{R}[\mathbf{x}]/\mathbf{I}$ is needed, where $\mathbf{I} = \langle h_i(\mathbf{x})\rangle_{i=1}^{l}$ is the ideal generated by equality constraints. The computations can be effectively done in $\mathbb{R}[\mathbf{x}]/\mathbf{I}$ after computing the Gröbner basis for $\mathbf{I}$ [26].

3.2 Multidimensional nonlinear control systems, performance optimization and Gröbner bases

Forsman and Glad [18] considered the following constrained optimization problem. Given a multivariate polynomial $V(\mathbf{x})$, $\mathbf{x}^T = (x_1 \; x_2 \; \ldots \; x_n)$, and m multivariate polynomial equations

$$c_1(\mathbf{x}) = c_2(\mathbf{x}) = \ldots = c_m(\mathbf{x}) = 0,$$

it is required to determine a maximum value $V(\mathbf{x}_0)$, where $\mathbf{x}_0$ belongs to the real algebraic variety formed by the m polynomial equations above. Forsman also showed that, without loss of generality, inequations can also be handled in the same framework by using the Rabinowitsch trick (embracing auxiliary variables). A condition on a solution $\mathbf{x} = \mathbf{x}_0$ to the optimization problem (when the Jacobian matrix of $\mathbf{c}$ at $\mathbf{x}$ is of full rank) is known to be

$$c_1 = c_2 = \ldots = c_m = 0, \quad \nabla V - \boldsymbol{\lambda}^T \nabla \mathbf{c} = 0,$$

where $\mathbf{c}^T = (c_1 \; c_2 \; \ldots \; c_m)$ and $\boldsymbol{\lambda}^T = (\lambda_1 \; \lambda_2 \; \ldots \; \lambda_m)$ is the vector of Lagrange multipliers. Note that the preceding equation is a square system of $n + m$ scalar equations in the $n + m$ unknowns $x_1, x_2, \ldots, x_n, \lambda_1, \lambda_2, \ldots, \lambda_m$ and hence, generically, there are a finite number of solutions. This problem may also be solved by the theory of Gröbner bases. The following problem, solved in [23, pp. 315–316] by classical methods, is tackled with Gröbner basis theory. Note that in the example below, there are inequalities in the constraints, which could be converted to equations by adding slack variables. Alternatively, discrete-time nonlinear systems described as constrained polynomial systems in a state-space model and subject to (semialgebraic) inequality constraints in the joint input-state space can, indeed, be tackled using Gröbner bases.

Example 3.5 Consider the problem

$$\begin{array}{ll} \text{minimize} & 2x_1^2 + 2x_1x_2 + x_2^2 - 10x_1 - 10x_2 \\ \text{subject to} & x_1^2 + x_2^2 \leq 5 \\ & 3x_1 + x_2 \leq 6. \end{array}$$

The first order (Karush-Kuhn-Tucker) necessary conditions for optimality in addition to the constraints are

$$\begin{aligned} 4x_1 + 2x_2 - 10 + 2\lambda_1 x_1 + 3\lambda_2 &= 0, \\ 2x_1 + 2x_2 - 10 + 2\lambda_1 x_2 + \lambda_2 &= 0, \\ \lambda_1, \lambda_2 &\geq 0, \\ \lambda_1(x_1^2 + x_2^2 - 5) &= 0, \\ \lambda_2(3x_1 + x_2 - 6) &= 0, \end{aligned}$$

where λ_1, λ_2 are Lagrange multipliers.

An inequality constraint $g_i(x) \leqslant 0$ is said to be active at the feasible point x if $g_i(x) = 0$ and inactive if $g_i(x) < 0$. The equality constraint $h_i(x) = 0$ is always considered active by convention. The active constraints at the feasible point restrict the domain of feasibility in the neighborhood of x, while the inactive constraints have no effect on the neighborhood. It is for this reason, attention is restricted towards active constraints while studying the properties of the optimum point.

To find the solution we consider various combinations of active constraints and check the sign of the corresponding Lagrange multipliers. We set one, two or no constraints to be active for this example. Assuming that the first constraint is active while the second is inactive yields

$$\begin{aligned} 4x_1 + 2x_2 - 10 + 2\lambda_1 x_1 &= 0, \\ 2x_1 + 2x_2 - 10 + 2\lambda_1 x_2 &= 0, \\ x_1^2 + x_2^2 - 5 &= 0. \end{aligned}$$

Computing the Gröbner bases for this system of polynomial equations with respect to lexicographical ordering and ordering $x_1 \succ x_2 \succ \lambda_1$ gives

$$\begin{aligned}\lambda_1^4 + 6\lambda_1^3 + \lambda_1^2 - 4\lambda_1 - 4 &= 0,\\ 5x_2 - 3\lambda_1^3 - 16\lambda_1^2 + 6\lambda_1 + 3 &= 0,\\ x_1 + x_2\lambda_1 + x_2 - 5 &= 0.\end{aligned}$$

The solution now could be easily inferred to be $x_1 = 1, x_2 = 2$, $\lambda_1 = 1$ (the other values of λ_1 being either negative or complex are not considered). Note that the second equality constraint is also satisfied at the solution point giving $3x_1 + x_2 = 5$. The value of the function at the solution point is -20. Considering only the second equation as active give a negative λ_2. At the minimum point, the Lagrange multiplier must be positive; so this solution is discarded. Finally, when both constraints are considered to be active, the Gröbner bases contains the polynomial 1 indicating, by Nullstellensatz, that no solution exists.

When all inequality constraints are active use of Gröbner bases is less straightforward in comparison to the case of equality constraints. Various combinations of inequality constraints are considered active and then the techniques developed for equality constraints is applied to solve the problem. The sign of Lagrange multipliers must be looked into. It is mandatory that at a regular point an optimum is attained only if all the Lagrange multipliers are of same sign. If the Lagrange multipliers are all positive then it is a minimum point and a maximum point if all the multipliers are negative. Besides the sign of Lagrange multipliers, the other inactive constraints must also be satisfied at the optimum point, i. e. it must not violate the inequalities. This process becomes a somewhat computation intensive as the number of inequality constraints increase.

The following problem, solved in [23, pp. 301–302] by classical methods, is tackled with Gröbner basis theory.

Example 3.6 Consider the problem on constructing a cardboard box of maximum volume with given surface area.

$$\begin{array}{ll}\text{maximize} & V = x_1x_2x_3\\ \text{subject to} & h = x_1x_2 + x_2x_3 + x_1x_3 - c/2 = 0,\end{array}$$

where x_1, x_2, x_3 are the variable dimensions of the box and c is a constant for fixed area. The first order necessary conditions gives us

$$\begin{aligned}f_1 &= \frac{\partial V}{\partial x_1} - \lambda\frac{\partial h}{\partial x_1} = x_2x_3 - \lambda(x_2 + x_3),\\ f_2 &= \frac{\partial V}{\partial x_2} - \lambda\frac{\partial h}{\partial x_2} = x_1x_3 - \lambda(x_1 + x_3),\\ f_3 &= \frac{\partial V}{\partial x_3} - \lambda\frac{\partial h}{\partial x_3} = x_1x_2 - \lambda(x_1 + x_2).\end{aligned}$$

Consider the ideal generated by the preceding three polynomials along with the two other polynomials

$$f_4 = x_1x_2x_3 - d,$$
$$f_5 = x_1x_2 + x_2x_3 + x_1x_3 - c/2,$$

where the real-valued parameter d is viewed to be an additional variable in f_4, which is introduced to directly obtain the optimal value. Computing the Gröbner basis of the resulting ideal $I = \langle f_1, f_2, f_3, f_4, f_5\rangle$ in lexicographic order with the ranking $x_1 \succ x_2 \succ x_3 \succ \lambda \succ d$ one gets the system of polynomials

$$g_1 = c\lambda - 3d,$$
$$g_2 = 216d^2 - c^3 \Rightarrow d = \sqrt{c^3/6^3},$$
$$g_3 = x_3 - 2\lambda,$$
$$g_4 = x_2 - 2\lambda,$$
$$g_5 = x_1 - 2\lambda.$$

Since the system of polynomial equations are in triangular form it could be easily inferred that the maximum value of the function is $\sqrt{c^3/6^3}$ and is attained at the point $x_1 = x_2 = x_3 = \sqrt{c/6}$.

Robustness of polynomial nonlinear systems with structured uncertainty. Consider the autonomous nonlinear system with structured uncertainty parameter vector $\boldsymbol{\eta}^T = (\eta_1\ \eta_2\ \dots\ \eta_q)$, characterized by the state-space vector differential equation

$$\dot{\mathbf{x}} = \mathbf{f}(\mathbf{x}, \boldsymbol{\eta}),$$

where $\mathbf{x}^T = (x_1\ x_2\ \dots\ x_n)$ and the components f_i of $\mathbf{f}$ are polynomials in both the components of $\mathbf{x}$ and $\boldsymbol{\eta}$. For a given Lyapunov function $V(\mathbf{x})$ at a nominal value of the parameter vector, the question regarding the subspace of parameter values that guarantee stability using $V(\mathbf{x})$ can be answered by Gröbner basis theory [17].

Example 3.7 Consider the system

$$\dot{x}_1 = -x_1 + \alpha x_1^2 x_2,$$
$$\dot{x}_2 = -x_2,$$

along with the Lyapunov function

$$V = 3x_1^2 + 4x_1x_2 + 4x_2^2.$$

Define

$$\Omega_v(\eta) = \{x \in \mathbb{R}^n \mid \dot{V}(x, \eta) \le -V\}$$

and

$$\mathbb{B}_d = \{x \in \mathbb{R}^n \mid V(x) \le d\}.$$

The parameter vector η here is just the scalar α. The robust performance problem is the following: Given V, for which η do we have $\mathbb{B}_d \subseteq \Omega_v(\eta)$?

By tangency condition, $\nabla V + \lambda \nabla(\dot{V}) = 0$, where $\dot{V}$ is the derivative of the Lyapunov function along the system trajectory. This derivative is calculated by the relation

$$\dot{V}(\mathbf{x}) = \sum_{j=1}^{n} \frac{\partial V}{\partial x_j}(\mathbf{x})\frac{dx_j}{dt},$$

where $\mathbf{x} = (x_1\ x_2\ \ldots\ x_n)^T$ is the state vector and $n = 2$ in this example. Here, the parameter d is fixed and η is varied. Assume that the value of d is 1 and let λ denote the Lagrange multiplier. Consider the ideal $\mathbf{I} = \langle f_1, f_2, f_3, f_4\rangle$, where

$$f_1 = V - d = 3x_1^2 + 4x_1x_2 + 4x_2^2 - 1,$$

$$f_2 = \dot{V} = 4x_1^2x_2^2\alpha - 8x_2^2 + 6x_2x_1^3\alpha - 8x_2x_1 - 6x_1^2,$$

$$f_3 = \frac{\partial V}{\partial x_1} + \lambda\frac{\partial \dot{V}}{\partial x_1} = +8x_2^2x_1\lambda\alpha + 18x_2x_1^2\lambda\alpha - 8x_2\lambda + 4x_2 - 12x_1\lambda + 6x_1,$$

$$f_4 = \frac{\partial V}{\partial x_2} + \lambda\frac{\partial \dot{V}}{\partial x_2} = +8x_2x_1^2\lambda\alpha - 16x_2\lambda + 8x_2 + 6x_1^3\lambda\alpha - 8x_1\lambda + 4x_1.$$

The Gröbner basis of $\mathbf{I} = \langle f_1, f_2, f_3, f_4\rangle$ with respect to lexicographic order and ranking $x_1 \succ x_2 \succ \lambda \succ \alpha$ gives a triangular system of four polynomials, the first of which is given below, for the sake of brevity:

$$g_1(\alpha) = -8192\alpha^2 - 1536\alpha^3 + 147\alpha^4 - 2\alpha^5.$$

From g_1 we can infer the interval of α within which exponential stability is guaranteed. Let us call this value α_{new}. The stability problem for it, i. e.

$$\Omega(\alpha) = \{x \in \mathbb{R}^n \mid \dot{V}(x, \eta) \le 0\}$$

gives another α in a different interval. Let us call it α_{old}. In general, $\alpha_{\text{new}} \subset \alpha_{\text{old}}$ or the connected region corresponding to exponential stability is a subset of the region in which the system is stable. Remember this analysis is dependent on the choice of the Lyapunov function. The obtained bounds are therefore not absolute in nature. It is possible for a different Lyapunov function to give better bounds and hence assurance of a more robust performance.

A concept central in real algebraic geometry (like Hilbert's Nullstellensatz in complex algebraic geometry) is stated next.

Theorem 3.8 (Stengle's Positivstellensatz (1974)) *Given polynomials* $\{f_1, \ldots, f_r\}$, $\{g_1, \ldots, g_k\}$ *and* $\{h_1, \ldots, h_l\}$ *in* $\mathbf{x} = (x_1, \ldots, x_n)$, *the following are equivalent*

1.

$$\left\{ \mathbf{x} \in \mathbb{R}^n \;\middle|\; \begin{array}{l} f_i(\mathbf{x}) \ge 0, i = 1, 2, \ldots, r \\ g_i(\mathbf{x}) \ne 0, i = 1, 2, \ldots, k \\ h_i(\mathbf{x}) = 0, i = 1, 2, \ldots, l \end{array} \right\}$$

is the empty set.

2. *There exist polynomials $f \in$ (cone generated by $\{f_1, \dots, f_r\}$), $g \in$ (cone generated by $\{g_1, \dots, g_k\}$) and $h \in$ (cone generated by $\{h_1, \dots, h_l\}$) such that*

$$f + g^2 + h = 0.$$

Comments

1. The multiplicative monoid M generated by $\{g_i\}_{i=1}^k$ is the set of all finite products of g_i's including 1, e. g.

$$M(g_1, g_2) = \left\{ g_1^{k_1}, g_2^{k_2} \;\middle|\; k_1, k_2 \in \mathbb{Z}_+ \cup \{0\} \right\}.$$

2. The cone generated by $\{f_i\}_{i=1}^r$ is

$$P(f_1, \dots, f_r) = \left\{ s_0 + \sum_{i=1}^{l} s_i b_i \;\middle|\; l \in \mathbb{Z}_+, s_i \in \Sigma_n, b_i \in M(f_1, \dots, f_r) \right\},$$

where Σ_n denotes the set of SOS polynomials in n variables. Note that $f_i^2 s_i \in \Sigma_n$ as well.

3. Positivstellensatz gives a characterization of the *infeasibility* of polynomial equations and inequalities over the reals.

The Positivstellensatz (P-satz) is useful because it provides a characterization of the infeasibility (refutation) of a system of polynomial equations and inequalities in conjunction with polynomial SOS and is beginning to be used in control theory [20].

4 Gröbner bases may not always produce the result sought

Let $R = K[z_1, z_2, z_3]$. The next example illustrates a fact in algebraic geometry that the maximal set of linearly independent elements of the Gröbner basis for a module generated by the columns of a multivariate polynomial matrix C using any lexicographical ordering may not form a minimal generating set of the column space of C. Its consequence in systems theory is pointed out.

Example 4.1 (Chalie Charoenlarpnopparut (2000)) It is known that

$$C = \left(\begin{array}{c|cc} z_1 z_2^2 z_3 & 0 & -z_1^2 z_2^2 - 1 \\ z_1^2 z_3^2 + z_3 & -z_3 & -z_1^3 z_3 - z_1 \end{array} \right) = \hat{G}(A_1 | B_1)$$

(note that the columns of $\hat{G}$ generate the same module as the columns of C) has zero coprime reduced minors and has a GCLF of the submatrices A_1, B_1. The 2×2 minors are

$$m_1 = -z_1 z_2^2 z_3^2, \quad m_2 = -z_1^2 z_2^2 z_3 - z_3, \quad m_3 = z_1^2 z_3^2 + z_3.$$

So, the reduced minors, obtained after extracting the gcd $d = z_3$, have no common zeros. To wit, a factorization is

$$\hat{G} = \begin{pmatrix} -z_1^2 z_2^2 - 1 & z_1 z_2^2 z_3 \\ -z_1 & z_3 \end{pmatrix}, \quad A_1 = \begin{pmatrix} z_1^3 z_2^2 z_3^2 \\ z_1^4 z_2^2 z_3 + z_1^2 z_3 + 1 \end{pmatrix},$$

$$B_1 = \begin{pmatrix} -z_1 z_2^2 z_3 & -z_1^4 z_2^2 z_3 + 1 \\ -z_1^2 z_2^2 - 1 & -z_1^3 (z_1^2 z_2^2 + 1) \end{pmatrix}.$$

Using degree reverse lexicographical ordering $z_1 \succ z_2 \succ z_3$, the matrix G whose columns are the reduced Gröbner basis vectors of the module generated by the columns of C are (using SINGULAR)

$$G = \begin{pmatrix} 0 & z_3 & z_1^2 z_2^2 + 1 \\ z_3 & 0 & z_1 \end{pmatrix}$$

The columns of $\hat{G}$ generate the same module as the columns of C. However, $\hat{G}$ cannot be computed from G by applying the algorithmic theory of Gröbner bases because it can be shown that no proper linearly independent subset of the columns of G can generate the column space of C. Thus, the maximal set of linearly independent elements of the reduced Gröbner basis for module M may only generate a proper submodule of M. Applications of Gröbner basis of modules over polynomial rings in wavelets and filter bank design with examples has been considered in [12] and [22].

Can a factorization be found by another algorithm? The answer to this question was provided in the affirmative by Wang and Kwong recently [33]. Let s_i be the ith row of $\begin{pmatrix} C \\ -z_3 I_3 \end{pmatrix}$, since $d = z_3$. They use CoCoA under the default module term ordering (Deg Revlex and ToPos) to get three generators of a syzygy module of s_i $(i = 1, \ldots, 5)$:

$$f_1 = \begin{pmatrix} 0 & z_3 & z_1^2 z_3^2 + z_3 & -z_3 & -z_1^3 z_3 - z_1 \end{pmatrix},$$

$$f_2 = \begin{pmatrix} z_3 & -z_1 z_2^2 z_3 & -z_1^3 z_2^2 z_3^2 & z_1 z_2^2 z_3 & z_1^4 z_2^2 z_3 - 1 \end{pmatrix},$$

$$f_3 = \begin{pmatrix} z_1 & -z_1^2 z_2^2 - 1 & -z_1^4 z_2^2 z_3 - z_1^2 z_3 - 1 & z_1^2 z_2^2 + 1 & z_1^5 z_2^2 + z_1^3 \end{pmatrix}.$$

Let M be the module generated by f_1, f_2, f_3. Using a CoCoA command "Minimalized (M)", they find that M has f_2 and f_3 as generators ($C \in \mathbb{R}^{l \times m}[z_1, z_2, z_3], l < m$).
Factorization:

$$C = \begin{pmatrix} z_1^2 z_2^2 + 1 & z_1 z_2^2 z_3 \\ z_1 & z_3 \end{pmatrix} \cdot \begin{pmatrix} -z_1^3 z_2^2 z_3^2 & z_1 z_2^2 z_3 & z_1^4 z_2^2 z_3 - 1 \\ z_1^4 z_2^2 z_3 + z_1^2 z_3 + 1 & -z_1^2 z_2^2 - 1 & -z_1^5 z_2^2 - z_1^3 \end{pmatrix}.$$

Limitation: Does not work if the system of generators of the syzygy module does not have ℓ (2 in this case) elements. Otherwise, more research needed in constructive (algorithmic algebra) n-variate factorization problems, $n > 2$.

5 Bihermitian forms and SOS

Theorem 5.1 *A bihermitian form $H(x, y)$ is representable as*

$$H(x, y) = \sum_{u=1}^{m} \sum_{v=1}^{m} (x^* A^{uv} x) \bar{y}_u y_v, \tag{5.1}$$

where the overbar and star superscript denote, respectively, complex conjugate and complex conjugate transpose, and $x^* = (\bar{x}_1, \bar{x}_2, \ldots, \bar{x}_n)$ and $A^{uv} = ((a_{ij}^{uv})) \in \mathbb{C}^{n \times n}$ are complex matrices such that $(A^{uv})^* = A^{vu}$. Furthermore, if A_{ij} represents a matrix of order $m \times m$ with a_{ij}^{uv} as its element in the (u, v)th position, $H(x, y)$ can be equivalently expressed as

$$H(x, y) = \sum_{i=1}^{n} \sum_{j=1}^{n} (y^* A_{ij} y) \bar{x}_i x_j. \tag{5.2}$$

With the above notation, a bihermitian form $H(x, y)$ is a Hermitian form in x for each y and a Hermitian form in y for each x. A bihermitian form $H(x, y)$ is called nonnegative definite Hermitian (nndH) if $H(x, y)$ is nonnegative for each pair of values for x and y. The fact that every bihermitian form H can be represented as in (5.1) and the dual representation (5.2) has been proved elsewhere by the author.

Consider the linear map

$$L = \phi : S \in M_n := \mathbb{C}^{n \times n} \longrightarrow L(S) \in \mathbb{C}^{m \times m}$$

acting on the space of nonnegative definite Hermitian (nndH) matrices. For such a linear operator $L : S$ nndH $\longrightarrow L(S)$ nndH when do there exist matrices $V_k \in \mathbb{C}^{n \times m}$ so that $\forall S$ nndH, there exists a finite sum of congruences representation for $L(S)$ [27]

$$L(S) = \sum_{k=1}^{p} V_k^* S V_k.$$

That the above does not hold over the field of complex numbers follows from the counterexample given next.

Counterexample. Consider $L(S) = (\text{tr } S)I - S$ so that $L : S$ nndH $\longrightarrow L(S)$ is nndH. Hypothesize, $L(S) = \sum_{k=1}^{p} V_k^* S V_k$. Choose $S = \mathbf{x}\mathbf{x}^*$, where $\mathbf{x}$ is a column-vector representation of a complex m-tuple such that $\mathbf{x}^*\mathbf{x} = 1$, but could be arbitrary otherwise. Then,

$$L(\mathbf{x}\mathbf{x}^*) = I - \mathbf{x}\mathbf{x}^* \text{ and}$$

$$\mathbf{x}\mathbf{x}^* L(\mathbf{x}\mathbf{x}^*) = \sum_k \mathbf{x}\mathbf{x}^* V_k^* \mathbf{x}\mathbf{x}^* V_k = 0,$$

$$\text{tr} \sum_k \mathbf{x}\mathbf{x}^* V_k^* \mathbf{x}\mathbf{x}^* V_k = \sum_k \mathbf{x}^* V_k^* \mathbf{x}\mathbf{x}^* V_k \mathbf{x} = 0$$

$$\Rightarrow \mathbf{x}^* V_k \mathbf{x} = 0, \forall k \quad \textbf{Contradiction}.$$

The feasibility or infeasibility of sum-of-squares representation of distinguished classes of forms can be linked to the study of linear transformations which map square matrices into square matrices and preserve hermitian symmetry and positivity.

For compactness of notation, we denote $\mathbb{C}^{n\times n}$ by M_n, the algebra of $n \times n$ complex matrices. $E_{jk} \in M_n$ is the $n\times n$ matrix with 1 at (j,k)-component and zeros elsewhere. $M_n(M_m) = M_m \otimes M_n$ is the collection of all $n\times n$ block matrices with $m\times m$ matrices as entries. Clearly, $M_n(M_m)$ belongs to $\mathbb{C}^{mn\times mn}$.

Definition 5.2 $\phi : M_n \to M_m$ is positive semidefinite (called, positive, for the sake of brevity in the present context) if ϕ maps the set of positive semidefinite matrices into itself (similarly $\phi : M_n \to M_p$). ϕ is p-positive definite if $\phi \otimes I_p$ is positive on $M_m \otimes M_p$, where $\phi \otimes I_p : M_p(M_n) \to M_p(M_m)$ is given by

$$\phi \otimes I_p((A_{jk}))_{1\leq j,k\leq p} = \phi((A_{jk}))_{1\leq j,k\leq p}.$$

ϕ is completely positive if ϕ is p-positive for every p, i.e. $\phi \otimes I_p$ is positive for all integers p.

Example 5.3 For each $n \times m$ matrix V the map $L = \phi : M_n \to M_m$ with $A \in M_n \to V^*AV \in M_m$ is completely positive [32].

The results below are proved in [13].

Fact 5.4 Let $\phi : M_n \to M_m$. Then ϕ is completely positive iff ϕ is of the form $\sum_i V_i^* A V_i$ for all $A \in M_n$, where V_i are $n \times m$ matrices.

An equivalent result is:

Let $\phi : M_n \to M_m$. Then ϕ is completely positive if $(\phi(E_{jk}))_{1\leq j,k\leq n}$ is positive.

Comment. Each linear map $\phi : M_n \to M_n$ is determined by its values on $E_{jk} \in M_n, 1 \leq j,k \leq n$. Hence ϕ is completely determined by the single element

$$(\phi(E_{jk}))_{1\leq j,k\leq n} \in M_n(M_m).$$

Completely positive maps and not positive maps provide the answer to SOS representation.

6 Conclusions

Several problems in multidimensional systems theory require tools from complex algebraic geometry (zero sets of multivariate polynomials in the complex polydomain, stabilization etc) as well as real algebraic geometry which is the study of real number solutions to algebraic equations with real number coefficients. Real algebraic geometry usage in multidimensional systems occurs when considering, for example, either a set of multivariate polynomial equations and inequalities in robust stability assessment of uncertain linear and nonlinear control systems or optimization of a multivariate polynomial cost function subject to equality and nonequality constraints. Such problems may also be solved by applying the Karush-Kuhn-Tucker conditions (a generalization of the method of Lagrange multipliers) in nonlinear programming or the method of cylindrical algebraic decomposition in elementary decision algebra [4]. The theory of Gröbner bases provide a viable alternative and is being increasingly explored for suitability in solving an increasing number of problems in multidimensional systems and signal processing. This paper provides a snapshot of the progress made to date.

A resurgence of interest in the implementation of algorithmic algebra algorithms based on elementary decision algebra took place after the emergence of software like QEPCAD (Quantifier Elimination by Partial Cylindrical Algebraic Decomposition). The problems tackled include the output stabilization problem, the simultaneous stabilization problem, the robust multiobjective problem, for which no general solutions exist. However, software packages like QEPCAD were found to be suitable only for tackling problems of modest complexity. Almost the same appears to be true, in general, for obtaining solutions to problems using software for generating Gröbner bases but the possibility for use over a variety of different coefficient domains, and for different variable and term orderings offers considerable flexibility. The time and memory required to calculate a Gröbner basis depend very much on the variable ordering, term ordering, and on which variables are regarded as constants.

Though the sum of squares decomposition is a sufficient but not necessary condition for multivariate polynomial nonnegativity, its bijective correspondence with completely positive linear maps that map square matrices to square matrices, preserving hermitian symmetry and positivity uncover the additional constraint on positive linear maps for such decompositions to materialize. A representation theorem for bihermitian forms is stated but not proved here. The proof will appear elsewhere and is currently available from the author. The linear maps associated with the two matrices in the biquadratic (bihermitian) form representation are required to be completely positive for the form to be representable as a sum of squares of bilinear forms.

Ongoing research efforts that are promising include the work on parametrized compactly supported wavelets with several vanishing moments [29], where a single parameter and lower order filters were considered because of the complexity of the expressions in the Gröbner basis. That was the case also in other applications like [8] and [9]. Nevertheless, the theory of Gröbner basis is effective in the study of parametrized systems in both signal processing and control. In fact the two parameter counterpart of the results in [29] is under study and the results will be reported later.

Acknowledgements. The author thanks Kamlesh Nahata for working out the three examples on constrained optimization and performance analysis of nonlinear systems using Gröbner bases. Thanks are also due to Nilesh Ahuja for his ready help with the computer processing of certain portions of this documentation. My special thanks go to the reviewer for his very careful reading of the paper and his valuable comments for improving the clarity and completeness of the presentation. Also, thanks are due to Dr.

Zhiping Lin and Kay Dimler for pointing out some additional typographical errors in the revised version.

Bibliography

[1] W. W. Adams and P. Loustanou, *An Introduction to Gröbner Bases*, vol. 3, American Mathematical Society, Providence, R.I., 1994.

[2] B. D. O. Anderson, N. K. Bose, and E. I. Jury, *Output feedback stabilization and related problems-solutions via decision methods*, IEEE Transactions on Automatic Control 20 (1975), pp. 53–56.

[3] D. Asimov, *There's no space like home*, The Sciences 35 (1995), pp. 20–25.

[4] N. K. Bose, *Applied Multidimensional Systems Theory*, Van Nostrand Reinhold Co., N.Y., 1982.

[5] N. K. Bose (ed.), *Multidimensional Systems Theory: Progress, Directions and Open Problems in Multidimensional Systems*, D. Reidel Publishing Company, Dordrecht, Holland, 1985.

[6] N. K. Bose, B. Buchberger, and J. P. Guiver, *Multidimensional Systems Theory and Applications*, Kluwer Academic Publishers, Dordrecht, The Netherlands, 2003.

[7] N. K. Bose and C. Charoenlarpnopparut, *Multivariate matrix factorization: new results*, Mathematical Theory of Networks and Systems, Proceedings of the MTNS-98 Symposium (Padova, Italy) (A. Beghi, L. Finesso, and G. Picci, eds.), Il Poligrafo, 1998, pp. 97–100.

[8] ______, *Minimax controller design using rate feedback*, Circuits, Systems and Signal Processing 18 (1999), pp. 17–25.

[9] ______, *Minimax controller using rate feedback: latest results*, Proceedings of the International Federation Automatic Control G-2e-20-4 (1999), pp 525–530.

[10] B. Buchberger, *An Algorithm for Finding a Basis for the Residue Class Ring of a Zero-dimensional Polynomial Ideal (German)*, Ph.D. thesis, Univ. Of Innsbruck (Austria), Math. Inst., 1965.

[11] B. Buchberger and F. Winkler (eds.), *Gröbner Bases and Applications*, London Mathematical Society Lecture Notes Series, vol. 251, Cambridge, Cambridge University Press, 1998, Proceedings of the International Conference "33 Years of Gröbner Bases".

[12] C. Charoenlarpnopparut and N. K. Bose, *Multidimensional FIR filter bank design using Gröbner Bases*, IEEE Transactions on Circuits and Systems II 46 (1999), pp. 1475–1486.

[13] M. D. Choi, *Completely positive linear maps on complex matrices*, Linear Algebra and its Applications 10 (1975), pp. 285–290.

[14] D. A. Cox, J. Little, and D. O'Shea, *Using Algebraic Geometry*, Springer, New York, 1998.

[15] C. N. Delzell, *A constructive, continuous solution to Hilbert's 17th problem, and other results in semi-algebraic geometry*, Ph.D. thesis, Stanford University, Math. Dept., 1980.

[16] S. V. Dushin and S. V. Chmutov, *Gaydar's formula for the greatest common divisor of several polynomials*, Communications of the Moscow Mathematical Society (1993), pp. 171–172.

[17] K. Forsman, *Optimization, Stability and Cylindrical Decomposition*, Tech. report, Linköping University, Linköping, Sweden, 1993.

[18] K. Forsman and S. Glad, *Constructive algebraic geometry in nonlinear control*, Proceedings of the 29th IEEE Conference on Decision and Control (Honolulu), vol. 5, 1990, pp. 2825–2827.

[19] J. P. Guiver and N. K. Bose, *Polynomial matrix primitive factorization over arbitrary coefficient field and related results*, IEEE Transactions on Circuits and Systems 29 (1982), pp. 649–657.

[20] D. Henrion and A. Garulli (eds.), *Positive Polynomials in Control*, Springer-Verlag Berlin Heidelberg, 2005.

[21] T. Y. Lam, *Serre's Problem on Projective Modules*, Springer-Verlag Berlin Heidelberg, 2006.

[22] Z. Lin, L. Xu, and Q. Wu, *Applications of Gröbner bases to signal and image processing: a survey*, Linear Algebra and its Applications 391 (2004), pp. 169–202.

[23] D. G. Luenberger, *Linear and Nonlinear Programming, 2nd. ed.*, Addison-Wesley Publishing Company, Reading, MA, 1984.

[24] U. Oberst, *Multidimensional constant linear systems*, Acta Applicandae Mathematicae (1990), pp. 1–175.

[25] P. A. Parrilo, *Semidefinite programming relaxations for semialgebraic problems*, Mathematical Programming Ser. B 96 (2003), pp. 293–320.

[26] ______, *Exploiting Algebraic Structure in Sum of Squares Programs*, Positive polynomials in control (D. Henrion and A. Garulli, eds.), Springer, 2005, pp. 181–194.

[27] J. de Pillis, *Linear transformations which preserve hermitian and positive semidefinite transformations*, Pacific Journal of Mathematics (1967), pp. 129–137.

[28] J. L. Rabinowitsch, *Zum Hilbertschen Nullstellensatz*, Mathematische Annalen 102 (1929), pp. 520–521.

[29] G. Regensburger and O. Scherzer, *Symbolic Computation for Moments and Filter Coeffcients of Scaling Functions*, Annals of Combinatorics 9 (2005), pp. 223–243.

[30] H. H. Rosenbrock, *State-Space and Multivariable Theory*, John Wiley & Sons Inc., New York, NY, 1970.

[31] N. Z. Shor, *Class of global minimum bounds of polynomial functions*, Cybernetics 23 (1987), pp. 731–734.

[32] W. Stinespring, *Positive functions on $\mathbb{C}^*$-algebras*, Proceedings of the American Mathematical Society 6 (1955), pp. 211–216.

[33] M. Wang and C. P. Kwong, *On multivariate polynomial matrix factorization problem*, Mathematics of Control, Signals and Systems 17 (2005), pp. 297–311.

[34] L. Xu, O. Saito, and K. Abe, *Output feedback stabilizability and stabilization algorithms for 2-D systems*, Multidimensional Systems and Signal Processing 5 (1994), pp. 41–60.

[35] J. Q. Ying, *On the strong stabilizability of MIMO n-dimensional linear systems*, SIAM Journal on Control and Optimization 38 (2000), pp. 313–335.

[36] J. Q. Ying, L. Xu, and Z. Lin, *A computational method for determining strong stabilizability of n-D systems*, Journal of Symbolic Computation 27 (1999), pp. 479–499.

[37] E. Zerz, *Topics in Multidimensional Linear Systems Theory*, Lecture Notes in Control and Information Sciences 256, Springer-Verlag London Limited, London, Great Britain, 2000.

Author information

N. K. Bose, Department of Electrical Engineering,
Pennsylvania State University,
University Park, PA 16203, USA.
Email: `nkb1@psu.edu`

Radon Series Comp. Appl. Math **3**, 23–106

Applications of the Quillen-Suslin theorem to multidimensional systems theory

Anna Fabiańska and Alban Quadrat

Key words. Constructive versions of the Quillen-Suslin theorem, Lin-Bose conjectures, multidimensional linear systems, flat systems, (weakly) doubly coprime factorizations of rational transfer matrices, decomposition of linear functional systems, symbolic computation.

AMS classification. 13C10, 13P99, 93B40, 47A65, 93B40, 93C23

1 Introduction

In 1784, Monge studied the integration of certain *underdetermined* nonlinear systems of ordinary differential equations, namely, systems containing more unknown functions than differentially independent equations ([29]). He showed how the solutions of these systems could be parametrized by means of a certain number of arbitrary functions of the independent variable. This problem was called *the Monge problem* and it was studied by famous mathematicians such as Hadamard, Hilbert, Cartan and Goursat. In particular, motivated by problems coming from linear elasticity theory, Hadamard considered the case of linear ordinary differential equations and Goursat investigated underdetermined systems of partial differential equations. We refer the reader to [29] for a historical account on the Monge problem and for the main references.

Within the *algebraic analysis* approach ([2, 21, 28, 33]), the Monge problem was recently studied for underdetermined systems of linear partial differential equations in [21, 33, 42, 43, 44] and for linear functional systems in [5, 6] (e. g., differential time-delay systems, discrete systems). Depending on the algebraic properties of a certain module M defined over a ring D of functional operators and intrinsically associated with the linear functional system, we can prove or disprove the existence of different kinds of parametrizations of the system (i. e., minimal parametrizations, non-minimal parametrizations, chains of successive parametrizations). Constructive algorithms for checking these algebraic properties (i. e., torsion, existence of torsion elements, torsion-free, reflexive, projective, stably free, free) and computing the different parametrizations were recently developed in [5, 42, 43, 44], implemented in the package OREMODULES ([5, 6]) and illustrated on numerous examples coming from mathematical physics and control theory ([5, 6]). Finally, we proved in [5, 42, 43, 44] how the Monge problem gave answers for the search of *potentials* in mathematical physics and *image representations* in control theory ([39, 40, 62, 63]).

The last results show that the Monge problem is constructively solved for certain classes of linear functional systems up to a last but important point: we can check whether or not a linear functional system admits injective parametrizations but we are generally not able to compute one even if some heuristic methods were presented in [5, 42, 43]. Indeed, the existence of injective parametrizations for a linear functional system was proved to be equivalent to the freeness of the corresponding module M. In the case of a linear functional system with constant coefficients, the corresponding ring D of functional operators is a commutative polynomial ring over a field k of constants. Using the famous Quillen-Suslin theorem ([52, 55]), also known as Serre's conjecture ([23, 24]), we then know that free D-modules are projective ones. Using Gröbner or Janet bases ([5, 10, 42]), we can check whether or not a module over a commutative polynomial ring is projective. See [3, 10, 20] and the references therein for introductions to Janet and Gröbner bases. Hence, we can constructively prove the existence of an injective parametrization for a linear functional system. However, we need to use a constructive version of the Quillen-Suslin theorem ([15, 19, 22, 26, 27, 35, 58, 59]) to get injective parametrizations of the corresponding system.

The main purpose of this paper is to recall a general algorithm for computing bases of a free module over a commutative polynomial ring, give four new applications of the Quillen-Suslin theorem to mathematical systems theory and demonstrate the implementation of the QUILLENSUSLIN package ([12]) developed in the computer algebra system MAPLE. To our knowledge, the QUILLENSUSLIN package is the first package available which performs bases computations of free modules over a commutative polynomial ring with rational and integer coefficients and is dedicated to different applications coming from the mathematical systems theory.

More precisely, the plan of the paper is the following one. In the second section, we recall how the structural properties of linear functional systems can be constructively studied within the algebraic analysis approach as well as different results on the Monge problem. A constructive version of the Quillen-Suslin theorem, which is the main tool we use in the paper, is presented in the third section and the implementation is illustrated on many examples in the appendix of the paper. We also describe some heuristic methods that highly simplify the computation of a basis of a free module over polynomial rings in certain special cases. The constructive version of the Quillen-Suslin theorem and, in particular the patching procedure, gives us the opportunity to make a new observation concerning linear functional systems which admit injective parametrizations also called *flat multidimensional systems* in mathematical systems theory. In the fourth section, we prove that a flat multidimensional system is algebraically equivalent to a 1-D flat linear system obtained by setting all but one functional operator to zero in the system matrix. This result gives an answer to a natural question on flat multidimensional systems. In particular, we prove that every flat differential time-delay system is algebraically equivalent to the differential system without delays, namely, the system obtained by setting to zero all the time-delay amplitudes. In the fifth section, we consider a generalization of Serre's conjecture. We recall that Serre's conjecture, also known as the Quillen-Suslin theorem, can be expressed in the language of matrices as follows: every matrix R over a commutative polynomial ring $D = k[x_1, \ldots, x_n]$ whose maximal minors generate D (*unimodular matrix*) can be completed to a square invertible matrix over D (i. e., its determinant is a non-zero element of the field k). The generalization, stated by Lin and Bose in [25] and first proved by Pommaret in [41] by means of algebraic analysis, can be formulated as the possibility of completing a matrix R whose maximal minors divided by their greatest common divisor d generate D to a square polynomial matrix whose determinant equals d. Serre's conjecture is then the special case where $d = 1$. Using the Quillen-Suslin theorem, we give a constructive algorithm for computing such a completion. Using the possibility of computing bases of a free module in our implementation QUILLENSUSLIN, this algorithm has been implemented in this package. In the sixth section, we study the existence of (weakly) left-/right-coprime factorizations of rational transfer matrices using recent results developed in [48]. We give algorithms for computing such factorizations using the constructive versions of the Quillen-Suslin theorem. These results constructively solve open questions in the literature of multidimensional linear systems (see [60, 61] and the references therein). Finally, we show that the constructive Quillen-Suslin theorem also plays an important role in the decomposition problem of linear functional systems studied in the literature of symbolic computation. See [8] and the references therein for more details. The main idea is to transform the system matrix into an equivalent block-triangular or a block-diagonal form ([8, 9]).

The different algorithms presented in the paper have been implemented in the package QUILLENSUSLIN based on the library INVOLUTIVE ([3]) (an OREMODULES ([6]) version will soon be available). The appendix illustrates the main procedures of the QUILLENSUSLIN package on different examples taken from the literature ([19, 22, 37, 58]). The package QUILLENSUSLIN also contains a completion algorithm for unimodular matrices over Laurent polynomial rings described in [34, 37]. See also [1] for a recent algorithm. In [37], Park explains the importance and the meaning of the completion problem of unimodular matrices over Laurent polynomial rings to signal processing and gives an algorithm for translating this problem into a polynomial case. Park's results can also be used for computing flat outputs of *δ-flat* multidimensional linear systems ([30, 31]). See [5] for another constructive algorithm and [6] for illustrations on different explicit examples.

Notation. In what follows, we shall denote by k a field, $D = k[x_1, \ldots, x_n]$ a commutative polynomial ring with coefficients in k, $D^{1\times p}$ the D-module formed by the row vectors of length p with entries in D and $D^{q\times p}$ the set of $q \times p$-matrices with entries in D. $\mathcal{F}$ will always denote a D-module. We denote by R^T the transpose of the matrix R and by I_p the $p \times p$ identity matrix. Finally, the symbol $\triangleq$ means "by definition".

2 A module-theoretic approach to systems theory

Let $D = k[x_1, \dots, x_n]$ be a commutative polynomial ring over a field k and $R \in D^{q \times p}$. We recall that a matrix R is said to have *full row rank* if the *first syzygy module* of the D-module $D^{1 \times q} R$ formed by the D-linear combinations of the rows of R, namely,

$$\ker_D(.R) \triangleq \{\lambda \in D^{1 \times p} \mid \lambda R = 0\},$$

is reduced to 0. In other words, $\lambda R = 0$ implies $\lambda = 0$, i. e., the rows of R are D-linearly independent.

The following definitions of *primeness* are classical in systems theory.

Definition 2.1 [32, 59, 63] Let $D = \mathbb{R}[x_1, \dots, x_n]$ be a commutative polynomial ring, $R \in D^{q \times p}$ a full row rank matrix, J the ideal generated by the $q \times q$ minors of R and $V(J)$ the algebraic variety defined by

$$V(J) = \{\xi \in \mathbb{C}^n \mid P(\xi) = 0, \ \forall \ P \in J\}.$$

1. R is called *minor left-prime* if $\dim_{\mathbb{C}} V(J) \leq n - 2$, i. e., the greatest common divisor of the $q \times q$ minors of R is 1.
2. R is called *weakly zero left-prime* if $\dim_{\mathbb{C}} V(J) \leq 0$, i. e., the $q \times q$ minors of R may only vanish simultaneously in a finite number of points of $\mathbb{C}^n$.
3. R is called *zero left-prime* if $\dim_{\mathbb{C}} V(J) = -1$, i. e., the $q \times q$ minors of R do not vanish simultaneously in $\mathbb{C}^n$.

The previous classification plays an important role in multidimensional systems theory. See [32, 59, 63] and the references therein for more details.

The purpose of this section is twofold. We first recall how we can generalize the previous classification for general multidimensional linear systems, i. e., systems which are not necessarily defined by full row rank matrices. We also explain the duality existing between the *behavioural approach* to multidimensional systems ([32, 39, 62, 63]) and the *module-theoretic one* ([42, 43, 44]). See also [62] for a nice introduction.

In what follows, D will denote a commutative polynomial ring with coefficients in a field k. In particular, we shall be interested in commutative polynomial rings of functional operators such as partial differential operators, differential time-delay operators or shift operators. Let us consider a matrix $R \in D^{q \times p}$ and a D-module $\mathcal{F}$, namely

$$\forall \ f_1, f_2 \in \mathcal{F}, \quad \forall \ a_1, a_2 \in D : \quad a_1 f_1 + a_2 f_2 \in \mathcal{F}.$$

If we define the following D-morphism, namely, D-linear map,

$$\begin{array}{rcl} .R : D^{1 \times q} & \xrightarrow{.R} & D^{1 \times p}, \\ \lambda = (\lambda_1 \ \dots \ \lambda_q) & \longmapsto & (.R)(\lambda) = \lambda R, \end{array}$$

where $D^{1\times p}$ denotes the D-module of row vectors of length p with entries in D, then the cokernel of the D-morphism $.R$ is defined by

$$M = D^{1\times p}/(D^{1\times q}\, R).$$

The D-module M is said to be presented by R or simply *finitely presented* ([5, 53]). Moreover, we can also define the *system* or *behaviour* as follows:

$$\ker_{\mathcal{F}}(R.) \triangleq \{\eta \in \mathcal{F}^p \mid R\,\eta = 0\}.$$

As it was noticed by Malgrange in [28], the D-module M and the system $\ker_{\mathcal{F}}(R.)$ are closely related. As this relation will play an important role in what follows, we shall explain it in details. In order to do that, let us first introduce a few classical definitions of homological algebra. We refer the reader to [53] for more details.

Definition 2.2 1. A sequence $(M_i, d_i : M_i \longrightarrow M_{i-1})_{i\in\mathbb{Z}}$ of D-modules M_i and D-morphisms $d_i : M_i \longrightarrow M_{i-1}$ is *a complex* if we have

$$\forall\, i \in \mathbb{Z}, \quad \operatorname{im} d_i \subseteq \ker d_{i-1}.$$

We denote the previous complex by

$$\ldots \xrightarrow{d_{i+2}} M_{i+1} \xrightarrow{d_{i+1}} M_i \xrightarrow{d_i} M_{i-1} \xrightarrow{d_{i-1}} \ldots \tag{2.1}$$

2. The *defect of exactness of the complex (2.1) at M_i* is defined by

$$H(M_i) = \ker d_i / \operatorname{im} d_{i+1}.$$

3. The complex (2.1) is said to be *exact at M_i* if we have

$$H(M_i) = 0 \quad \Longleftrightarrow \quad \ker d_i = \operatorname{im} d_{i+1}.$$

4. The complex (2.1) is *exact* if

$$\forall\, i \in \mathbb{Z}, \quad \ker d_i = \operatorname{im} d_{i+1}.$$

5. The complex (2.1) is a *split exact sequence* if (2.1) is exact and if there exist D-morphisms $s_i : M_{i-1} \longrightarrow M_i$ satisfying the following conditions:

$$\forall\, i \in \mathbb{Z}, \quad \begin{cases} s_{i+1} \circ s_i = 0, \\ s_i \circ d_i + d_{i+1} \circ s_{i+1} = id_{M_i}. \end{cases}$$

6. A *finite free resolution* of a D-module M is an exact sequence of the form

$$0 \longrightarrow D^{1\times p_m} \xrightarrow{.R_m} \ldots \xrightarrow{.R_2} D^{1\times p_1} \xrightarrow{.R_1} D^{1\times p_0} \xrightarrow{\pi} M \longrightarrow 0, \tag{2.2}$$

where $p_i \in \mathbb{Z}_+ = \{0, 1, 2, \ldots\}$, $R_i \in D^{p_i \times p_{i-1}}$, and the D-morphism $.R_i$ is defined by

$$\begin{aligned} .R_i : D^{1\times p_i} &\longrightarrow D^{1\times p_{i-1}} \\ \lambda &\longmapsto (.R_i)(\lambda) = \lambda\, R_i. \end{aligned}$$

The next classical result of homological algebra will play a crucial role in what follows.

Theorem 2.3 [53] *Let $\mathcal{F}$ be a D-module, M a D-module and (2.2) a finite free resolution of M. Then, the defects of exactness of the following complex*

$$\ldots \xleftarrow{R_3.} \mathcal{F}^{p_2} \xleftarrow{R_2.} \mathcal{F}^{p_1} \xleftarrow{R_1.} \mathcal{F}^{p_0} \longleftarrow 0, \tag{2.3}$$

where the D-morphism $R_i. : \mathcal{F}^{p_{i-1}} \longrightarrow \mathcal{F}^{p_i}$ is defined by

$$\forall \eta \in \mathcal{F}^{p_{i-1}}, \quad (R_i.)(\eta) = R_i\, \eta,$$

only depend on the D-modules M and $\mathcal{F}$. Up to an isomorphism, we denote these defects of exactness by

$$\begin{cases} \operatorname{ext}_D^0(M, \mathcal{F}) \cong \ker_{\mathcal{F}}(R_1.), \\ \operatorname{ext}_D^i(M, \mathcal{F}) \cong \ker_{\mathcal{F}}(R_{i+1}.)/(R_i\, \mathcal{F}^{p_i}), \quad i \geq 1. \end{cases}$$

Finally, we have $\operatorname{ext}_D^0(M, \mathcal{F}) = \hom_D(M, \mathcal{F})$, *where* $\hom_D(M, \mathcal{F})$ *denotes the D-module of D-morphisms from M to $\mathcal{F}$.*

We refer the reader to Example 5.3 for explicit computations of $\operatorname{ext}_D^i(N, D)$, $i \geq 0$.

Coming back to the D-module M, we have the following beginning of a finite free resolution of M:

$$\begin{array}{ccccc} D^{1\times q} & \xrightarrow{.R} & D^{1\times p} & \xrightarrow{\pi} & M \longrightarrow 0, \\ \lambda & \longmapsto & \lambda R & & \end{array} \tag{2.4}$$

where π denotes the D-morphism which sends elements of $D^{1\times p}$ to their residue classes in M. If we "apply the left-exact contravariant functor" $\hom_D(\cdot, \mathcal{F})$ to (2.4) (see [53] for more details), by Theorem 2.3, we obtain the following exact sequence:

$$\begin{array}{ccccc} \mathcal{F}^q & \xleftarrow{R.} & \mathcal{F}^p & \longleftarrow & \hom_D(M, \mathcal{F}) \longleftarrow 0. \\ R\eta & \longleftarrow\!\shortmid & \eta & & \end{array}$$

This implies the following important isomorphism ([28]):

$$\ker_{\mathcal{F}}(R.) = \{\eta \in \mathcal{F}^p \mid R\,\eta = 0\} \cong \hom_D(M, \mathcal{F}). \tag{2.5}$$

For more details, see [5, 28, 32, 44, 62] and the references therein. In particular, (2.5) gives an intrinsic characterization of the $\mathcal{F}$-solutions of the system $\ker_{\mathcal{F}}(R.)$. It only depends on two mathematical objects:

1. The finitely presented D-module M which algebraically represents the linear functional system.
2. The D-module $\mathcal{F}$ which represents the "functional space" where we seek the solutions of the system.

If D is now a ring of functional operators (e. g., differential operators, time-delay operators, difference operators), then the issue of understanding which $\mathcal{F}$ is suitable for a particular linear system has long been studied in functional analysis and is still nowadays a very active subject of research. It does not seem that constructive algebra and symbolic computation can propose new methods to handle this functional analysis problem. However, they are very useful for classifying $\hom_D(M, \mathcal{F})$ by means of the algebraic properties of the D-module M. Indeed, a large classification of the properties of modules is developed in module theory and homological algebra. See [53] for more information. Let us recall a few of them.

Definition 2.4 [53] Let D be a commutative polynomial ring with coefficients in a field k and M a finitely presented D-module. Then, we have

1. M is said to be *free* if it is isomorphic to $D^{1\times r}$ for a non-negative integer r, i. e.,
$$M \cong D^{1\times r}, \quad r \in \mathbb{Z}_+ = \{0, 1, 2 \ldots\}.$$
2. M is said to be *stably free* if there exist two non-negative integers r and s such that
$$M \oplus D^{1\times s} \cong D^{1\times r}.$$
3. M is said to be *projective* if there exist a D-module P and non-negative integer r such that
$$M \oplus P \cong D^{1\times r}.$$
4. M is said to be *reflexive* if the canonical map
$$\varepsilon_M : M \longrightarrow \hom_D(\hom_D(M, D), D),$$
defined by
$$\forall\, m \in M, \forall\, f \in \hom_D(M, D) : \; \varepsilon_M(m)(f) = f(m),$$
is an isomorphism, where $\hom_D(M, D)$ denotes the D-module of D-morphisms from M to D.
5. M is said to be *torsion-free* if the submodule of M defined by
$$t(M) = \{m \in M \mid \exists\, 0 \neq P \in D : \; P\, m = 0\}$$
is reduced to the zero module. $t(M)$ is called the *torsion submodule* of M and the elements of $t(M)$ are the *torsion elements* of M.
6. M is said to be *torsion* if $t(M) = M$, i. e., every element of M is a torsion element.

Let $K = Q(D) = k(x_1, \ldots, x_n)$ be the *quotient field* of D ([53]) and M a finitely presented D-module. We call the *rank* of M over D, denoted by $\text{rank}_D(M)$, the dimension of the K-vector space $K \otimes_D M$ obtained by extending the scalars of M from D to K, i. e.,
$$\text{rank}_D(M) = \dim_K(K \otimes_D M).$$

We can check that if M is a torsion D-module, we then have $K \otimes_D M = 0$, a fact implying $\mathrm{rank}_D(M) = 0$. See [53] for more details.

Let us recall a few results about the notions previously introduced in Definition 2.4.

Theorem 2.5 [53] *Let $D = k[x_1, \ldots, x_n]$ be a commutative polynomial ring with coefficients in a field k. We have the following results:*

1. *We have the implications among the previous concepts:*
$$\textit{free} \Longrightarrow \textit{stably free} \Longrightarrow \textit{projective} \Longrightarrow \textit{reflexive} \Longrightarrow \textit{torsion-free}.$$
2. *If $D = k[x_1]$, then D is a* principal ideal domain *– namely, every ideal of D is* principal, *i. e., it can be generated by one element of D – and every finitely generated torsion-free D-module is free.*
3. (Serre theorem [10]) *Every projective module over D is stably free.*
4. (Quillen-Suslin theorem [52, 55]) *Every projective module over D is free.*

The famous Quillen-Suslin theorem will play an important role in what follows. We refer to [23, 24] for the best introductions nowadays available on this subject.

The next theorem gives some characterizations of the definitions given in Definition 2.4.

Theorem 2.6 [5, 33, 44] *Let $D = k[x_1, \ldots, x_n]$ be a commutative polynomial ring over a field k, $R \in D^{q \times p}$ and the finitely presented D-modules*
$$M = D^{1\times p}/(D^{1\times q}\, R), \quad N = D^{1\times q}/(D^{1\times p}\, R^T).$$
We then have the equivalences between the first two columns of Figure 2.1.

Combining the results of Theorem 2.6 and the Quillen-Suslin theorem (see 4 of Theorem 2.5), we then obtain a way to check whether or not a finitely presented D-module M has some torsion elements or is torsion-free, reflexive, projective, stably free or free. We point out that the explicit computation of $\mathrm{ext}^i_D(N, D)$ can always be done using Gröbner or Janet bases. See [5, 42, 43] for more details and for the description of the corresponding algorithms. We also refer the reader to [4, 6] for the library OREMODULES in which the different algorithms were implemented as well as to the large library of examples of OREMODULES which illustrates them. Finally, see also [3, 10, 20] and the references therein for an introduction to Gröbner and Janet bases.

Remark 2.7 The D-module $N = D^{1\times q}/(D^{1\times p}\, R^T)$ is called the *transposed module* of $M = D^{1\times p}/(D^{1\times q}\, R)$ even if N depends on M only up to a *projective equivalence* ([45]), namely, if $M = D^{1\times r}/(D^{1\times s}\, R')$ and $N' = D^{1\times s}/(D^{1\times r}\, R'^T)$, then there exist two projective D-modules P and P' such that $N \oplus P \cong N' \oplus P'$ ([53]). However, for every D-module $\mathcal{F}$, we have $\mathrm{ext}^i_D(N \oplus P, \mathcal{F}) \cong \mathrm{ext}^i_D(N, \mathcal{F}) \oplus \mathrm{ext}^i_D(P, \mathcal{F})$ and, for $i \geq 1$, $\mathrm{ext}^i_D(P, \mathcal{F}) = 0$ because P is a projective D-module ([53]). Hence, we then get $\mathrm{ext}^i_D(N, \mathcal{F}) \cong \mathrm{ext}^i_D(N', \mathcal{F})$, for $i \geq 1$. Hence, the results of Theorem 2.6 do not depend on the choice of a presentation of M, i. e., on R. In what follows, we shall sometimes denote N by $T(M)$.

In order to explain why the definitions given in Definition 2.4 extend the concepts of primeness defined in Definition 2.1, we first need to introduce some more definitions ([2]).

Definition 2.8 1. If M is a non-zero finitely presented D-module, then the *grade* $j_D(M)$ of M is defined by

$$j_D(M) = \min \{i \geq 0 \mid \mathrm{ext}^i_D(M, D) \neq 0\}.$$

2. If M is a non-zero finitely presented D-module, the *dimension* $\dim_D(M)$ of M is defined by

$$\dim_D(M) = \mathrm{Kdim}(D/\sqrt{\mathrm{ann}_D(M)}),$$

where Kdim denotes the *Krull dimension* ([53]) and

$$\mathrm{ann}_D(M) = \{a \in D \mid a\, M = 0\},$$
$$\sqrt{\mathrm{ann}_D(M)} = \{a \in D \mid \exists\, l \in \mathbb{Z}_+ :\ a^l\, M = 0\}.$$

We are now in position to state an important result.

Theorem 2.9 [2, 33] *If M is a non-zero finitely presented $D = k[x_1, \ldots, x_n]$-module, where k is a field containing $\mathbb{Q}$, we then have*

$$j_D(M) + \dim_D(M) = n.$$

Let us suppose that R has full row rank and let us consider the finitely presented D-module $M = D^{1\times p}/(D^{1\times q}\, R)$. Using the notations of Definition 2.1 and the fact that

$$\dim_D(N) = \dim_{\mathbb{C}} V(J),$$

where $N = T(M) = D^{1\times q}/(D^{1\times p}\, R^T)$ is then a torsion D-module, i. e., it satisfies $\mathrm{ext}^0_D(M, D) = \hom_D(M, D) = 0$, by Theorem 2.9, we then obtain

$$j_D(N) = n - \dim_{\mathbb{C}} V(J) \geq 1.$$

Hence, by Theorems 2.6 and 2.9, we obtain that R is minor left-prime (resp., zero left-prime) iff the D-module M is torsion-free (resp., projective, i. e., free by the Quillen-Suslin theorem stated in 4 of Theorem 2.5). See [44] for more details and the extension of these results to the case of non-commutative rings of differential operators.

We finally obtain the table given in Figure 2.1 which sums up the different results previously obtained. We note that the last two columns of this table only hold when the matrix R has full row rank.

To finish, we explain what the system interpretations of the definitions given in Definition 2.4 are. In particular, these interpretations solve the Monge problem stated in the introduction of the paper. In order to do that, we also need to introduce a few more definitions (see, e. g., [53]).

Module M	$\mathbf{ext}^i_D(N, D)$	$\dim_D(N)$	**Primeness**
With torsion	$t(M) \cong \text{ext}^1_D(N, D)$	$n-1$	$\emptyset$
Torsion-free	$\text{ext}^1_D(N, D) = 0$	$n-2$	Minor left-prime
Reflexive	$\text{ext}^i_D(N, D) = 0$, $i = 1, 2$	$n-3$	
…	…	…	…
…	$\text{ext}^i_D(N, D) = 0$, $1 \leq i \leq n-1$	0	Weakly zero left-prime
Projective	$\text{ext}^i_D(N, D) = 0$, $1 \leq i \leq n$	-1	Zero left-prime

Figure 2.1 Classification of some module properties

Definition 2.10 1. A D-module $\mathcal{F}$ is called *injective* if, for every D-module M, and, for all $i \geq 1$, we have $\text{ext}^i_D(M, \mathcal{F}) = 0$.

2. A D-module $\mathcal{F}$ is called *cogenerator* if, for every D-module M, we have

$$\hom_D(M, \mathcal{F}) = 0 \quad \Longrightarrow \quad M = 0.$$

Roughly speaking, an injective cogenerator is a space rich enough for seeking solutions of linear systems of the form $R\eta = 0$, where $R \in D^{q \times p}$ is any matrix and $\eta \in \mathcal{F}^p$. In particular, using (2.5), if $\mathcal{F}$ is a cogenerator D-module and $M \neq 0$, then $\hom_D(M, \mathcal{F}) \neq 0$, meaning that the corresponding system $\ker_{\mathcal{F}}(R.)$ is not trivial. Finally, if $\mathcal{F}$ is an injective cogenerator D-module, then we can prove that any complex of the form (2.3) is exact at $\mathcal{F}^{p_i}$, $i \geq 1$, if and only if the corresponding complex (2.2) is exact. See [32, 39, 62] and the references therein for more details.

The following result proves that there always exists an injective cogenerator.

Theorem 2.11 [53] *An injective cogenerator D-module $\mathcal{F}$ exists for every ring D.*

Let us give important examples of injective cogenerator modules.

Example 2.12 If Ω is an open convex subset of $\mathbb{R}^n$, then the space $C^\infty(\Omega)$ (resp., $\mathcal{D}'(\Omega)$) of smooth real functions (resp., real distributions) on Ω is an injective cogenerator module over the ring $\mathbb{R}[\partial_1, \ldots, \partial_n]$ of differential operators with coefficients in $\mathbb{R}$, where we have denoted by $\partial_i = \partial / \partial x_i$ ([32, 28, 39]).

Example 2.13 Let k be a field, $\mathcal{F} = k^{\mathbb{Z}^n_+}$ be the set of sequences with values in k and $D = k[x_1, \ldots, x_n]$ be the ring of *shift operators*, namely,

$$\forall\, f \in \mathcal{F},\ i = 1, \ldots, n, \quad (x_i\, f)(\mu) = f(\mu + 1_i),$$

where $\mu = (\mu_1, \ldots, \mu_n) \in \mathbb{Z}^n_+$ and $\mu + 1_i = (\mu_1, \ldots, \mu_{i-1}, \mu_i + 1, \mu_{i+1}, \ldots, \mu_n)$. Then, $\mathcal{F}$ is an injective D-module ([32, 62]).

We have the following important corollary of Theorem 2.6 which solves the Monge problem in the case of linear functional systems with constant coefficients. See [64] and the references therein and the introduction of the paper.

Corollary 2.14 [5, 42] *Let $\mathcal{F}$ be an injective cogenerator $D = k[x_1, \ldots, x_n]$-module, $R \in D^{q \times p}$ and $M = D^{1 \times p}/(D^{1 \times q}\, R)$. Then, we have the following results:*

1. *There exists $Q_1 \in D^{q_1 \times q_2}$, where $p = q_1$, such that we have the exact sequence*

$$\mathcal{F}^q \xleftarrow{R.} \mathcal{F}^{q_1} \xleftarrow{Q_1.} \mathcal{F}^{q_2},$$

i. e., $\ker_{\mathcal{F}}(R.) = Q_1\, \mathcal{F}^{q_2}$, iff the D-module M is torsion-free.

2. *There exist $Q_1 \in D^{q_1 \times q_2}$ and $Q_2 \in D^{q_2 \times q_3}$ such that we have the exact sequence*

$$\mathcal{F}^q \xleftarrow{R.} \mathcal{F}^{q_1} \xleftarrow{Q_1.} \mathcal{F}^{q_2} \xleftarrow{Q_2.} \mathcal{F}^{q_3},$$

i. e., $\ker_{\mathcal{F}}(R.) = Q_1\, \mathcal{F}^{q_2}$ and $\ker_{\mathcal{F}}(Q_1.) = Q_2\, \mathcal{F}^{q_3}$, iff the D-module M is reflexive.

3. *There exists a chain of n successive parametrizations, namely, for $i = 1, \ldots, n$, there exist $Q_i \in D^{q_i \times q_{i+1}}$ such that we have the following exact sequence*

$$\mathcal{F}^{q} \xleftarrow{R.} \mathcal{F}^{q_1} \xleftarrow{Q_1.} \ldots \xleftarrow{Q_{n-1}.} \mathcal{F}^{q_n} \xleftarrow{Q_n.} \mathcal{F}^{q_{n+1}},$$

i. e., $\ker_{\mathcal{F}}(R.) = Q_1\, \mathcal{F}^{q_2}$ *and* $\ker_{\mathcal{F}}(Q_i.) = Q_{i+1}\, \mathcal{F}^{q_{i+1}}$, $i = 1, \ldots, n-1$, *iff the D-module M is projective.*

4. *There exist $Q \in D^{p \times m}$ and $T \in D^{m \times p}$ such that $T\,Q = I_m$ and the sequence*

$$\mathcal{F}^{q} \xleftarrow{R.} \mathcal{F}^{p} \xleftarrow{Q.} \mathcal{F}^{m} \longleftarrow 0, \tag{2.6}$$

is exact, i. e., $\ker_{\mathcal{F}}(R.) = Q\, \mathcal{F}^{m}$, *and iff the D-module M is free.*

We refer the reader to [5, 42, 43, 44, 49, 50] for the solutions of the Monge problem for different classes of linear functional systems with variables coefficients such as partial differential, differential time-delay or difference equations.

The matrices Q_i defined in Corollary 2.14 are called *parametrizations* ([5, 42, 43, 44]). Indeed, from 1 of Corollary 2.14, if M is torsion-free, then there exists a matrix of operators $Q_1 \in D^{q_1 \times q_2}$ which satisfies $\ker_{\mathcal{F}}(R.) = Q_1\, \mathcal{F}^{q_2}$. This means that every solution $\eta \in \mathcal{F}^p$ satisfying $R\,\eta = 0$ is of the form $\eta = Q_1\,\xi$ for a certain $\xi \in \mathcal{F}^{q_2}$. In the *behaviour approach* ([40]), the parametrization is called an *image representation* of $\ker_{\mathcal{F}}(R.)$ ([39, 62, 63]). We point out that the parametrizations Q_i are obtained by computing $\operatorname{ext}^i_D(N, D)$ (see Theorem 2.6). Hence, checking whether or not a D-module is torsion-free, reflexive or projective gives the corresponding successive parametrizations. We refer to [5, 42, 43, 44] for more details, the extension of the previous results to non-commutative algebras of functional operators and the implementation of the corresponding algorithms in the library OREMODULES. Finally, the matrix Q defined in 4 of Corollary 2.14 is called an *injective parametrization* of $\ker_{\mathcal{F}}(R.)$ as every $\mathcal{F}$-solution of $\ker_{\mathcal{F}}(R.)$ has the form $\eta = Q\,\xi$ for a certain $\xi \in \mathcal{F}^m$ and we have

$$\xi = (T\,Q)\,\xi = T\,\eta,$$

i. e., ξ is uniquely defined by $\eta \in \ker_{\mathcal{F}}(R.)$. At this stage, it is important to point out that no general algorithm has been developed to get injective parametrizations when the D-module M is free. It is the main purpose of this paper to constructively study this question and to apply the computation of injective parametrizations to some open questions appearing in mathematical systems theory.

Finally, we point out that, if M is a free D-module, then there always exist matrices $Q \in D^{p\times m}$ and $T \in D^{m\times p}$ such that, for every D-module $\mathcal{F}$, we have the exact sequence (2.6). Indeed, let us recall two standard arguments of homological algebra ([53]).

Proposition 2.15 *1. Let us consider the following short exact sequence:*

$$M' \xrightarrow{f} M \xrightarrow{g} M'' \longrightarrow 0.$$

If M'' is a projective D-module, then the previous exact sequence splits (see 5 of Definition 2.2).

2. Let $\mathcal{F}$ be a D-module. The functor $\hom_D(\cdot, \mathcal{F})$ *transforms split exact sequences of D-modules into split exact sequences of D-modules.*

By 1 of Proposition 2.15, we obtain that $D^{1\times q} \xrightarrow{.R} D^{1\times p} \xrightarrow{.Q} D^{1\times m} \longrightarrow 0$ is a splitting exact sequence and applying the functor $\hom_D(\cdot, \mathcal{F})$ to it, by 2 of Proposition 2.15, we obtain the splitting exact sequence (2.6). Hence, the assumption that $\mathcal{F}$ is an injective cogenerator D-module is only important for the converse implication of 6 of Corollary 2.14.

Explicit examples of computation of parametrizations can be found in [5, 6, 42, 43, 44] as well in the OREMODULES large library of examples ([4]). We refer the reader to these references and to section 4 for the computation of injective parametrizations. However, let us give a simple example in order to illustrate the previous results.

Example 2.16 Let us consider the ring $D = \mathbb{Q}[\partial_1, \partial_2, \partial_3]$ of differential operators with rational coefficients ($\partial_i = \partial/\partial x_i$), the matrix $R = (\partial_1\ \partial_2\ \partial_3)$ defining the so-called *divergent operator* in $\mathbb{R}^3$ and the finitely presented D-module $M = D^{1\times 3}/(D\,R)$. Let us check whether or not the D-module M has some torsion elements or is torsion-free, reflexive or projective, i. e., free by the Quillen-Suslin theorem. In order to do that, we define the D-module $N = D/(D^{1\times 3}\,R^T)$. A finite free resolution of N can easily be computed by means of Gröbner or Janet bases. We obtain the following exact sequence

$$0 \longrightarrow D \xrightarrow{.P_3} D^{1\times 3} \xrightarrow{.P_2} D^{1\times 3} \xrightarrow{.R^T} D \xrightarrow{\sigma} N \longrightarrow 0,$$

where σ denotes the canonical projection onto N and

$$P_2 = \begin{pmatrix} 0 & -\partial_3 & \partial_2 \\ \partial_3 & 0 & -\partial_1 \\ -\partial_2 & \partial_1 & 0 \end{pmatrix}, \quad P_3 = R.$$

We note that P_2 corresponds to the so-called *curl operator* whereas R^T is the *gradient operator*. Then, the defects of exactness of the following complex

$$0 \longleftarrow D \xleftarrow{.P_3^T} D^{1\times 3} \xleftarrow{.P_2^T} D^{1\times 3} \xleftarrow{.R} D \longleftarrow 0 \tag{2.7}$$

are defined by

$$\begin{cases} \mathrm{ext}^0_D(N,D) \cong \ker_D(.R), \\ \mathrm{ext}^1_D(N,D) \cong \ker_D(.P_2^T)/(D\,R), \\ \mathrm{ext}^2_D(N,D) \cong \ker_D(.P_3^T)/(D^{1\times 3}\,P_2^T), \\ \mathrm{ext}^3_D(N,D) \cong D/(D^{1\times 3}\,P_3^T). \end{cases}$$

Using the fact that R has full row rank, we obtain that $\mathrm{ext}^0_D(N,D) \cong \ker_D(.R) = 0$, which is equivalent to say that N is a torsion D-module. Now computing the syzygy modules $\ker_D(.P_2^T)$ and $\ker_D(.P_3^T)$ by means of Gröbner or Janet bases, we obtain that

$$\ker_D(.P_2^T) = D\,R, \quad \ker_D(.P_3^T) = D^{1\times 3}\,P_2^T,$$

which shows that $\mathrm{ext}^1_D(N,D) = \mathrm{ext}^2_D(N,D) = 0$. Finally, we can easily check that 1 does not belong to the ideal $I = D\,\partial_1 + D\,\partial_2 + D\,\partial_3$ of D, and thus, we have

$$\mathrm{ext}^3_D(N,D) \cong D/I \neq 0.$$

Using Theorem 2.6, we obtain that M is a reflexive but not a projective, i. e., not a free D-module. This last fact can also be checked as R has full row rank and the dimension $\dim_D(N)$ is 0 as the corresponding system is defined by the gradient operator, namely,

$$\partial_i\,y = 0, \quad i = 1, 2, 3,$$

whose solution is a constant, i. e., the solution of the system only depends on "a function of zero independent variables". Hence, by Theorem 2.9, we obtain that $j_D(N) = 3$, meaning that the first non-zero $\mathrm{ext}^i_D(N,D)$ has index 3. By Theorem 2.6, we then get that M is a reflexive D-module but not a projective one.

Finally, if we consider the D-module $\mathcal{F} = C^\infty(\Omega)$, where Ω is an open convex subset of $\mathbb{R}^3$, using Example 2.12, we obtain that $\mathcal{F}$ is an injective cogenerator D-module. Hence, if we apply the functor $\hom_D(\cdot, \mathcal{F})$ to the complex (2.7), we then obtain the following exact sequence:

$$\mathcal{F} \xrightarrow{P_3^T.} \mathcal{F}^3 \xrightarrow{P_2^T.} \mathcal{F}^3 \xrightarrow{R.} \mathcal{F} \longrightarrow 0.$$

We find again the classical results in mathematical physics that the smooth solutions on an open convex subset of $\mathbb{R}^3$ of the divergence operator are parametrized by the curl operator and the solutions of the curl operator are parametrized by the gradient operator.

The only point left open is to constructively compute injective parametrizations of linear functional systems defining free modules over a commutative polynomial ring D. Indeed, checking the vanishing of the $\mathrm{ext}^i_D(N,D)$, we generally obtain a successive chain of n parametrizations but not an injective one. In the case of linear systems of partial differential equations with polynomial or rational coefficients, we have recently solved this problem in [49, 50, 51] using a constructive proof of a famous result in non-commutative algebra due to Stafford. However, the same technique cannot be used if we want an injective parametrization Q of $\ker_{\mathcal{F}}(R.)$ to have only constant coefficients. The main purpose of this paper is to solve this problem using a constructive proof of the Quillen-Suslin theorem and to show some applications of this result to mathematical systems theory.

3 The Quillen-Suslin theorem

Since Quillen and Suslin independently proved Serre's conjecture stating that projective modules over commutative polynomial rings with coefficients in a field are free, some algorithmic versions of the proof have been proposed in the literature in order to constructively compute bases of free modules ([15, 19, 26, 27, 35, 56, 57, 58, 59]). We refer the interested reader to Lam's nice books [23, 24] concerning Serre's conjecture.

3.1 Projective and stably free modules

In module theory, it is well known that a finitely presented $D = k[x_1, \dots, x_n]$-module $M = D^{1\times p}/(D^{1\times q}\, R)$, where k is a field and $R \in D^{q\times p}$, admits a finite free resolution. This is a result is due to Hilbert ([10]). Moreover, if k is a *computable field*, we can even construct a finite free resolution of M using Gröbner or Janet basis ([3, 10, 20]).

A classical result due to Serre proves that every projective D-module is stably free (a stably free module always being a projective D-module). See [10, 23, 24] for more details. In [49, 51], a constructive proof of this result was given and the corresponding algorithm was implemented in OREMODULES. Let us recall these useful results.

Proposition 3.1 [49, 51] *Let M be a D-module defined by the finite free resolution*

$$0 \longrightarrow D^{1\times p_m} \xrightarrow{.R_m} \dots \xrightarrow{.R_2} D^{1\times p_1} \xrightarrow{.R_1} D^{1\times p_0} \xrightarrow{\pi} M \longrightarrow 0. \tag{3.1}$$

1. *If $m \geq 3$ and there exists $S_m \in D^{p_{m-1}\times p_m}$ such that $R_m\, S_m = I_{p_m}$, then we have the following finite free resolution of M*

$$0 \longrightarrow D^{1\times p_{m-1}} \xrightarrow{.T_{m-1}} D^{1\times (p_{m-2}+p_m)} \xrightarrow{.T_{m-2}} D^{1\times p_{m-3}} \xrightarrow{.R_{m-3}} \dots \xrightarrow{\pi} M \longrightarrow 0, \tag{3.2}$$

with the following notations:

$$\begin{cases} T_{m-1} = (R_{m-1} \quad S_m) \in D^{p_{m-1}\times (p_{m-2}+p_m)}, \\ T_{m-2} = \begin{pmatrix} R_{m-2} \\ 0 \end{pmatrix} \in D^{(p_{m-2}+p_m)\times p_{m-3}}. \end{cases}$$

2. *If $m = 2$ and there exists $S_2 \in D^{p_1\times p_2}$ such that $R_2\, S_2 = I_{p_2}$, then we have the following finite free resolution of M*

$$0 \longrightarrow D^{1\times p_1} \xrightarrow{.T_1} D^{1\times (p_0+p_2)} \xrightarrow{\tau} M \longrightarrow 0, \tag{3.3}$$

with the notations $T_1 = (R_1 \quad S_2) \in D^{p_1 \times (p_0+p_2)}$ *and*

$$\begin{aligned} \tau = \pi \oplus 0 : D^{1\times(p_0+p_2)} &\longrightarrow M \\ \lambda = (\lambda_1 \; \lambda_2) &\longmapsto \tau(\lambda) = \pi(\lambda_1). \end{aligned}$$

Remark 3.2 We note that Proposition 3.1 holds for every ring D.

Let $R \in D^{q\times p}$ and let us suppose that the D-module $M = D^{1\times p}/(D^{1\times q} R)$ is projective (using the results summed up in Figure 2.1, we can constructively check this result). Using 1 of Proposition 2.15, we obtain that the exact sequence (3.1) splits (see 5 of Definition 2.2), and thus, there exists $S_m \in D^{p_{m-1}\times p_m}$ such that $R_m S_m = I_{p_m}$. Repeating inductively the same method with the new finite free resolution of M, we can assume that we have the finite free resolution of M:

$$0 \longrightarrow D^{1\times p'_3} \xrightarrow{.R'_3} D^{1\times p'_2} \xrightarrow{.R'_2} D^{1\times p_1} \xrightarrow{.R_1} D^{1\times p_0} \xrightarrow{\pi} M \longrightarrow 0.$$

As M is a projective D-module, by 1 of Proposition 2.15, the previous exact sequence splits and thus, there exists a matrix $S'_3 \in D^{p'_2\times p'_3}$ satisfying $R'_3 S'_3 = I_{p'_3}$. By 1 of Proposition 3.1, we then get the finite free resolution of M:

$$0 \longrightarrow D^{1\times p'_2} \xrightarrow{.(R'_2 \quad S'_3)} D^{1\times(p_1+p'_3)} \xrightarrow{.(R_1^T \quad 0^T)^T} D^{1\times p_0} \xrightarrow{\pi} M \longrightarrow 0.$$

Let us denote by $T_1 = (R_1^T \quad 0^T)^T$. Again, as M is a projective D-module, by 1 of Proposition 2.15, the previous exact sequence splits and there exists $S'_2 \in D^{(p_1+p'_3)\times p'_2}$ such that $(R'_2 \quad S'_3)\, S'_2 = I_{p'_2}$. Using 2 of Proposition 3.1, we obtain the following finite free presentation of the D-module $M' = D^{1\times(p_0+p'_2)}/(D^{1\times(p_1+p'_3)}\,(T_1 \quad S'_2))$

$$0 \longrightarrow D^{1\times(p_1+p'_3)} \xrightarrow{.(T_1 \quad S'_2)} D^{1\times(p_0+p'_2)} \xrightarrow{\pi'} M' \longrightarrow 0,$$

where π' denotes the standard projection onto M' and $\tau : M' \longrightarrow M$ is defined by $\tau(m) = \pi(\lambda_1)$, for all $\lambda = (\lambda_1 \; \lambda_2) \in D^{1\times(p_0+p'_2)}$ satisfying $m = \pi'(\lambda)$. Moreover, 2 of Proposition 2.15 says that τ is an isomorphism, i. e., $M' \cong M$, a fact that can be also directly checked. We then obtain the following result.

Corollary 3.3 *Let* $D = k[x_1, \ldots, x_n]$ *be a commutative polynomial ring over a field* k *and* $R \in D^{q\times p}$. *If the* D*-module* $M = D^{1\times p}/(D^{1\times q} R)$ *is projective, then there exists a full row rank matrix* $R' \in D^{q'\times p'}$ *such that*

$$M \cong D^{1\times p'}/(D^{1\times q'} R'). \tag{3.4}$$

We refer to Example 5.5 for an illustration of Corollary 3.3. See also [49, 50, 51].

We note that $\operatorname{rank}_D(M) = \operatorname{rank}_D(M') = p' - q'$. Finally, we have the following short exact sequence of D-modules

$$0 \longrightarrow D^{1\times q'} \xrightarrow{.R'} D^{1\times p'} \xrightarrow{\pi'} M' \longrightarrow 0,$$

and using the fact that $M' \cong M$ and M is a projective D-module, by 1 of Proposition 2.15, we obtain that the previous exact sequence splits and we then get ([5, 53])

$$M' \oplus D^{1\times q'} \cong D^{1\times p'},$$

which, by 2 of Definition 2.4, shows that $M \cong M'$ is a stably free D-module.

Corollary 3.4 (Serre) [10, 23, 24] *Let M be a finitely generated projective $D = k[x_1, ..., x_n]$-module. Then, M is stably free.*

3.2 Stably free and free modules

Let M be a stably free module over $D = k[x_1, \ldots, x_n]$, where k is a field. Using Corollary 3.3, we can always suppose that M has the form $M = D^{1\times p}/(D^{1\times q}\, R)$, where $R \in D^{q\times p}$ admits a right-inverse $S \in D^{p\times q}$. We note that R has then full row rank ($\lambda\, R = 0 \Rightarrow \lambda = \lambda\, R\, S = 0$). Let us characterize when M is a free D-module.

In order to do that, we first need to introduce a definition.

Definition 3.5 Let D be a ring. The *general linear group* $\mathrm{GL}_p(D)$ is defined by

$$\mathrm{GL}_p(D) = \{U \in D^{p\times p} \mid \exists\, V \in D^{p\times p} :\, U\, V = V\, U = I_p\}.$$

An element $U \in \mathrm{GL}_p(D)$ is called a *unimodular matrix*.

In the case where $D = k[x_1, \ldots, x_n]$, we note that $U \in \mathrm{GL}_p(D)$ iff the determinant $\det U$ of U is invertible in D, i. e., is a non-zero element of k. The following result holds for every ring D.

Lemma 3.6 *Let $R \in D^{q\times p}$ be a matrix which admits a right-inverse over D. Then, the D-module $M = D^{1\times p}/(D^{1\times q}\, R)$ is free if and only if there exists $U \in \mathrm{GL}_p(D)$ such that $R\, U = (I_q \quad 0)$.*

Indeed, let us suppose that there exists $U \in \mathrm{GL}_p(D)$ such that $R\, U = (I_q \quad 0)$ and let us denote by $J = (I_q \quad 0) \in D^{q\times p}$. We easily check that $D^{1\times p}/(D^{1\times q}\, J) = D^{1\times (p-q)}$. Moreover, using the fact that $R\, U = J$ and $U \in \mathrm{GL}_p(D)$, we obtain the following commutative exact diagram

$$\begin{array}{ccccccccc}
 & & 0 & & 0 & & & & \\
 & & \downarrow & & \downarrow & & & & \\
0 & \longrightarrow & D^{1\times q} & \xrightarrow{.R} & D^{1\times p} & \xrightarrow{\pi} & M & \longrightarrow & 0 \\
 & & \| & & \downarrow .U & & & & \\
0 & \longrightarrow & D^{1\times q} & \xrightarrow{.J} & D^{1\times p} & \xrightarrow{\kappa} & D^{1\times (p-q)} & \longrightarrow & 0, \\
 & & \downarrow & & \downarrow & & & & \\
 & & 0 & & 0 & & & &
\end{array}$$

which proves that $M \cong D^{1\times (p-q)}$, i. e., M is a free D-module of rank $p - q$.

Conversely, let us suppose that $M \cong D^{1\times(p-q)}$. Then, combining the isomorphism $\psi : M \longrightarrow D^{1\times(p-q)}$ and the short exact sequence

$$0 \longrightarrow D^{1\times q} \xrightarrow{.R} D^{1\times p} \xrightarrow{\pi} M \longrightarrow 0,$$

we obtain the following exact sequence:

$$0 \longrightarrow D^{1\times q} \xrightarrow{.R} D^{1\times p} \xrightarrow{\psi\circ\pi} D^{1\times(p-q)} \longrightarrow 0.$$

If we consider the matrix which corresponds to the D-morphism $\psi \circ \pi$ in the canonical bases of $D^{1\times p}$ and $D^{1\times(p-q)}$, we then obtain a matrix $Q \in D^{p\times(p-q)}$ such that

$$\forall\, \lambda \in D^{1\times p}: \quad (\psi\circ\pi)(\lambda) = \lambda\, Q.$$

By 1 of Proposition 2.15, the previous exact sequence splits, i. e., we have

$$0 \longrightarrow D^{1\times q} \underset{\xleftarrow{.S}}{\xrightarrow{.R}} D^{1\times p} \underset{\xleftarrow{.T}}{\xrightarrow{.Q}} D^{1\times(p-q)} \longrightarrow 0,$$

or, equivalently, there exists a matrix $T \in D^{(p-q)\times p}$ such that the following Bézout identities hold (see [5, 42, 48, 53] for more details):

$$\begin{pmatrix} R \\ T \end{pmatrix} (S \quad Q) = \begin{pmatrix} I_q & 0 \\ 0 & I_{p-q} \end{pmatrix}, \quad (S \quad Q) \begin{pmatrix} R \\ T \end{pmatrix} = I_p.$$

In particular, we obtain that there exists a matrix $U = (S \quad Q) \in \mathrm{GL}_p(D)$ satisfying

$$R\,U = (I_q \quad 0).$$

Finally, the family $\{\pi(T_i)\}_{1\leq i\leq p-q}$ forms a basis of the free D-module M, where T_i denotes the ith row of $T \in D^{(p-q)\times p}$.

We are now in position to state the famous Quillen-Suslin theorem ([23, 24, 53]).

Theorem 3.7 (Quillen-Suslin theorem) [52, 55] *Let A be a principal ideal domain (e. g., a field k) and $D = A[x_1,\ldots,x_n]$ a polynomial ring with coefficients in A. Moreover, let $R \in D^{q\times p}$ be a matrix which admits a right-inverse $S \in D^{p\times q}$, i. e., $R\,S = I_q$. Then, there exists a unimodular $U \in \mathrm{GL}_p(D)$ satisfying*

$$R\,U = (I_q \quad 0). \tag{3.5}$$

Using Lemma 3.6 and Theorem 3.7, we obtain the following important corollary.

Corollary 3.8 (Quillen-Suslin) [52, 55] *Let A be a principal ideal domain (e. g., a field k) and $D = A[x_1,\ldots,x_n]$. Then, every stably free D-module is free.*

Moreover, the problem of finding a basis of a free finitely generated D-module M can be reformulated in terms of matrices as follows:

Problem 1 *Let $R \in D^{q \times p}$ be a matrix which admits a right-inverse over D. Find a matrix $U \in \mathrm{GL}_p(D)$ such that $R\,U = (I_q \quad 0)$.*

The previous problem is equivalent to completing R to a square invertible matrix

$$U^{-1} = \begin{pmatrix} R \\ T \end{pmatrix} \in D^{p \times p}.$$

The Quillen-Suslin theorem states that Problem 1 has always a solution over a polynomial ring $D = A[x_1, \ldots, x_n]$ with coefficients in a principal ring A and, in particular, in a field k. In what follows, an algorithm which computes such a matrix U will be called a *QS-algorithm*.

Let us consider a matrix $R \in D^{q \times p}$ which admits a right-inverse over D and let us denote by R_i the ith row of R. We note that the row $R_1 \in D^{1 \times p}$ admits a right-inverse over D. Let us suppose that we can find a matrix $U_1 \in \mathrm{GL}_p(D)$ satisfying $R_1\,U = (1\;0\;\ldots\;0)$. Then, we have

$$R\,U_1 = \begin{pmatrix} 1 & 0 \\ \star & R_2 \end{pmatrix},$$

where $R_2 \in D^{(q-1) \times (p-1)}$ and $\star$ denotes a certain element of $D^{(q-1) \times 1}$. Hence, we restrict our considerations to the new matrix R_2, which can easily be shown to admit a right-inverse over D, and reduce Problem 1 to the following one:

Problem 2 *Let $R \in D^{1 \times p}$ be a row vector which admits a right-inverse over D. Find a matrix $U \in \mathrm{GL}_p(D)$ such that $R\,U = (1\;0\;\ldots\;0)$.*

The purpose of the next paragraphs is to recall a QS-algorithm solving Problem 2 over a commutative polynomial ring $D = k[x_1, \ldots, x_n]$ over a computable field k (for instance, $k = \mathbb{Q}$). This algorithm was implemented in the package QUILLENSUSLIN ([12]). See also the appendix. We also point out that a QS-algorithm has also been implemented in QUILLENSUSLIN for the case $D = \mathbb{Z}[x_1, \ldots, x_n]$. Even though there are some differences in the constructive proofs of the Quillen-Suslin theorem developed in [15, 19, 23, 26, 35, 56, 57, 59], we note that our algorithm is based on the same main idea, i. e., it proceeds by induction on the number of variables x_i in $D = k[x_1, \ldots, x_n]$. Each inductive step of the general QS-algorithm reduces the problem to the case with one variable less. A more global and interesting approach has recently been developed in [27, 58] which needs to be studied with care in the future.

3.3 Solution of Problem 2 in some special cases

Although the tedious inductive method, which will be explained in the next section, cannot generally be avoided, there are cases where simpler and faster heuristic methods can be used. We shall first consider such cases.

3.3.1 Matrices over a principal ideal domain D

We first consider the special case of matrices over a principal ideal domain D (e. g., $D = k[x_1]$, where k a field, $\mathbb{Z}$). Let $R \in D^{q \times p}$ be a matrix which admits a right-inverse over D. Then, computing the *Smith canonical form* of R ([40]), we obtain two matrices $F \in \mathrm{GL}_q(D)$ and $G \in \mathrm{GL}_p(D)$ satisfying

$$R = F\,(I_q \quad 0)\,G.$$

If we denote by $r = p - q$, $G = (G_1^T \quad G_2^T)^T$, where $G_1 \in D^{q \times p}$, $G_2 \in D^{r \times p}$ and $G^{-1} = (H_1 \quad H_2) \in D^{p \times p}$, where $H_1 \in D^{p \times q}$, $H_2 \in D^{p \times r}$, then we get $R = F\,G_1$, i. e., $G_1 = F^{-1}\,R$, and thus, we get

$$\begin{aligned}
\begin{pmatrix} F^{-1}\,R \\ G_2 \end{pmatrix} (H_1 \quad H_2) = I_p \quad &\Rightarrow \quad \begin{pmatrix} F^{-1} & 0 \\ 0 & I_r \end{pmatrix} \begin{pmatrix} R \\ G_2 \end{pmatrix} (H_1 \quad H_2) = I_p \\
&\Rightarrow \quad \begin{pmatrix} R \\ G_2 \end{pmatrix} (H_1 \quad H_2) \begin{pmatrix} F^{-1} & 0 \\ 0 & I_r \end{pmatrix} = I_p \\
&\Rightarrow \quad \begin{pmatrix} R \\ G_2 \end{pmatrix} (H_1\,F^{-1} \quad H_2) = I_p,
\end{aligned}$$

which solves Problem 1 as we can take $U = (H_1\,F^{-1} \quad H_2) \in \mathrm{GL}_p(D)$ and $T = G_2$.

3.3.2 $(p-1) \times p$ matrices over an arbitrary commutative ring D

Let us consider the case of a matrix $R \in D^{(p-1) \times p}$ which admits a right-inverse over a commutative ring D. If we denote by m_i the $(p-1) \times (p-1)$ minor of R obtained by removing the ith column of R, then, using the fact R admits right-inverse, we get that the family $\{m_i\}_{1 \leq i \leq p}$ satisfies a Bézout identity $\sum_{i=1}^{p} n_i\,m_i = 1$ for certain $n_i \in D$ and $i = 1, \ldots, p$. Let us denote by

$$V = \begin{pmatrix} & R & \\ (-1)^{p+1}\,n_1 & \ldots & (-1)^{2\,p}\,n_p \end{pmatrix} \in D^{p \times p}.$$

Expand the determinant of V along the last row, using the Laplace's formula, we then get $\det V = 1$. Hence, if we denote by $U \in D^{p \times p}$ the inverse of the matrix V, we then obtain $R\,U = (I_{p-1} \quad 0)$, which solves Problem 1.

3.3.3 $1 \times p$ rows over an arbitrary commutative ring D

We now consider Problem 2, i. e., the case of a row vector $f = (f_1 \ \ldots \ f_p) \in D^{1 \times p}$ which admits a right-inverse over an arbitrary commutative ring D.

Remark 3.9 (Special form of the row) 1. We note that if one of the components of f is an invertible element of D, we can then transform the row f into $(1\ 0\ \ldots\ 0)$

by means of trivial elementary operations. For instance, if $f_1^{-1} \in D$, then the matrix defined by

$$W = \begin{pmatrix} f_1^{-1} & 0 \\ 0 & I_{p-1} \end{pmatrix}$$

satisfies $\det W = f_1^{-1} \in D$ and $f\,W = (1 \; f_2 \; \ldots \; f_p)$. Then, simple elementary operations transform $f\,W$ into the vector $(1\,0\,\ldots\,0)$.

2. Another simple case is when two components of f generate D. Let us suppose that there exist h_1 and $h_2 \in D$ such that we have the Bézout identity $f_1 h_1 + f_2 h_2 = 1$ and let us define the following matrix:

$$W = \begin{pmatrix} h_1 & -f_2 & 0 \\ h_2 & f_1 & 0 \\ 0 & 0 & I_{p-2} \end{pmatrix}.$$

We easily check that $\det W = 1$ and $f\,W = (1\;0\;f_3\;\ldots\;f_p)$. Then, we can reduce $f\,W$ to $(1\,0\,\ldots\,0)$ by means of elementary operations.

3. If the ith component of f is 0 or the ideal generated by the elements $f_1, \ldots, f_{i-1}, f_{i+1}, \ldots, f_p$ is already D, then we can follow an idea analogous to the one developed in [49, 51]. Let us suppose that $i = 1$, i.e., f_1 is a redundant component in the sense that $(f_2, \ldots, f_p) = D$. Then, there exist $h_2, \ldots, h_p \in D$ satisfying the Bézout equation $\sum_{i=2}^{p} f_i\,h_i = 1$. Then, the matrix

$$W = \begin{pmatrix} 1 & & & \\ (1-f_1)\,h_2 & 1 & & \\ \vdots & & \ddots & \\ (1-f_1)\,h_p & & & 1 \end{pmatrix}$$

satisfies $f\,W = (1\;f_2\;\ldots\;f_p)$ and $\det W = 1$. We can now reduce $f\,W$ to $(1\,0\,\ldots\,0)$ by means of elementary operations.

In particular, this strategy is always successful when the length p of the row f exceeds the *stable range* of the ring D. We note that the stable range of $D = \mathbb{R}[x_1, \ldots, x_n]$ is equal to $n+1$. We refer the reader to [49, 51] for more details.

We note that all the conditions given in Remark 3.9 can be checked using Gröbner or Janet bases.

The matrix U can also be easily computed in cases where a right-inverse g of the row f has a special form.

Remark 3.10 (Special form of the right-inverse) Let $g \in D^{p\times 1}$ be a right-inverse of the unimodular row $f \in D^{1\times p}$, i.e., $fg = 1$.

1. Let us suppose that one of the entries of a right-inverse g of f, say g_1, is invertible in D. Then, the following matrix

$$W = \begin{pmatrix} g_1 & & & \\ g_2 & 1 & & \\ \vdots & & \ddots & \\ g_p & & & 1 \end{pmatrix}$$

satisfies $\det W = g_1$ and $f\,W = (1\ f_2\ \ldots\ f_p)$. As g_1 is an invertible element of D, then W is a unimodular matrix and $f\,W$ can easily be transformed into $(1\ 0\ \ldots\ 0)$ by means of elementary operations.

2. If two components g_1, g_2 of g generate the whole ring D, then there exist elements $h_1, h_2 \in D$ such that $g_1\,h_1 + g_2\,h_2 = 1$. Then, the matrix defined by

$$W = \begin{pmatrix} g_1 & -h_2 & & & \\ g_2 & h_1 & & & \\ g_3 & & 1 & & \\ \vdots & & & \ddots & \\ g_p & & & & 1 \end{pmatrix}$$

satisfies $\det W = 1$ and $f\,W = (1\ \star\ f_3\ \ldots\ f_p)$, where $\star$ denotes a certain element of D. We can then reduce $f\,W$ to $(1\ 0\ \ldots\ 0)$ by means of elementary operations.

Finally, we also note that if $f \in D^{1\times p}$ admits a right-inverse g over D for which any of the heuristic methods explained in Remark 3.9 may be used for g^T, then a unimodular matrix V having g^T as a first row can be easily computed. Then, the product $f\,V^T = (1\ \star\ \ldots\ \star)$ can be reduced to the first standard basis vector by elementary column operations.

For instance, let us illustrate 1 of Remark 3.10. In some of the illustrating examples, we shall also use the notation $D = k[z_1,\ldots,z_n]$ as these examples come from the control theory and signal processing literatures where z_i is commonly used. The independent variables z_i, $i = 1,\ldots,n$, usually denote the variables appearing in the discrete Laplace transform.

Example 3.11 Let us consider $D = \mathbb{Q}[z_1, z_2, z_3]$ and the following row vector:

$$R = (z_1^2\,z_2^2 + 1 \quad z_1^2\,z_3 + 1 \quad z_1\,z_2^2\,z_3).$$

We can easily check that R admits the following right-inverse $S = (-z_1^2\,z_3 \quad 1 \quad z_1^3)^T$. As the second component of S is invertible over D, we can apply 1 of Remark 3.10 in order to find a unimodular matrix U over D which satisfies $R\,U = (1\ 0\ 0)$. Let us define the following elementary matrices:

$$U_1 = \begin{pmatrix} 0 & 1 & 0 \\ 1 & 0 & 0 \\ 0 & 0 & 1 \end{pmatrix}, \quad U_2 = \begin{pmatrix} 1 & 0 & 0 \\ -z_1^2\,z_3 & 1 & 0 \\ z_1^3 & 0 & 1 \end{pmatrix}.$$

We then have $R\,(U_1\,U_2) = (1 \quad z_1^2\,z_2^2 + 1 \quad z_1\,z_2^2\,z_3)$. Finally, if we denote by

$$U_3 = \begin{pmatrix} 1 & -z_1^2\,z_2^2 - 1 & -z_1\,z_2^2\,z_3 \\ 0 & 1 & 0 \\ 0 & 0 & 1 \end{pmatrix},$$

we then have $R\,U = (1\;0\;0)$, where the unimodular $U = U_1\,U_2\,U_3$ is defined by

$$U = \begin{pmatrix} -z_1^2\,z_3 & z_1^4\,z_2^2\,z_3 + z_1^2\,z_3 + 1 & z_1^3\,z_2^2\,z_3^2 \\ 1 & -z_1^2\,z_2^2 - 1 & -z_1\,z_2^2\,z_3 \\ z_1^3 & -z_1^3\,(z_1^2\,z_2^2 + 1) & -z_1^4\,z_2^2\,z_3 + 1 \end{pmatrix}. \tag{3.6}$$

3.4 A QS-algorithm for commutative polynomial rings

Over an arbitrary commutative ring A, not every row admitting a right-inverse can be completed to a unimodular matrix over A. The module-theoretic interpretation of this result is that, over certain rings, there exist stably free modules which are not free. For instance, using a classical topological theorem on vector fields on the sphere $\mathbb{S}_2(\mathbb{R})$, we can prove that the row vector $R = (x_1\;x_2\;x_3) \in D^{1\times 3}$ with entries in the commutative ring $D = \mathbb{R}[x_1, x_2, x_3]/(x_1^2 + x_2^2 + x_3^2 - 1)$, which admits the right-inverse R^T, cannot be completed to a unimodular matrix over D. For more details, see [24].

However, it is always possible over a polynomial ring with coefficients on a field k or a principal ideal domain A. See Quillen-Suslin theorem (Theorem 3.7). We shortly describe a QS-algorithm which has recently been implemented in a package called QUILLENSUSLIN ([12]). See the appendix for more details. In what follows, we shall only consider a commutative polynomial ring $D = k[x_1, \ldots, x_n]$ over a field k even if the extension of the algorithms exists when k is replaced by a principal ideal ring A. For instance, the case of $A = \mathbb{Z}$ has also been implemented in QUILLENSUSLIN. Let $f \in D^{1\times p}$ be a row vector which admits a right-inverse g over D. When no method explained in section 3.3 can be applied to f, we then need to consider a general algorithm. However, we point out that most of the examples we know do not require the general algorithm as the previous heuristic methods are generally enough to get the result.

The QS-algorithm proceeds by induction on the number n of the independent variables x_i of the ring $D = k[x_1, \ldots, x_n]$. Each inductive step, which simplifies the problem to the case of a polynomial ring containing one variable less, consists of three main parts:

1. Finding a normalized component in the last variable of the polynomial ring.
2. Computing finitely many local solutions of Problem 2 over certain local rings (local loop).
3. Patching/glueing the local solutions together in order to obtain a global one.

3.4.1 Normalization step

The next lemma is essential for Horrocks' theorem which is used in the local loop.

Lemma 3.12 [53, 57] *Let us consider a polynomial* $a \in k[y_1, \ldots, y_n]$ *and let us denote by* $m = \deg(a) + 1$, *where* $\deg(a)$ *denotes the total degree of* a. *Using the following invertible transformation*

$$\begin{cases} x_n = y_n, \\ x_i = y_i - y_n^{m^{n-i}}, \quad 1 \leq i \leq n-1, \end{cases} \Leftrightarrow \begin{cases} y_n = x_n, \\ y_i = x_i + x_n^{m^{n-i}}, \quad 1 \leq i \leq n-1, \end{cases}$$

we obtain $a(y_1, \ldots, y_n) = r\, b(x_1, \ldots, x_n)$, *where* $0 \neq r \in k$ *and* b *is a* monic polynomial *in* x_n *with coefficients in the ring* $E = k[x_1, \ldots, x_{n-1}]$, *namely, the leading coefficient of* $b \in E[x_n]$ *is* 1.

In the case where k is an infinite field, we can achieve the same result by using only a linear transformation whose coefficients are appropriately chosen ([57, 59]). The normalization step can also be generalized to the case $D = A[x_1, \ldots, x_n]$, where A is a principal ideal domain. See [56] for more details.

3.4.2 Local loop

In the second step, we need to compute a finite number of *local solutions* of Problem 2 over a local ring A, namely, a ring A which has only one maximal ideal, i. e., a proper ideal $\mathcal{M}$ of A which is not properly contained in any ideal of A other than A itself. In order to do that, we use the so-called Horrocks' theorem. Let us recall it.

Theorem 3.13 [53, 57] *Let* B *be a local ring and* f *a row vector which admits a right-inverse over* $B[y]$. *If one of the components* f_i *of* f *is monic, then there exists a unimodular matrix* V *over* $B[y]$ *such that* f *is the first row of* V *or, equivalently, such that* $f\, V^{-1} = (1\ 0\ \ldots\ 0)$.

Horrocks' theorem can easily be implemented using, for instance, the approaches developed in [26, 53, 59]. In particular, the implementation in QUILLENSUSLIN of this theorem follows [53]. If $\mathcal{M}$ is a maximal ideal of D, we then denote by $D_{\mathcal{M}}$ the local ring, which is a standard *localization* of D with respect to the multiplicative closed subset $S = D \backslash \mathcal{M}$ of D, namely, $D_{\mathcal{M}} = \{a/b \mid a \in D,\ b \notin \mathcal{M}\}$ ([53]).

We can now give the first main part of the general algorithm ([26, 59]).

Algorithm 3.14

- **Input:** Let $D = k[x_1, \ldots, x_n]$ and $f \in D^{1 \times p}$ a row vector which admits a right-inverse over D and a monic component in the last variable x_n.
- **Output:** A finite number of maximal ideals $\{\mathcal{M}_i\}_{i \in I}$ of $E = k[x_1, \ldots, x_{n-1}]$ and unimodular matrices $\{H_i\}_{i \in I}$ over the ring $E_{\mathcal{M}_i}[x_n]$ which satisfy the relation $f\, H_i = (1\ 0\ \ldots\ 0)$, and such that the ideal generated by the denominators of the matrices H_i, $i \in I$, generates the ring E.

1. Let $\mathcal{M}_1$ be an arbitrary maximal ideal of the ring E. Using Horrocks' theorem, compute a unimodular matrix H_1 over $E_{\mathcal{M}_1}[x_n]$ which satisfies the relation $f\, H_1 = (1\ 0\ \ldots\ 0)$.

2. Let $d_1 \in E$ be the denominator of H_1 and J the ideal in E generated by d_1. Set $i = 1$.
3. While $J \neq E$, do:
   (i) For $i := i + 1$, compute a maximal ideal $\mathcal{M}_i$ of E such that $J \subset \mathcal{M}_i$.
   (ii) Using Horrocks' theorem, compute a matrix H_i over the ring $E_{\mathcal{M}_i}[x_n]$ such that $\det H_i$ is invertible in $E_{\mathcal{M}_i}[x_n]$ and $f\,H_i = (1\ 0\ \ldots\ 0)$.
   (iii) Let d_i be the denominator of the matrix H_i and consider the ideal $J = (d_1, \ldots, d_i)$.
4. Return $\{\mathcal{M}_i\}_{i\in I}$, $\{H_i\}_{i\in I}$ and $\{d_i\}_{i\in I}$.

The local loop stops when all the denominators d_i generate E. As the ring E is noetherian ([53]), the number of the local solutions, i. e., the cardinal of the set I, is finite.

3.4.3 Patching

To obtain a polynomial solution of Problem 1, we use the following lemma.

Lemma 3.15 [26] *Let $f \in D^{1\times p}$ be a row vector which admits a right-inverse over $D = k[x_1, \ldots, x_n]$ and U a unimodular matrix over $k[x_1, \ldots, x_{n-1}]_{\mathcal{M}}[x_n]$, where $\mathcal{M}$ is a certain maximal ideal of $E = k[x_1, \ldots, x_{n-1}]$, which satisfies $f\,U = (1\ 0\ \ldots\ 0)$. Let us denote by $d \in E$ the denominator of U. Then, the matrix defined by*

$$\Delta(x_n, z) = U(x_1, \ldots, x_n)\,U^{-1}(x_1, \ldots, x_{n-1}, x_n + z) \in (E_{\mathcal{M}}[x_n, z])^{p\times p}$$

is such that

$$\forall\, z \in D: \quad f(x_1, \ldots, x_n)\,\Delta(x_n, z) = f(x_1, \ldots, x_{n-1}, x_n + z), \tag{3.7}$$

and its denominator is d^{α} with $0 \leq \alpha \leq p$.

(3.7) is clear as the identity $f(x_1, \ldots, x_n)\,U(x_1, \ldots, x_n) = (1\ 0\ \ldots\ 0)$ implies that $f(x_1, \ldots, x_n + z)\,U(x_1, \ldots, x_n + z) = (1\ 0\ \ldots\ 0)$ and we then have

$$f(x_1, \ldots, x_n + z) = f(x_1, \ldots, x_n)\,U(x_1, \ldots, x_n)\,U(x_1, \ldots, x_n + z)^{-1}.$$

Moreover, using the standard formula $U^{-1} = (\det U)^{-1}\,\mathrm{adj}(U)$, where $\mathrm{adj}(U)$ denotes the adjugate of U, we can also prove that the common denominator of $\Delta(x_n, z)$ is d^{α}, where $0 \leq \alpha \leq p$.

Let $\{\mathcal{M}_i\}_{i\in I}$, $\{H_i\}_{i\in I}$ and $\{d_i\}_{i\in I}$ be the output of Algorithm 3.14, where I is a finite set. Let us set $I = \{1, \ldots, l\}$. The ideal of $E = k[x_1, \ldots, x_{n-1}]$ defined by $\{d_i\}_{i\in I}$ generates E. Hence, there exists $c_i \in E$, $i \in I$, such that the Bézout identity holds:

$$\sum_{i=1}^{l} c_i\, d_i^p = 1.$$

Let us define the following matrices

$$\Delta_i(x_n, z) = H_i(x_1, \ldots, x_n)\, H_i^{-1}(x_1, \ldots, x_{n-1}, x_n + z), \quad i = 1, \ldots, l,$$

and, in order to simplify the notations, we denote by $\tilde{f}(x_n)$ the function $f(x_1, \ldots, x_n)$. Then, we have

$$\begin{aligned}
&\tilde{f}(x_n)\, \Delta_1(x_n, (a_n - x_n)\, c_1\, d_1^p) = \tilde{f}(x_n + (a_n - x_n)\, c_1\, d_1^p),\\
&\tilde{f}(x_n + (a_n - x_n)\, c_1\, d_1^p)\, \Delta_2(x_n + (a_n - x_n)\, c_1\, d_1^p, (a_n - x_n)\, c_2\, d_2^p) =\\
&\qquad\qquad \tilde{f}\left(x_n + (a_n - x_n)\left(\textstyle\sum_{i=1}^{2} c_i\, d_i^p\right)\right),\\
&\qquad\qquad \vdots\\
&\tilde{f}\left(x_n + (a_n - x_n)\left(\textstyle\sum_{i=1}^{l-1} c_i\, d_i^p\right)\right)\\
&\qquad \Delta_l\left(x_n + (a_n - x_n)\left(\textstyle\sum_{i=1}^{l-1} c_i\, d_i^p\right), (a_n - x_n)\, c_l\, d_l^p\right) = \tilde{f}(a_n).
\end{aligned}$$

Finally, we can prove that we have $\Delta_i(x_n, d_i^p\, z) \in \mathrm{GL}_p(D)$, $i = 1, \ldots, l$, ([26]) and

$$\begin{aligned}
U_1 = {}& \Delta_1(x_n, (a_n - x_n)\, c_1\, d_1^p)\, \Delta_2(x_n + (a_n - x_n)\, c_1\, d_1^p, (a_n - x_n)\, c_2\, d_2^p)\\
&\ldots \Delta_l\left(x_n + (a_n - x_n)\left(\sum_{i=1}^{l-1} c_i\, d_i^p\right), (a_n - x_n)\, c_l\, d_l^p\right) \in \mathrm{GL}_p(D).
\end{aligned}$$

The previous computations then show that $f(x_1, \ldots, x_n)\, U_1 = f(x_1, \ldots, x_{n-1}, a_n)$.

We can now state the main result.

Theorem 3.16 [26, 53, 57, 59] *Let $f \in D^{1\times p}$ be a row vector which admits a right-inverse over the ring $D = k[x_1, \ldots, x_n]$. Then, for every $a \in k$, there exists a matrix $U \in \mathrm{GL}_p(D)$ such that*

$$f(x_1, \ldots, x_n)\, U(x_1, \ldots, x_n) = f(x_1, \ldots, x_{n-1}, a).$$

Let us now consider a row vector $f(x_1, \ldots, x_n) \in D^{1\times p}$ which admits a right-inverse $g(x_1, \ldots, x_n) \in D^{p\times 1}$. Applying inductively Theorem 3.16 to $f(x_1, \ldots, x_n)$ for the values $a_2, \ldots, a_n \in k$, we obtain $U_1, \ldots, U_{n-1} \in \mathrm{GL}_p(D)$ such that

$$\begin{aligned}
f(x_1, \ldots, x_n)\, U_1 &= f(x_1, \ldots, x_{n-1}, a_n),\\
f(x_1, \ldots, x_{n-i}, a_{n-i+1}, \ldots, a_n)\, U_{i+1} &= f(x_1, \ldots, x_{n-i-1}, a_{n-i}, \ldots, a_n),
\end{aligned}$$

for $i = 1, \ldots, n-2$. Hence, we get $f(x_1, \ldots, x_n)\,(U_1 \ldots U_{n-1}) = f(x_1, a_2, \ldots, a_n)$ and we have simplified Problem 2 to the case of a row vector $f(x_1, a_2, \ldots, a_n)$ over a principal ideal domain $k[x_1]$ which admits a right-inverse $g(x_1, a_2, \ldots, a_n)$ over $k[x_1]$. Using the first result of section 3.3, we can find a matrix $U_n \in \mathrm{GL}_p(D)$ such that

$$f(x_1, a_2, \ldots, a_n)\, U_n(x_1) = (1\, 0 \ldots 0) \Leftrightarrow U_n^{-1}(x_1) = \begin{pmatrix} f(x_1, a_2, \ldots, a_n) \\ \star \end{pmatrix}.$$

Hence, Problem 2 is then solved if we take $U = U_1 \dots U_n \in \mathrm{GL}_p(D)$. We also note that it is generally simpler to take the particular values $a_2 = \dots = a_n = 0$.

Now, let us find a matrix U' satisfying $f(x_1, \dots, x_n)\, U' = f(a_1, \dots, a_n)$, where $a_1 \in k$. Let us define by $U_n'(x_1) = U_n(x_1)\, U_n^{-1}(a_1) \in \mathrm{GL}_p(D)$. Then, we have

$$f(x_1, a_2, \dots, a_n)\, U_n'(x_1) = (1\, 0 \dots 0) \begin{pmatrix} f(a_1, a_2, \dots, a_n) \\ \star \end{pmatrix} = f(a_1, a_2, \dots, a_n).$$

Hence, the matrix $U' = U_1 \dots U_{n-1}\, U_n' \in \mathrm{GL}_p(D)$ satisfies

$$f(x_1, \dots, x_n)\, U' = f(a_1, \dots, a_n).$$

Let us illustrate the QS-algorithm on a simple example.

Example 3.17 Let us consider the commutative polynomial ring $D = \mathbb{Q}[x_1, x_2]$ and the row vector $R = (x_1\, x_2^2 + 1 \quad 3\, x_2/2 + x_1 - 1 \quad 2\, x_1\, x_2) \in D^{1\times 3}$. We can check that $S = (1 \quad 0 \quad -x_1/2)^T$ is a right-inverse of R, a fact implying that the D-module $M = D^{1\times 3}/(D\, R)$ is projective, and thus, free by the Quillen-Suslin theorem. Let us compute a matrix $U \in \mathrm{GL}_3(D)$ such that $R\, U = (1\ 0\ 0)$. As the first component of S is 1, we can easily find such a matrix U using the heuristic methods explained in section 3.3. However, let us illustrate the main algorithm previously described.

We first note that R contains the normalized component $3\, x_2/2 + x_1 - 1$ over the ring $D = E[x_2]$, where $E = \mathbb{Q}[x_1]$. The second step consists in computing certain local solutions. Let us consider the maximal ideal $\mathcal{M}_1 = (x_1)$ of E. Using an effective version of Horrocks' theorem, we obtain that

$$H_1 = \frac{1}{d_1} \begin{pmatrix} 4 & -2\,(3\, x_1 + 2\, x_2 - 2) & 4\, x_1\,(3\, x_1 - 2) \\ 2\, x_1\,(3\, x_1 - 2\, x_2 - 2) & 4\,(x_1\, x_2^2 + 1) & -4\, x_1\,(3\, x_1^2\, x_2 - 2\, x_1\, x_2 + 2) \\ 0 & 0 & 9\, x_1^3 - 12\, x_1^2 + 4\, x_1 + 4 \end{pmatrix},$$

where $d_1 = 9\, x_1^3 - 12\, x_1^2 + 4\, x_1 + 4 \notin \mathcal{M}_1$. We can check that $\det H_1 = 4/d_1$, i.e., $H_1 \in \mathrm{GL}_3(E_{\mathcal{M}_1}[x_2])$, and $R\, H_1 = (1\ 0\ 0)$, showing that H_1 is a local solution.

The ideal $J = (d_1)$ is strictly contained in E. Therefore, we consider another maximal ideal $\mathcal{M}_2$ such that $J \subseteq \mathcal{M}_2$. We can take $\mathcal{M}_2 = (9\, x_1^3 - 12\, x_1^2 + 4\, x_1 + 4)$. Using an effective version of Horrocks' theorem, we obtain the matrix

$$H_2 = \frac{1}{d_2} \begin{pmatrix} 0 & 0 & 4\, x_1\,(3\, x_1 - 2) \\ 8\, x_1 & -8\, x_1\, x_2 & -4\, x_1\,(3\, x_1^2\, x_2 - 2\, x_1\, x_2 + 2) \\ -4 & 2\,(3\, x_1 + 2\, x_2 - 2) & 9\, x_1^3 - 12\, x_1^2 + 4\, x_1 + 4 \end{pmatrix},$$

where $d_2 = 4\, x_1\,(3\, x_1 - 2) \notin \mathcal{M}_2$. We then have $\det H_2 = -1/(x_1\,(3\, x_1 - 2))$, i.e., $H_2 \in \mathrm{GL}_3(E_{\mathcal{M}_2}[x_2])$ and $R\, H_2 = (1\ 0\ 0)$. We can check that the ideal $(d_1, d_2) = E$ as we have the Bézout identity $c_1\, d_1 + c_2\, d_2 = 1$, where $c_1 = 1/4$ and $c_2 = -(3\, x_1 - 2)/16$.

The matrix $\Delta_i(x_2, -c_1\, d_1\, x_2)$ is defined by

$$\begin{pmatrix} (9\,x_1^4/4 - 3\,x_1^3 + x_1^2)\,x_2^2 + (3\,x_1^2/2 - x_1)\,x_2 + 1 & -x_2 & -2\,x_1\,x_2 \\ -(18\,x_1^4 - 24\,x_1^3 + 8\,x_1^2)\,x_1\,x_2^3/8 + (27\,x_1^5 - 54\,x_1^4 + 36\,x_1^3 - 20\,x_1^2 + 8\,x_1)\,x_1\,x_2^2/8 - x_1\,x_2 & x_1\,x_2^2 + (-3\,x_1^2/2 + x_1)\,x_2 + 1 & 2\,x_1^2\,x_2^2 - x_1^2\,(3\,x_1 - 2)\,x_2 \\ 0 & 0 & 1 \end{pmatrix}.$$

We can check that $R(x_1, x_2)\,\Delta_i(x_2, -c_1\, d_1\, x_2) = R(x_1, x_2 - c_1\, d_1\, x_2)$ as well as $\Delta_1(x_2, -c_1\, d_1\, x_2) \in \mathrm{GL}_3(D)$. Moreover, the matrix $\Delta_2(x_2 - c_1\, d_1\, x_2, -c_2\, d_2\, x_2)$ is defined by

$$\begin{pmatrix} 1 & 0 & 0 \\ 0 & (3\,x_1^2/2 - x_1)\,x_2 + 1 & x_1^2\,(3\,x_1 - 2)\,x_2 \\ (9\,x_1^2 - 12\,x_1 + 4)\,x_1\,x_2/8 & (-3\,x_1 + 2)\,x_2/4 & (-3\,x_1^2/2 + x_1)\,x_2 + 1 \end{pmatrix}.$$

We can also check that $R(x_1, x_2 - c_1\, d_1\, x_2)\,\Delta_2(x_2 - c_1\, d_1\, x_2, -c_2\, d_2\, x_2) = R(x_1, 0)$ and $\Delta_2(x_2 - c_1\, d_1\, x_2, -c_2\, d_2\, x_2) \in \mathrm{GL}_3(D)$. Defining the matrix

$$U_1 = \Delta_1(x_2, -c_1\, d_1\, x_2)\,\Delta_2(x_2 - c_1\, d_1\, x_2, -c_2\, d_2\, x_2) \in \mathrm{GL}_3(D),$$

we then get $R(x_1, x_2)\,U_1(x_1, x_2) = R(x_1, 0) = (1 \quad 3\,x_1/2 - 1 \quad 0)$.

Finally, if we denote by

$$U_2 = \begin{pmatrix} 1 & -3\,x_1/2 + 1 & 0 \\ 0 & 1 & 0 \\ 0 & 0 & 1 \end{pmatrix} \in \mathrm{GL}_3(D),$$

then, the matrix $R(x_1, 0)$ is then equivalent to $(1\ 0\ 0)$, i.e., $R(x_1, 0)\,U_2 = (1\ 0\ 0)$. Hence, if we define the matrix $U = U_1\,U_2 \in \mathrm{GL}_3(D)$, i.e.,

$$\begin{pmatrix} (3\,x_1^2/2 - x_1)\,x_2 + 1 & (-9\,x_1^3/4 + 3\,x_1^2 - x_1 - 1)\,x_2 - 3\,x_1/2 + 1 & -2\,x_1\,x_2 \\ (-3\,x_1^3/2 + x_1^2)\,x_2^2 - x_1\,x_2 & (9\,x_1^4/4 - 3\,x_1^3 + x_1^2 + x_1)\,x_2^2 + (3\,x_1^2/2 - x_1)\,x_2 + 1 & 2\,x_1^2\,x_2^2 \\ (9\,x_1^2 - 12\,x_1 + 4)\,x_1\,x_2/8 & (-27\,x_1^4/16 + 27\,x_1^3/8 - 9\,x_1^2/4 - x_1/4 + 1/2)\,x_2 & (-3\,x_1^2/2 + x_1)\,x_2 + 1 \end{pmatrix},$$

we finally obtain $R\,U = (1\ 0\ 0)$.

In the third point of section 3.3, we saw that the case of a matrix $R \in D^{q \times p}$ admitting a right-inverse over D can be solved by applying q times Theorem 3.16 on certain row vectors obtained during the process having smaller and smaller lengths. Hence, we obtain the following corollary.

Corollary 3.18 [26, 53, 57, 59] *Let $R \in D^{q \times p}$ be a matrix which admits a right-inverse over D. Then, for all $a_1, \ldots, a_n \in k$, there exists $U \in \mathrm{GL}_p(D)$ such that*

$$R(x_1, \ldots, x_n)\, U(x_1, \ldots, x_n) = R(a_1, \ldots, a_n).$$

We note that as the matrix $R(a_1, \ldots, a_n)$ has full row rank over a field k, there always exists a right-inverse $V \in k^{p \times q}$ such that $R(a_1, \ldots, a_n)\, V = I_q$. Hence, we obtain that $R\,(U\,V) = (I_q \quad 0)$, which also solves Problem 1. Another possibility is to first obtain a matrix $W \in \mathrm{GL}_p(D)$ such that $R(x_1, \ldots, x_n)\, W = R(x_1, a_2, \ldots, a_n)$ and then compute a Smith canonical form of $R(x_1, a_2, \ldots, a_n)$ as we did for the row vector case.

Remark 3.19 In [37], it was shown how a certain transformation maps a matrix $\overline{R}$ with entries in a Laurent polynomial ring $\overline{D} = k[x_1, \ldots x_n, x_1^{-1}, \ldots, x_n^{-1}]$, where k is a field, and which admits a right-inverse over D to a matrix R with entries in $D = k[x_1, \ldots, x_n]$ and which admits a right-inverse over D. Hence, we can use a QS-algorithm to solve Problems 2 and 1 over D. See [37] for more details. See also [9] for explicit examples. Finally, a new algorithm has recently been developed in [1].

3.4.4 Computation of bases of free modules

If $R \in D^{q \times p}$ is a matrix which admits a right-inverse over D, then, in section 3.2, we showed that a basis of the free D-module $M = D^{1 \times p}/(D^{1 \times q}\, R)$ is defined by $\{\pi(T_i)\}_{1 \le i \le (p-q)}$, where $\pi : D^{1 \times p} \longrightarrow M$ denotes the canonical projection onto M and T_i is the ith row of the matrix $T \in D^{(p-q) \times p}$ defined by

$$U^{-1} = \begin{pmatrix} R \\ T \end{pmatrix} \in \mathrm{GL}_p(D).$$

Example 3.20 Let us consider again Example 3.17. If we consider $d_i = \partial/\partial x_i$ instead of x_i, i. e., $D = \mathbb{Q}[d_1, d_2]$, $R = (d_1\, d_2^2 + 1 \quad 3\, d_2/2 + d_1 - 1 \quad 2\, d_1\, d_2) \in D^{1 \times 3}$, denote by $x = (x_1, x_2, x_3)$ and choose $\mathcal{F} = C^\infty(\mathbb{R}^3)$, we then obtain that the underdetermined linear system of PDEs $\ker_{\mathcal{F}}(R.) = \{y = (y_1\ y_2\ y_3)^T \in \mathcal{F}^3 \mid R\, y = 0\}$, namely,

$$d_1\, d_2^2\, y_1(x) + y_1(x) + \frac{3}{2}\, d_2\, y_2(x) + d_1\, y_2(x) - y_2(x) + 2\, d_1\, d_2\, y_3(x) = 0,$$

admits the parametrization $(y_1(x)\ y_2(x)\ y_3(x))^T = Q\,(z_1(x)\ z_2(x))^T$, where Q is the matrix of differential operators formed by the last two columns of the matrix U defined in Example 3.17 and $z = (z_1 \quad z_2)^T$ is any arbitrary element of $\mathcal{F}^2$, i. e.,

$$\begin{cases} y_1 = (-\frac{9}{4}\, d_1^3 + 3\, d_1^2 - d_1 - 1)\, d_2\, z_1 - \frac{3}{2}\, d_1\, z_1 + z_1 - 2\, d_1\, d_2\, z_2, \\ y_2 = \left(\frac{9}{4}\, d_1^4 - 3\, d_1^3 + d_1^2 + d_1\right)\, d_2^2\, z_1 + \left(\frac{3}{2}\, d_1^2 - d_1\right)\, d_2\, z_1 + z_1 + 2\, d_1^2\, d_2^2\, z_2, \\ y_3 = \left(-\frac{27}{16}\, d_1^4 + \frac{27}{8}\, d_1^3 - \frac{9}{4}\, d_1^2 - \frac{1}{4}\, d_1 + \frac{1}{2}\right)\, d_2\, z_1 + \left(-\frac{3}{2}\, d_1^2 + d_1\right)\, d_2\, z_2 + z_2. \end{cases}$$

Finally, if we denote by $T \in D^{2\times 3}$ the matrix formed by the last two rows of the matrix U^{-1}, namely,

$$\begin{pmatrix} d_1\, d_2 & 1 & 0 \\ \frac{1}{4}\,(3\,d_1^2 - 2\,d_1)\,d_2^2 + \frac{1}{8}\,(-9\,d_1^3 + 12\,d_1^2 - 4\,d_1)\,d_2 & \frac{1}{4}\,(3\,d_1 - 2)\,d_2 & \frac{1}{2}\,(3\,d_1^2 - 2\,d_1)\,d_2 \end{pmatrix},$$

we then have $T\,Q = I_2$, i. e., the parametrization Q of $\ker_{\mathcal{F}}(R.)$ is injective.

Now, if $M = D^{1\times p}/(D^{1\times q}\,R)$ is a projective D which is defined by a non full row rank matrix $R \in D^{q'\times p'}$, then, using Proposition 3.1, we first compute a full row rank matrix $R' \in D^{q'\times p'}$ satisfying

$$M \cong M' = D^{1\times p'}/(D^{1\times p'}\,R'),$$

and we then apply the previous QS-algorithm to $R' \in D^{q'\times p'}$ to obtain $U \in \mathrm{GL}_{p'}(D)$ such that $R'\,U = (I_{q'} \quad 0)$. Let $S' \in D^{p'\times q'}$, $Q' \in D^{p'\times(p'-q')}$, $T' \in D^{(p'-q')\times p'}$ be the matrices defined by

$$U = (S' \quad Q'), \quad U^{-1} = \begin{pmatrix} R' \\ T' \end{pmatrix}.$$

Then, we have the following split exact sequence:

$$0 \longrightarrow D^{1\times q'} \underset{\overset{.S'}{\longleftarrow}}{\overset{.R'}{\longrightarrow}} D^{1\times p'} \underset{\overset{.T'}{\longleftarrow}}{\overset{.Q'}{\longrightarrow}} D^{1\times(p'-q')} \longrightarrow 0. \tag{3.8}$$

We now need to precisely describe the isomorphism between M and M' in order to get a basis of M from one of M'. In order to do that, we take the same notations as the ones used at the end of section 3.1, namely, $R_1 = R$, $T_1 = (R_1^T \quad 0^T)^T$, $R' = (T_1 \quad S_2')$, $p_0 = p$, $p_1 = q$, $q' = p_1 + p_3'$, $p' = p_0 + p_2'$. We first easily check that we have the following commutative exact diagram

$$\begin{array}{ccccccc} D^{1\times p_1} & \xrightarrow{.R_1} & D^{1\times p_0} & \xrightarrow{\pi} & M & \longrightarrow 0 \\ \uparrow .X & & \uparrow .I_{p_0} & & \uparrow \mathrm{id}_M & \\ D^{1\times(p_1+p_3')} & \xrightarrow{.T_1} & D^{1\times p_0} & \xrightarrow{\pi} & M & \longrightarrow 0, \end{array}$$

where $X = (I_q^T \quad 0^T)^T$. Moreover, we also have the commutative exact diagram

$$\begin{array}{ccccccc} D^{1\times(p_1+p_3')} & \xrightarrow{.T_1} & D^{1\times p_0} & \xrightarrow{\pi} & M & \longrightarrow 0 \\ \uparrow .Z & & \uparrow .Y & & \uparrow \sigma & \\ D^{1\times(p_1+p_3')} & \xrightarrow{.R'} & D^{1\times(p_0+p_2')} & \xrightarrow{\pi'} & M' & \longrightarrow 0, \end{array}$$

where $Y = (I_{p_0}^T \quad 0^T)^T$, $Z = (I_{p_1}^T \quad 0^T)^T$ and the isomorphism σ is defined by

$$\forall\, m' = \pi'(\lambda), \quad \lambda = (\lambda_1\ \lambda_2) \in D^{1\times(p_0+p_2')}, \quad \sigma(m') = \pi(\lambda_1).$$

Combining the two commutative exact diagrams, we then obtain the following one:

$$\begin{array}{ccccccc} D^{1\times p_1} & \xrightarrow{.R_1} & D^{1\times p_0} & \xrightarrow{\pi} & M & \longrightarrow 0 \\ \uparrow .(Z\;X) & & \uparrow .Y & & \uparrow \sigma & \\ D^{1\times (p_1+p_3')} & \xrightarrow{.R'} & D^{1\times (p_0+p_2')} & \xrightarrow{\pi'} & M' & \longrightarrow 0. \end{array}$$

Hence, if we denote by $\{f_i\}_{1\leq i\leq (p'-q')}$ the standard basis of $D^{1\times (p'-q')}$, using (3.8), we then obtain that $\{\sigma(\pi'(f_i\,T')) = \pi(f_i\,(T'\,Y))\}_{1\leq i\leq (p'-q')}$ is a basis of M, i. e., a basis of M is defined by taking the residue classes of the rows of $(T'\,Y)\in D^{(p'-q')\times p_0}$.

We can check that the D-morphism $\sigma^{-1}: M \longrightarrow M'$ is defined by

$$\forall\, m = \pi(\lambda), \quad \lambda\in D^{1\times p_0}, \quad \sigma^{-1}(m) = \tau'(\lambda\,Y^T).$$

Then, using (3.8), we obtain the following split exact sequence

$$D^{1\times q} \underset{\xleftarrow{.S}}{\xrightarrow{.R}} D^{1\times p} \underset{\xleftarrow{.(T'\,Y)}}{\xrightarrow{.(Y^T\,Q')}} D^{1\times (p'-q')} \longrightarrow 0,$$

where $S\in D^{p\times q}$ is a *generalized inverse* of R, i. e., S satisfies $R\,S\,R = R$ ([42]). If we denote by $T' = (T_1'\quad T_2')$, where $T_1'\in D^{(p'-q')\times p}$ and $T_2'\in D^{(p'-q')\times p_2}$ and $Q' = ((Q_1')^T\quad (Q_2')^T)^T$, where $Q_1'\in D^{p\times (p'-q')}$ and $Q_2'\in D^{p_2\times (p'-q')}$, we then get

$$Y^T\,Q' = Q_1', \quad T'\,Y = T_1',$$

i. e., we need to select the first p columns of T' and the first p rows of Q'.

Remark 3.21 If the free D-module $M = D^{1\times p}/(D^{1\times q}\,R)$ is defined by the finite free resolution (3.1), where $R_1 = R$, $p_0 = p$ and $p_1 = q$, we point out that we only apply once the QS-algorithm to the matrix R' in order to obtain a basis of M contrary to the algorithm developed in [26] where the QS-algorithm is applied m times. Hence, our algorithm is generally more efficient than the one developed in [26].

If $\mathcal{F}$ is a D-module, then applying the functor $\hom_D(\cdot,\mathcal{F})$ to the previous split exact sequence, by 2 of Proposition 2.15, we then obtain the following split exact sequence:

$$\mathcal{F}^q \underset{\xrightarrow{S.}}{\xleftarrow{R.}} \mathcal{F}^p \underset{\xrightarrow{T_1'.}}{\xleftarrow{Q_1'.}} \mathcal{F}^{(p'-q')} \longleftarrow 0.$$

The system $\ker_{\mathcal{F}}(R.)$ admits the injective parametrization Q_1', namely

$$\ker_{\mathcal{F}}(R.) = Q_1'\,\mathcal{F}^{(p'-q')}, \quad T_1'\,Q_1' = I_{p'-q'}.$$

Remark 3.22 Let us now explain how we can compute a basis of free D-modules $\mathrm{im}_D(.R)$, $\ker_D(.R)$ and $\mathrm{coim}_D(.R) \triangleq D^{1\times q}/\ker_D(.R)$, where $R\in D^{q\times p}$.

1. A basis of $\mathrm{im}_D(.R) = D^{1\times q} R$ can be obtained as follows: we first compute the first syzygy D-module of $\mathrm{im}_D(.R)$ and we obtain a matrix $R_2 \in D^{r\times q}$ satisfying $\ker_D(.R) = D^{1\times r} R_2$. Let us denote by $M_2 = D^{1\times q}/(D^{1\times r} R_2) \cong D^{1\times q} R$. Using the method previously described, we can compute a basis of the free D-module M_2. We get $Q_2 \in D^{q\times l}$ and $T_2 \in D^{l\times q}$ such that we have the exact split sequence
$$D^{1\times r} \underset{\xleftarrow{.S_2}}{\xrightarrow{.R_2}} D^{1\times q} \underset{\xleftarrow{.T_2}}{\xrightarrow{.Q_2}} D^{1\times l} \longrightarrow 0,$$
where $S_2 \in D^{q\times r}$ denotes a generalized inverse of R_2. A basis of $D^{1\times q} R$ is then given by the D-linearly independent rows of the matrix $T_2 R \in D^{l\times p}$ and we have $D^{1\times q} R = D^{1\times l} (T_2 R)$.
2. Using the same notations as before, we have $\ker_D(.R) = D^{1\times r} R_2$ and a basis of the free D-module $\ker_D(.R)$ can then be obtained by computing a basis of $D^{1\times r} R_2$ as it was shown in the previous point.
3. Using again the same notations as in the first point, we get
$$\mathrm{coim}_D(.R) = D^{1\times q}/\ker_D(.R) = D^{1\times q}/(D^{1\times r} R_2),$$
and a basis of $\mathrm{coim}_D(.R)$ can be computed using the general method previously described in this section.

To finish, all the algorithms presented in this section were implemented in the package QUILLENSUSLIN ([12]). See the appendix for more details and examples.

4 Flat multidimensional linear systems

4.1 Computation of flat outputs of flat multidimensional systems

Our first motivation to study and implement constructive versions of the Quillen-Suslin theorem was the computation of *flat outputs* and injective parametrizations of *flat multidimensional linear systems* and, particularly, differential time-delay systems. The study of flat linear ordinary differential time-delay systems has recently been initiated in [18, 30]. As for nonlinear ordinary differential systems ([17]), this class of systems has interesting mathematical properties which can be used to do motion planning and tracking as shown in [30] and the references therein on explicit examples.

However, the theory of flat linear ordinary differential time-delay systems is still in its infancy and some concepts developed for nonlinear ordinary differential systems seem to have no counterparts for this second class of systems. In particular, for flat linear differential time-delay systems, we can wonder which kind of linear systems could play a similar role as the one played by the Brunovský systems for flat nonlinear systems ([17]). To answer this question, we first need to understand which kind of equivalence plays a similar role for differential time-delay linear systems as the one played by the *Lie-Bäcklund equivalence* for nonlinear differential systems ([17]). To our knowledge, these important questions have not been studied in the literature till now. This section aims at constructively answering these two questions.

As the differential time-delay systems is a particular class of multidimensional systems, we can define the concept of a flat multidimensional linear system in terms of the existence of an injective parametrization of the trajectories of the system ([5, 42, 62]).

Definition 4.1 Let $D = k[x_1, \ldots, x_n]$, $R \in D^{q \times p}$ and $\mathcal{F}$ a D-module. Then, the system $\ker_{\mathcal{F}}(R.)$ is called *flat* if there exist $Q \in D^{p \times m}$ and $T \in D^{m \times p}$ satisfying

$$\ker_{\mathcal{F}}(R.) = Q\,\mathcal{F}^m, \quad T\,Q = I_m.$$

In terms of the module-theoretic/behaviour approach recently developed for multidimensional linear systems ([5, 39, 32, 62, 63]), it means that the module M intrinsically associated with the multidimensional linear system is free over the commutative polynomial ring D of functional operators ([5, 16, 17, 30, 42]).

Proposition 4.2 [5] *Let $D = k[x_1, \ldots, x_n]$, $R \in D^{q \times p}$, $M = D^{1 \times p}/(D^{1 \times q}\,R)$ and $\mathcal{F}$ be an injective cogenerator D-module. Then, $\ker_{\mathcal{F}}(R.)$ is a flat system iff the D-module M is free. Moreover, the bases of the D-module M are then in a one-to-one correspondence with flat outputs of* $\ker_{\mathcal{F}}(R.)$.

Remark 4.3 Using the end of the section 2, we obtain that the condition that M is a free D-module is a sufficient condition for $\ker_{\mathcal{F}}(R.)$ to be a flat system.

Using Proposition 4.2 and the Quillen-Suslin theorem (see 4 of Theorem 2.5), we then get the following important corollary.

Corollary 4.4 *Let $D = k[x_1, \ldots, x_n]$, $R \in D^{q \times p}$, $M = D^{1 \times p}/(D^{1 \times q}\,R)$ and $\mathcal{F}$ be an injective cogenerator D-module. Then, $\ker_{\mathcal{F}}(R.)$ is a flat system iff the D-module M is projective.*

When R has a full row rank, then, using Theorem 2.6, a constructive test for flatness of multidimensional linear systems with constant coefficients consists in checking if the $q \times q$ minors of R do not simultaneously vanish on complex common zeros ([23, 59]). This last result can algorithmically be checked by computing a Gröbner or Janet basis of the ideal I of D generated by the $q \times q$ minors of R and check whether or not $1 \in I$. We can also check whether or not R admits a right-inverse over D ([4, 5, 42]).

In the general case, using Theorem 2.6, the projectiveness of M can constructively be obtained by verifying the vanishing of $\mathrm{ext}^i_D(N, D)$, for $i = 1, \ldots, n$, where N is the transposed D-module $N = D^{1 \times q}/(D^{1 \times p}\,R^T)$. Other possibilities are to compute the so-called *global dimension* of M ([53]) by means of Proposition 3.1 and Corollary 3.3 as it was shown in [49], check whether or not R admits a generalized inverse S over D, i. e., check for the existence of a matrix $S \in D^{p \times q}$ satisfying $R\,S\,R = R$ ([42]) or check some straightforward conditions on the so-called *Fitting ideals* of M as it is explained in [10].

However, we point out that, till now, there has been no easy way for obtaining the flat outputs of the system, i. e., the bases of the free D-module M. Hence, we are led to use constructive versions of the Quillen-Suslin theorem developed in the symbolic algebra community ([19, 26, 27, 35, 58]) for computing a basis of the free D-module M. It was our first main purpose for developing the package QUILLENSUSLIN ([12]). See the appendix for more details and examples.

Example 4.5 Let us consider the differential time-delay linear system (see [30])

$$\begin{cases} \dot{y}_1(t) - y_1(t-h) + 2\,y_1(t) + 2\,y_2(t) - 2\,u(t-h) = 0, \\ \dot{y}_1(t) + \dot{y}_2(t) - \dot{u}(t-h) - u(t) = 0. \end{cases} \tag{4.1}$$

Let us denote by $D = \mathbb{Q}\left[\frac{d}{dt}, \delta\right]$ the commutative ring of differential time-delay operators with rational contant coefficients, where $(d/dt)\,y(t) = \dot{y}(t)$ and $(\delta\,y)(t) = y(t-h)$, $h \in \mathbb{R}_+$. Let us also denote the matrix of functional operators defining (4.1) by

$$R = \begin{pmatrix} \frac{d}{dt} - \delta + 2 & 2 & -2\,\delta \\ \frac{d}{dt} & \frac{d}{dt} & -\frac{d}{dt}\,\delta - 1 \end{pmatrix} \in D^{2\times 3}.$$

Using the algorithms developed in [5, 42] and implemented in the package OREMODULES ([4]), we obtain that R admits a right-inverse over D defined by

$$S = \frac{1}{2}\begin{pmatrix} 0 & 0 \\ \frac{d}{dt}\,\delta + 2 & -2\,\delta \\ \frac{d}{dt} & -2 \end{pmatrix},$$

a fact proving that $M = D^{1\times 3}/(D^{1\times 2}\,R)$ is a projective, and thus, a free D-module by the Quillen-Suslin theorem (see 4 of Theorem 2.5).

Using a constructive version of the Quillen-Suslin theorem (see also the heuristic methods developed in [5, 42]), we obtain the following split exact sequence of D-modules

$$0 \longrightarrow D^{1\times 2} \underset{\overset{.S}{\longleftarrow}}{\overset{.R}{\longrightarrow}} D^{1\times 3} \underset{\overset{.T}{\longleftarrow}}{\overset{.Q}{\longrightarrow}} D \longrightarrow 0, \tag{4.2}$$

where $T = (1 \quad 0 \quad 0)$ and

$$Q = \frac{1}{2}\begin{pmatrix} 2 \\ -\frac{d^2}{dt^2}\,\delta + \frac{d}{dt}\,\delta^2 - \frac{d}{dt} + \delta - 2 \\ \frac{d}{dt}\,\delta - \frac{d^2}{dt^2} \end{pmatrix}.$$

Using the split exact sequence (4.2), we can check that we have

$$M = D^{1\times 3}/(D^{1\times 2}\,R) \cong (D^{1\times 3}\,Q) = D,$$

i. e., we find again that M is a free D-module of rank 1.

Now, if $\mathcal{F}$ is a D-module (e. g., $\mathcal{F} = C^\infty(\mathbb{R})$), by applying the functor $\hom_D(\cdot, \mathcal{F})$ to the split exact sequence (4.2), we then obtain the following split exact sequence of D-modules (see 2 of Proposition 2.15):

$$0 \longleftarrow \mathcal{F}^2 \underset{\xrightarrow{S.}}{\xleftarrow{R.}} \mathcal{F}^3 \underset{\xrightarrow{T.}}{\xleftarrow{Q.}} \mathcal{F} \longleftarrow 0.$$

Hence, for any D-module $\mathcal{F}$, we get that the system $\ker_{\mathcal{F}}(R.)$ defined by (4.1) is parametrized by the following injective parametrization:

$$\forall\, x_1 \in \mathcal{F}, \quad \begin{cases} y_1(t) = x_1(t), \\ y_2(t) = \dfrac{1}{2}\,(-\ddot{x}_1(t-h) + \dot{x}_1(t-2\,h) - \dot{x}_1(t) + x_1(t-h) - 2\,x_1(t)), \\ u(t) = \dfrac{1}{2}\,(\dot{x}_1(t-h) - \ddot{x}_1(t)). \end{cases} \tag{4.3}$$

We refer the reader to [49, 51] for a constructive algorithm for the computation of bases, and thus, of flat outputs of a class of linear systems defined by partial differential equations with polynomial or rational coefficients. See [50, 49] for an implementation of this algorithm in the package STAFFORD of the library OREMODULES.

Finally, we say that the $D = k[x_1, \ldots, x_n]$-module $M = D^{1\times p}/(D^{1\times q}\,R)$ is π-*free*, where $\pi \in D$, if the D_π-module $D_\pi \otimes_D M$ is free, where D_π denotes the localization $D_\pi = \{a/b \mid a \in D,\ b = \pi^i,\ i \in \mathbb{Z}_+\}$ of the ring D with respect to the multiplicatively closed subset $S_\pi = \{1,\ \pi,\ \pi^2,\ \ldots\}$ of D ([53]). By extension, we can define the concept of a π-flat system. See [5, 30, 31] for more details. Given a finitely presented $D = k[x_1, \ldots, x_n]$-module $M = D^{1\times p}/(D^{1\times q}\,R)$, constructive algorithms computing the corresponding polynomials π and basis of the free D_π-module $D_\pi \otimes_D M$ were given in [5] and implemented in the OREMODULES package ([4, 6]). However, we can also use Remark 3.19 to compute the corresponding basis in the case where $\pi = x_i$. We refer the reader to [14] for more details and examples.

4.2 Equivalences of flat multidimensional systems

Using a QS-algorithm, the purpose of this section is to prove that a flat multidimensional linear system with constant coefficients is algebraically equivalent to a linear controllable 1-D system obtained by setting all but one functional operator to 0 in the system matrix. In particular, the algebraic equivalence we use is the natural equivalence developed in module theory, namely, two multidimensional linear systems are said to be algebraically equivalent if their canonical associated modules are isomorphic over the underlying commutative polynomial ring of functional operators D. This equivalence is nothing else than the natural substitute to the Lie-Bäcklund equivalence for multidimensional linear systems. In the case of ordinary differential linear systems, we already know that Lie-Bäcklund transformations correspond to isomorphisms of the underlying modules (see e. g. [17] and the references therein). Finally, we prove that a flat differential time-delay linear system is algebraically equivalent to the controllable ordinary differential system without delays, namely, the system obtained by setting all the delay amplitudes to 0. This last system plays a similar role as the one played by the Brunovský canoncial form in the nonlinear case.

We have the following corollary of Theorem 3.16.

Corollary 4.6 *Let $D = k[x_1, \ldots, x_n]$, $R \in D^{q \times p}$ be a full row rank and $\mathcal{F}$ an injective cogenerator D-module. The flat multidimensional system $\ker_{\mathcal{F}}(R(x_1, \ldots, x_n).)$ is then D-isomorphic to a controllable 1-D linear system obtained by setting any functional operator to 0. For instance, the system $\ker_{\mathcal{F}}(R(x_1, \ldots, x_n).)$ is D-isomorphic to the system $\ker_{\mathcal{F}}(R(x_1, 0, \ldots, 0).)$ and the elements of $\ker_{\mathcal{F}}(R(x_1, \ldots, x_n).)$ are in a one-to-one correspondence with the ones of $\ker_{\mathcal{F}}(R(x_1, 0, \ldots, 0).)$.*

Proof. Using Proposition 4.2, we obtain that $M = D^{1\times p}/(D^{1\times q} R)$ is a free D-module. Using the fact that R has full row rank, by Theorem 3.16, there exists a matrix $U \in \mathrm{GL}_p(D)$ such that $RU = \overline{R}$, where $\overline{R} = R(x_1, 0, \ldots, 0)$. Therefore, we have the following commutative exact diagram

$$\begin{array}{ccccccccc}
 & & 0 & & 0 & & 0 & & \\
 & & \downarrow & & \downarrow & & \downarrow & & \\
0 & \longrightarrow & D^{1\times q} & \xrightarrow{.R} & D^{1\times p} & \xrightarrow{\pi} & M & \longrightarrow & 0 \\
 & & \| & & \downarrow .U & & \downarrow f & & \\
0 & \longrightarrow & D^{1\times q} & \xrightarrow{.\overline{R}} & D^{1\times p} & \xrightarrow{\kappa} & M' & \longrightarrow & 0, \\
 & & \downarrow & & \downarrow & & \downarrow & & \\
 & & 0 & & 0 & & 0 & &
\end{array}$$

where $\kappa : D^{1\times p} \longrightarrow M$ denotes the canonical projection onto M and the D-isomorphism $f : M \longrightarrow M'$ is defined by

$$\forall\, m = \pi(\lambda), \quad \lambda \in D^{1\times p}, \quad f(m) = \kappa(\lambda\, U).$$

Applying the functor $\hom_D(\cdot, \mathcal{F})$ to the previous commutative exact diagram and using the fact that horizontal exact sequences split because $M \cong M'$ is a free D-module, we then obtain the following commutative exact diagram:

$$\begin{array}{ccccccccc}
 & & 0 & & 0 & & 0 & & \\
 & & \uparrow & & \uparrow & & \uparrow & & \\
0 & \longleftarrow & \mathcal{F}^q & \xleftarrow{R.} & \mathcal{F}^p & \xleftarrow{\pi^\star} & \ker_{\mathcal{F}}(R.) & \longleftarrow & 0 \\
 & & \| & & \uparrow U. & & \uparrow f^\star & & \\
0 & \longleftarrow & \mathcal{F}^q & \xleftarrow{\overline{R}.} & \mathcal{F}^p & \xleftarrow{\kappa^\star} & \ker_{\mathcal{F}}(\overline{R}.) & \longleftarrow & 0. \\
 & & \uparrow & & \uparrow & & \uparrow & & \\
 & & 0 & & 0 & & 0 & &
\end{array}$$

The D-isomorphism $f^\star : \ker_{\mathcal{F}}(\overline{R}.) \longrightarrow \ker_{\mathcal{F}}(R.)$ is defined by

$$\forall\, \eta \in \ker_{\mathcal{F}}(\overline{R}.), \quad f^\star(\eta) = U\,\eta.$$

Hence, $f^\star$ induces a one-to-one correspondence between the elements of $\ker_{\mathcal{F}}(\overline{R}.)$ and those of $\ker_{\mathcal{F}}(R.)$ and $(f^\star)^{-1}$ is defined by

$$\forall\, \zeta \in \ker_{\mathcal{F}}(R.), \quad (f^\star)^{-1}(\zeta) = U^{-1}\,\zeta.$$

□

Using Corollary 3.3 and the end of the section 3.4, we can always reduce the case of a non full row rank matrix R to the case of a full row rank matrix R' and then apply Corollary 4.6 to R'.

Despite the fact that Corollary 4.6 is a straightforward consequence of the Quillen-Suslin theorem, its applications to flat multidimensional systems seem to be ignored. In particular, it shows that the Lie-Bäcklund equivalence in the nonlinear case needs to be replaced by the isomorphism equivalence in the multidimensional case. Moreover, the right substitute of the Brunovský linear system in the nonlinear case becomes the controllable 1-D linear linear system with constant coefficients obtained by setting all but one functional operator to 0.

Let us illustrate Corollary 4.6 on an example.

Example 4.7 Let us consider again the differential time-delay linear system defined by (4.1). In Example 4.5, we proved that the corresponding D-module M is free. It is well known that $\mathcal{F} = C^\infty(\mathbb{R})$ is not an injective D-module but, by Remark 4.3, the system $\ker_{\mathcal{F}}(R.)$ is flat as the D-module M is free. Hence, according to Corollary 4.6, the flat system (4.1) is algebraically equivalent to the following controllable ordinary differential linear system

$$\begin{cases} \dot{z}_1(t) + 2\,z_1(t) + 2\,z_2(t) = 0, \\ \dot{z}_1(t) + \dot{z}_2(t) - v(t) = 0, \end{cases} \tag{4.4}$$

i. e., the system obtained by setting δ to 0 in the matrix R. Using the constructive QS-algorithm to R, after a few computations, we obtain an invertible transformation which bijectively maps the trajectories of (4.1) to the ones of (4.4) is defined by

$$\begin{cases} y_1(t) = z_1(t), \\ y_2(t) = \dfrac{1}{2}\,(\dot{z}_1(t-2\,h) + z_1(t-h)) + z_2(t) + v(t-h), \\ u(t) = \dfrac{1}{2}\,\dot{z}_1(t-h) + v(t) \end{cases}$$
$$\Leftrightarrow \quad \begin{cases} z_1(t) = y_1(t), \\ z_2(t) = -\dfrac{1}{2}\,y_1(t-h) + y_2(t) - u(t-h), \\ v(t) = -\dfrac{1}{2}\,\dot{y}_1(t-h) + u(t). \end{cases} \tag{4.5}$$

Applying again Corollary 4.6 to (4.4), we get that the ordinary differential system (4.4) is equivalent to the purely algebraic system

$$\begin{cases} 2\,x_1(t) + 2\,x_2(t) = 0, \\ -w(t) = 0, \end{cases} \tag{4.6}$$

i. e., the system obtained by setting to δ and d/dt to 0 in R. Applying a QS-algorithm to R,we obtain that a transformation which bijectively maps the trajectories of (4.4) to the ones of (4.6) is defined by

$$\begin{cases} z_1(t) = x_1(t), \\ z_2(t) = x_2(t) - \frac{1}{2}\,\dot{x}_1(t), \\ v(t) = w(t) - \frac{1}{2}\,\ddot{x}_1(t) + \dot{x}_1(t) + \dot{x}_2(t) \end{cases} \Leftrightarrow \begin{cases} x_1(t) = z_1(t), \\ x_2(t) = z_2(t) + \frac{1}{2}\,\dot{z}_1(t), \\ w(t) = v(t) + \dot{z}_1(t) + \dot{z}_2(t). \end{cases} \tag{4.7}$$

Combining (4.5) and (4.7), we finally obtain a one-to-one correspondence between the solutions of (4.1) and (4.6).

We note that the solutions of (4.1) (resp., (4.4)) are parametrized by means of (4.5) (resp., (4.7)), where z_1, z_2 and v (resp., x_1, x_2 and w) are not arbitrary functions as they must satisfy (4.4) (resp., (4.6)). However, solving the algebraic system (4.6), we obtain that $x_2 = -x_1$ and $w = 0$. Substituting these values in (4.7) and the result into (4.5), we find that an injective parametrization of (4.1) is defined by (4.3).

Finally, we can check that an injective parametrization of (4.4) is obtained by setting $\delta = 0$ in the matrix of operators defining (4.3), i. e.,

$$\forall\,\psi \in \mathcal{F}, \quad \begin{cases} z_1(t) = \psi(t), \\ z_2(t) = -\frac{1}{2}\,(\ddot{\psi}(t) + 2\,\psi(t)), \\ v(t) = -\frac{1}{2}\,\ddot{\psi}(t). \end{cases}$$

Similarly, if we set δ and d/dt to 0 in the matrix of operators defining (4.3), we obtain the following injective parametrization of (4.6):

$$\forall\,\varphi \in \mathcal{F}, \quad \begin{cases} x_1(t) = \varphi(t), \\ x_2(t) = -\varphi(t), \\ w(t) = 0. \end{cases}$$

These last results can be obtained by applying the functor $(D/(D\,\delta)) \otimes_D \cdot$ (resp., $(D/\left(D\,\delta + D\,\frac{d}{dt}\right)) \otimes_D \cdot$) to the split exact sequence (4.2) to get the corresponding split exact sequence of $D/(D\,\delta)$-modules (resp., $D/\left(D\,\delta + D\,\frac{d}{dt}\right)$-modules) ([53]).

See [13, 14] for more examples.

Using Corollary 3.18, we can also set the different functional operators appearing in the system matrix of a flat multidimensional linear system to any particular value belonging to k. Applying this result to the class of flat differential time-delay linear systems, we show that a flat differential time-delay linear system is equivalent to the controllable ordinary differential linear system obtained by setting all the time-delay amplitudes to 0, i. e., to the corresponding ordinary differential system without delays.

Corollary 4.8 *Let $D = k\left[\frac{d}{dt}, \delta_1, \ldots, \delta_{n-1}\right]$ be the ring of differential incommensurable time-delay operators, namely, the amplitudes $h_i \in \mathbb{R}_+$ of the time-delay operator $(\delta_i\, y)(t) = y(t - h_i)$, $i = 1, \ldots, n-1$, are such that the $\mathbb{Q}$-vector space generated by $h_1, \ldots, h_{n-1}$ is n-dimensional. Let us consider $R \in D^{q\times p}$ which admits a right-inverse over D and $\mathcal{F}$ an injective cogenerator D-module. Then, the time-invariant flat differential time-delay linear system $\ker_{\mathcal{F}}(R\left(\frac{d}{dt}, \delta_1, \ldots, \delta_{n-1}\right).)$ is D-isomorphic to the controllable ordinary differential linear system $\ker_{\mathcal{F}}(R\left(\frac{d}{dt}, 1, \ldots, 1\right).)$ obtained by setting the amplitudes of all the delays to 0, i. e., it is equivalent to the linear system without delays. In particular, the elements of the system $\ker_{\mathcal{F}}(R\left(\frac{d}{dt}, \delta_1, \ldots, \delta_{n-1}\right).)$ are in a one-to-one correspondence with the ones of $\ker_{\mathcal{F}}(R\left(\frac{d}{dt}, 1, \ldots, 1\right).)$.*

Let us illustrate Corollary 4.8 on an example.

Example 4.9 Let us consider again the flat differential time-delay linear system defined by (4.1). Applying Corollary 4.8 on (4.1), we obtain that (4.1) is equivalent to the ordinary differential linear system obtained by substituting $h = 0$ into (4.1), i. e., by setting $\delta = 1$ in the matrix R defined in Example 4.5, namely

$$\begin{cases} \dot{z}_1(t) + z_1(t) + 2\, z_2(t) - 2\, v(t) = 0, \\ \dot{z}_1(t) + \dot{z}_2(t) - \dot{v}(t) - v(t) = 0. \end{cases} \tag{4.8}$$

Using a QS-algorithm, we then obtain that the following transformation

$$\begin{cases} z_1(t) = y_1(t), \\ z_2(t) = \dfrac{1}{2}\,(\dot{y}_1(t) - \dot{y}_1(t-h) + y_1(t) - y_1(t-h)) + y_2(t) + u(t) - u(t-h), \\ v(t) = \dfrac{1}{2}\,(\dot{y}_1(t) - \dot{y}_1(t-h)) + u(t), \end{cases} \tag{4.9}$$

whose inverse is defined by

$$\begin{cases} y_1(t) = z_1(t), \\ y_2(t) = -\dfrac{1}{2}\,(\dot{z}_1(t-h) - \dot{z}_1(t-2\,h) + z_1(t-h) - z_1(t)) + z_2(t) \\ \qquad + v(t-h) - v(t), \\ u(t) = \dfrac{1}{2}\,(\dot{z}_1(t-h) - \dot{z}_1(t)) + v(t), \end{cases}$$

bijectively maps the trajectories of (4.1) to the ones of (4.8). An injective parametrization of (4.8) can then be obtained by taking $h = 0$ in (4.3), i. e.,

$$\forall\, \psi \in \mathcal{F}, \quad \begin{cases} z_1(t) = \psi(t), \\ z_2(t) = -\dfrac{1}{2}\,(\ddot{\psi}(t) + \psi(t)), \\ v(t) = \dfrac{1}{2}\,(-\dddot{\psi}(t) + \dot{\psi}(t)). \end{cases}$$

See [13, 14] for more examples.

In the previous example, we note that the invertible transformations can easily be computed by hand but it is generally not the case for more complicated examples. Hence, we need to use an implementation of constructive versions of the Quillen-Suslin theorem for computing the invertible transformations and the injective parametrizations of flat multidimensional linear systems. Such an implementation has recently been done in the package QUILLENSUSLIN ([12]) which, with the library OREMODULES ([4]), allows us to effectively handle these difficult computations.

As for the flat nonlinear ordinary differential systems, using the fact that there is a one-to-one correspondence between the trajectories of the flat differential time-delay systems with those of the ordinary differential system without delays, we can use stabilizing controllers of the latter in order to stabilize the former. This approach echoes the Smith predictor method. We refer the reader to [13, 14] for more details and examples.

5 Pommaret's theorem of Lin-Bose's conjectures

The purpose of this section is to show how to use a QS-algorithm to constructively solve Pommaret's theorem of Lin-Bose's conjectures ([41]). Let us first recall this conjecture recently developed in the multidimensional systems theory which generalizes Serre's conjecture ([25]). Let us state a new problem.

Problem 3 *Let $D = k[x_1, \ldots, x_n]$ be a commutative polynomial ring with coefficients in a field k, $R \in D^{q\times p}$ a full row rank matrix and $M = D^{1\times p}/(D^{1\times q}\,R)$ the D-module finitely presented by R. We suppose that $M/t(M)$ is a free D-module. Does exist a full row rank matrix $R' \in D^{q\times p}$ satisfying*

$$M/t(M) = D^{1\times p}/(D^{1\times q}\,R')?$$

If so, compute such a matrix R'.

If we can solve Problem 3, we then have

$$t(M) = (D^{1\times q}\,R')/(D^{1\times q}\,R),$$

and using the fact that $D^{1\times q}\,R \subseteq D^{1\times q'}\,R'$, there exists $R'' \in D^{q\times q}$ such that

$$R = R''\,R'. \tag{5.1}$$

Let us denote by $r = p!/((p-q)!\,q!)$. The fact that $M/t(M)$ is a projective D-module implies that there is no common zero in the $q \times q$ minors $\{m'_i\}_{1 \leq i \leq r}$ of R', i. e., there exists a family $\{p_i\}_{1 \leq i \leq r}$ of elements of D satisfying the following Bézout identity:

$$\sum_{i=1}^{r} p_i\, m'_i = 1. \tag{5.2}$$

Now, using the fact that we have $m_i = (\det R'')\, m'_i$, for $i = 1, \ldots, r$, where the m_i denote the $q \times q$-minors of R, we obtain that the following inclusion of ideals of D:

$$\sum_{i=1}^{r} D\, m_i \subseteq (D\,(\det R''))\left(\sum_{i=1}^{r} D\, m'_i\right) = D\,(\det R'').$$

Multiplying (5.2) by $\det R''$, we obtain

$$\det R'' = \sum_{i=1}^{r} p_i\,(\det R'')\, m'_i = \sum_{i=1}^{r} p_i\, m_i,$$

which shows that $D\,(\det R'') \subseteq \sum_{i=1}^{r} D\, m_i$ and $\sum_{i=1}^{r} D\, m_i = D\,(\det R'')$. Hence, the greatest common divisor of the $q \times q$ minors $\{m_i\}_{1 \leq i \leq r}$ is then equal to $\det R''$.

Solving Problem 3 gives us a way to factorize R under the form $R = R''\, R'$, where $R' \in D^{q \times p}$ admits a right-inverse over D and $\det R''$ is the greatest common divisor of the $q \times q$ minors of R. The question of the possibility to achieve this factorization was first asked by Lin and Bose in [25] and solved by Pommaret in [41]. See also [36, 54, 60]. It was proved in [41] that this factorization problem is equivalent to Problem 3. The purpose of this paragraph is to give a general constructive algorithm which solves Problem 3, and thus, performs the corresponding factorization. The algorithm has recently been implemented in the package QUILLENSUSLIN.

Based on the Quillen-Suslin theorem, we first prove that a matrix R' satisfying Problem 3 always exists. We then show how to effectively compute it.

The fact that R has full row rank implies that we have the following exact sequence:

$$0 \longrightarrow D^{1\times q} \xrightarrow{.R} D^{1\times p} \xrightarrow{\pi} M \longrightarrow 0. \tag{5.3}$$

Let $N = D^{1\times q}/(D^{1\times p}\, R^T)$ be the transposed D-module of M (see Remark 2.7), according to Theorem 2.6, there exists $Q \in D^{q' \times p}$ such that

$$M/t(M) = D^{1\times p}/(D^{1\times q'}\, Q).$$

In particular, using the fact that $(D^{1\times q}\, R) \subseteq (D^{1\times q'}\, Q)$, there then exists a matrix $P \in D^{q \times q'}$ satisfying $R = P\, Q$. We refer the reader to [4] for the implementation of the corresponding algorithms in the library OREMODULES as well as the large library of examples which demonstrates these results.

Then, we have the following commutative exact diagram:

$$\begin{array}{ccccccc} & & & & 0 & & \\ & & & & \downarrow & & \\ & & 0 & & t(M) & & \\ & & \downarrow & & \downarrow i & & \\ 0 \longrightarrow & D^{1\times q} & \xrightarrow{.R} & D^{1\times p} & \xrightarrow{\pi} & M & \longrightarrow 0 \\ & \downarrow .P & & \| & & \downarrow \rho & \\ & D^{1\times q'} & \xrightarrow{.Q} & D^{1\times p} & \xrightarrow{\pi'} & M/t(M) & \longrightarrow 0. \\ & & & \downarrow & & \downarrow & \\ & & & 0 & & 0 & \end{array}$$

As, by hypothesis, the D-module $M/t(M)$ is projective, using 1 of Proposition 2.15, we obtain that the following exact sequence

$$0 \longrightarrow (D^{1\times q'}\, Q) \longrightarrow D^{1\times p} \xrightarrow{\pi'} M/t(M) \longrightarrow 0 \qquad (5.4)$$

splits and we obtain

$$D^{1\times p} \cong M/t(M) \oplus (D^{1\times q'}\, Q),$$

which shows that $D^{1\times q'}\, Q$ is a projective D-module. By the Quillen-Suslin theorem, we obtain that $D^{1\times q'}\, Q$ is then a free D-module.

Let us compute the rank of the free D-module $D^{1\times q'}\, Q$. Applying the exact functor $K \otimes_D \cdot$ to the short exact sequence (5.4), where $K = Q(D)$ denotes the quotient field of D ([53]), we obtain that

$$\operatorname{rank}_D(D^{1\times q'}\, Q) = p - \operatorname{rank}_D(M/t(M)).$$

See [53] for more details (*Euler characteristic*). Similarly with the two short exact sequences (5.3) and

$$0 \longrightarrow t(M) \xrightarrow{i} M \xrightarrow{\rho} M/t(M) \longrightarrow 0,$$

and, using the fact that $K \otimes_D t(M) = 0$ because $t(M)$ is a torsion D-module ([53]), we then get

$$\operatorname{rank}_D(M/t(M)) = \operatorname{rank}_D(M) = p - q.$$

Therefore, we obtain $\operatorname{rank}_D(D^{1\times q'}\, Q) = p - (p-q) = q$, which shows that $D^{1\times q'}\, Q$ is a free D-module of rank q, i. e., $(D^{1\times q'}\, Q) \cong D^{1\times q}$. Computing a basis of this free D-module, we obtain a full row rank matrix $R' \in D^{q\times p}$ satisfying

$$D^{1\times q'}\, Q = D^{1\times q}\, R', \qquad (5.5)$$

which implies that $M/t(M) = D^{1\times p}/(D^{1\times q}\, R')$ and we have the following finite free short resolution of $M/t(M)$:

$$0 \longrightarrow D^{1\times q} \xrightarrow{.R'} D^{1\times p} \xrightarrow{\pi'} M/t(M) \longrightarrow 0. \qquad (5.6)$$

We note that if Q has full row rank, we then can take $R' = Q$ and $q' = q$.

In order to compute the matrix $R' \in D^{q \times p}$ which satisfies (5.5), we need to compute a basis of the free D-module $D^{1 \times q'}\, Q$. Hence, we can use the first point of Remark 3.22 to compute a basis of the D-module $D^{1 \times q'}\, Q$.

Algorithm 5.1 • **Input:** A commutative polynomial ring $D = k[x_1, \ldots, x_n]$ over a computable field k, a full row rank matrix $R \in D^{q \times p}$ and the finitely presented D-module $M = D^{1 \times p}/(D^{1 \times q}\, R)$ such that $M/t(M)$ is a free D-module.

- **Output:** A full row rank matrix $R' \in D^{q \times p}$ such that

$$M/t(M) = D^{1 \times p}/(D^{1 \times q}\, R').$$

1. Transpose the matrix R and define the finitely presented D-module

$$N = D^{1 \times q}/(D^{1 \times p}\, R^T).$$

2. Compute the D-module $\operatorname{ext}_D^1(N, D)$. We obtain a matrix $Q \in D^{q' \times p}$ such that

$$M/t(M) = D^{1 \times p}/(D^{1 \times q'}\, Q).$$

3. Compute the first syzygy module $\ker_D(.Q)$ of $D^{1 \times q'}\, Q$.
4. If $\ker_D(.Q) = 0$, then Q has full row rank and exit the algorithm with $R' = Q$. Else, denote by $Q_2 \in D^{q_2' \times q'}$ a matrix satisfying $\ker_D(.Q) = D^{1 \times q_2'}\, Q_2$.
5. Compute a basis of the free D-module

$$L = D^{1 \times q'}/(D^{1 \times q_2'}\, Q_2).$$

In particular, we obtain a full row rank matrix $B \in D^{q \times q'}$ such that

$$L = \pi_2(D^{1 \times q}\, B),$$

where $\pi_2 : D^{1 \times q'} \longrightarrow L$ denotes the canonical projection onto L.
6. Return the full row rank matrix $R' = B\, Q \in D^{q \times p}$.

Remark 5.2 The computation of a basis of L gives two matrices $P_2 \in D^{q' \times q}$ and $B \in D^{q \times q'}$ such that we have the following split exact sequence:

$$\begin{array}{ccccccc}
 & & & & 0 & & \\
 & & & & \uparrow & & \\
D^{1 \times q_2'} & \xrightarrow{.Q_2} & D^{1 \times q'} & \xrightarrow{\pi_2} & L & \longrightarrow & 0 \\
 & & \| & & \uparrow \phi & & \\
 & & D^{1 \times q'} & \underset{\xleftarrow{.B}}{\xrightarrow{.P_2}} & D^{1 \times q} & \longrightarrow & 0, \\
 & & & & \uparrow & & \\
 & & & & 0 & &
\end{array}$$

where $\phi : D^{1\times q} \longrightarrow L$ denotes the corresponding isomorphism. We can now check that the matrix $R' = B\,Q$ has full row rank. Let $\lambda \in D^{1\times q}$ be such that $\lambda\,R' = 0$. Then, we get $(\lambda\,B)\,Q = 0$, i.e., $\lambda\,B \in \ker_D(.Q) = D^{1\times q_2'}\,Q_2$, and thus, there exists $\mu \in D^{1\times q_2'}$ such that $\lambda\,B = \mu\,Q_2$. Using the identity $B\,P_2 = I_q$, we then obtain

$$\lambda = (\lambda\,B)\,P_2 = \mu\,(Q_2\,P_2) = 0.$$

We illustrate Algorithm 5.1 on a simple example.

Example 5.3 Let us consider the differential time-delay model of a flexible rod with a force applied on one end developed in [30]:

$$\begin{cases} \dot y_1(t) - \dot y_2(t-1) - u(t) = 0, \\ 2\,\dot y_1(t-1) - \dot y_2(t) - \dot y_2(t-2) = 0. \end{cases} \tag{5.7}$$

Let us define the ring $D = \mathbb{Q}\left[\frac{d}{dt}, \delta\right]$ of differential time-delay operators with rational constant coefficients. The system matrix of (5.7) is defined by

$$R = \begin{pmatrix} \dfrac{d}{dt} & -\dfrac{d}{dt}\,\delta & -1 \\ 2\,\dfrac{d}{dt}\,\delta & -\dfrac{d}{dt}\,\delta^2 - \dfrac{d}{dt} & 0 \end{pmatrix} \in D^{2\times 3}.$$

Let $M = D^{1\times 3}/(D^{1\times 2}\,R)$ be the D-module associated with (5.7) and its transposed D-module $N = D^{1\times 2}/(D^{1\times 3}\,R^T)$. N admits the following finite free resolution

$$0 \longleftarrow N \xleftarrow{\sigma} D^{1\times 2} \xleftarrow{.R^T} D^{1\times 3} \xleftarrow{.R_2^T} D \longleftarrow 0,$$

where $R_2^T = \left(-\delta^2 - 1 \quad -2\,\delta \quad \frac{d}{dt}\,\delta^2 - \frac{d}{dt}\right)$. The defects of exactness of the complex

$$0 \longrightarrow D^{1\times 2} \xrightarrow{.R} D^{1\times 3} \xrightarrow{.R_2} D \longrightarrow 0$$

are then defined by

$$\begin{cases} \mathrm{ext}_D^0(N, D) = \ker_D(.R) = 0, \\ \mathrm{ext}_D^1(N, D) = \ker_D(.R_2)/(D^{1\times 2}\,R), \\ \mathrm{ext}_D^2(N, D) = D/(D^{1\times 3}\,R_2). \end{cases}$$

Computing the first syzygy module $\ker_D(.R_2)$ of $D^{1\times 2}\,R$, we obtain $\ker_D(.R_2) = D^{1\times 3}\,Q$, where the matrix Q is defined by

$$Q = \begin{pmatrix} -2\,\delta & \delta^2 + 1 & 0 \\ -\dfrac{d}{dt} & \dfrac{d}{dt}\,\delta & 1 \\ \dfrac{d}{dt}\,\delta & -\dfrac{d}{dt} & \delta \end{pmatrix} \in D^{3\times 3}. \tag{5.8}$$

We get $t(M) \cong (D^{1\times 3}\, Q)/(D^{1\times 2}\, R)$ and reducing the rows of Q with respect to $D^{1\times 2}\, R$, we obtain that the only non-trivial torsion element of M is defined by

$$\begin{cases} m = -2\,\delta\, y_1 + (\delta^2 + 1)\, y_2, \\ \dfrac{d}{dt}\, m = 0, \end{cases}$$

where y_1, y_2 and y_3 denote the residue classes of the standard basis of $D^{1\times 3}$ in M.

Following Algorithm 5.1, we compute the first syzygy module $\ker_D(.Q)$ and obtain $\ker_D(.Q) = D\, Q_2$, where

$$Q_2 = \begin{pmatrix} \dfrac{d}{dt} & -\,\delta & 1 \end{pmatrix} \in D^{1\times 3}. \tag{5.9}$$

We now have to compute a basis of the free D-module $L = D^{1\times 3}/(D\, Q_2)$. Using a constructive version of the Quillen-Suslin theorem, we obtain the split exact sequence

$$0 \longrightarrow D \underset{\underleftarrow{.S_2}}{\xrightarrow{.Q_2}} D^{1\times 2} \underset{\underleftarrow{.B}}{\xrightarrow{.P_2}} D \longrightarrow 0$$

with the following notations:

$$S_2 = (0 \;\; 0 \;\; 1)^T, \quad P_2 = \begin{pmatrix} -1 & 0 \\ 0 & 1 \\ \dfrac{d}{dt} & \delta \end{pmatrix}, \quad B = \begin{pmatrix} -1 & 0 & 0 \\ 0 & 1 & 0 \end{pmatrix}.$$

Computing $R' = B\, Q$, we obtain that the following full row rank matrix

$$R' = \begin{pmatrix} 2\,\delta & -\delta^2 - 1 & 0 \\ -\dfrac{d}{dt} & \dfrac{d}{dt}\,\delta & 1 \end{pmatrix} \in D^{2\times 3}$$

satisfies $D^{1\times 3}\, Q = D^{1\times 2}\, R'$. Finally, we have the factorization $R = R''\, R'$, where the R'' is defined by

$$R'' = \begin{pmatrix} 0 & -1 \\ \dfrac{d}{dt} & 0 \end{pmatrix},$$

and satisfies $\det R'' = d/dt$, where d/dt is the greatest common divisor of the 2×2 minors of R and is the functional operator which annihilates the torsion element m.

Using the fact that $M/t(M)$ is a free D-module of rank $p - q$, i. e., there exists an isomorphism

$$\psi : M/t(M) \longrightarrow D^{1\times (p-q)},$$

and the exact sequence (5.6), we then obtain the following exact sequence

$$0 \longrightarrow D^{1\times q} \xrightarrow{.R'} D^{1\times p} \xrightarrow{.P} D^{1\times (p-q)} \longrightarrow 0, \tag{5.10}$$

where $P \in D^{p\times(p-q)}$ is the matrix defining the morphism $\pi' \circ \psi$ in the standard bases of $D^{1\times p}$ and $D^{1\times(p-q)}$. As the exact sequence (5.10) ends with a free D-module, by 1 of Proposition 2.15, it splits, i. e., there exist $S \in D^{p\times q}$ and $T \in D^{(p-q)\times p}$ such that we have the following Bézout identities:

$$\begin{pmatrix} R' \\ T \end{pmatrix} (S \quad P) = \begin{pmatrix} I_q & 0 \\ 0 & I_{p-q} \end{pmatrix} = I_p, \tag{5.11}$$

$$(S \quad P) \begin{pmatrix} R' \\ T \end{pmatrix} = I_p. \tag{5.12}$$

Now, we have

$$\begin{pmatrix} R \\ T \end{pmatrix} = \begin{pmatrix} R''\, R' \\ T \end{pmatrix} = \begin{pmatrix} R'' & 0 \\ 0 & I_{p-q} \end{pmatrix} \begin{pmatrix} R' \\ T \end{pmatrix}$$

and using (5.11), we obtain that $\det((R'^T \quad T^T)^T) = 1$ and

$$\det \begin{pmatrix} R \\ T \end{pmatrix} = \det \begin{pmatrix} R'' & 0 \\ 0 & I_{p-q} \end{pmatrix} \det \begin{pmatrix} R' \\ T \end{pmatrix} = \det R''.$$

Finally, using the fact that we have proved that $\det R''$ is the greatest common divisor of the $q \times q$ minors of the matrix R, we then have solved the following problem.

Problem 4 *Let $R \in D^{q\times p}$ be a full row rank matrix such that the ideal $\sum_{i=1}^{r} D\, m_i$ of D generated by the $q \times q$ minors $\{m_i\}_{1\le i\le r}$ of the matrix R satisfies*

$$\sum_{i=1}^{r} D\, m_i = D\, d,$$

where d denotes the greatest common divisor of the $q \times q$ minors of the matrix R. Find a matrix $T \in D^{(p-q)\times p}$ such that we have

$$\det \begin{pmatrix} R \\ T \end{pmatrix} = d.$$

To our knowledge, such a problem was first stated by Bose and Lin in [25]. Let us give a constructive algorithm solving Problem 4.

Algorithm 5.4

- **Input:** A commutative polynomial ring $D = k[x_1, \ldots, x_n]$ over a computable field k, a full row rank matrix $R \in D^{q\times p}$ such that the ideal of D generated by the $q \times q$ minors $\{m_i\}_{1\le i\le r}$ of R satisfies $\sum_{i=1}^{r} D\, m_i = D\, d$, where d denotes the greatest common divisor of the $q \times q$ minors of R.

- **Output:** A matrix $T \in D^{(p-q)\times p}$ such that $\det \begin{pmatrix} R \\ T \end{pmatrix} = d$.

1. Transpose the matrix R and define the finitely presented D-module
$$N = D^{1\times q}/(D^{1\times p}\, R^T).$$
2. Compute the D-module $\mathrm{ext}^1_D(N, D)$. We obtain a matrix $Q \in D^{q'\times p}$ such that
$$M/t(M) = D^{1\times p}/(D^{1\times q'}\, Q).$$
3. Compute a basis of the free D-module $M/t(M) = D^{1\times p}/(D^{1\times q'}\, Q)$. In particular, we obtain a full row rank matrix $T \in D^{(p-q)\times p}$ such that $M/t(M) = \pi'(D^{1\times(p-q)}\, T)$, where $\pi' : D^{1\times p} \longrightarrow M/t(M)$ denotes the canonical projection on $M/t(M)$.
4. Return the matrix $U = (R^T \quad T^T)^T$ which satisfies $\det U = d$.

We illustrate Algorithm 5.4 on an example.

Example 5.5 We consider again the model of a flexible rod defined in (5.7). In Example 5.3, we have proved that $M/t(M) = D^{1\times 3}/(D^{1\times 3}\, Q)$, where the matrix Q is defined by (5.8). Let us compute a basis of the free D-module $M/t(M)$. The D-module $M/t(M)$ admits the following free resolution

$$0 \longrightarrow D \xrightarrow{.Q_2} D^{1\times 3} \xrightarrow{.Q} D^{1\times 3} \xrightarrow{\pi'} M/t(M) \longrightarrow 0,$$

where Q_2 is defined by (5.9). Using the fact that Q_2 admits the right-inverse S_2 defined by (5.3), we obtain the following minimal free resolution of $M/t(M)$

$$0 \longrightarrow D^{1\times 3} \xrightarrow{.\overline{Q}} D^{1\times 4} \xrightarrow{\pi'\oplus 0} M/t(M) \longrightarrow 0,$$

where the full row rank matrix $\overline{Q}$ is defined by $\overline{Q} = (Q^T \quad S_2^T)^T$.

Applying a constructive version of the Quillen-Suslin theorem to $\overline{Q}$, we then find that a basis of $M/t(M)$ is given by $(\pi' \oplus 0)(\overline{T})$, where $\overline{T}$ denotes the matrix

$$\overline{T} = \begin{pmatrix} 1 & \frac{1}{2}\,\delta & 0 & 0 \end{pmatrix}.$$

If we denote by T the matrix defined by the first three entries of $\overline{T}$, we then obtain a square matrix $U = (R^T \quad T^T)^T$ satisfying $\det U = d/dt$.

The explicit computation of $\mathrm{ext}^1_D(N, D)$ gives a matrix $R_{-1} \in D^{p\times m}$ which satisfies $\ker_D(.R_{-1}) = D^{1\times q'}\, Q$, i. e., such that we have the following exact sequence:

$$D^{1\times q'} \xrightarrow{.Q} D^{1\times p} \xrightarrow{.R_{-1}} D^{1\times m}.$$

A direct way to solve Problem 4 exists when the matrix R_{-1} admits a left-inverse $S_{-1} \in D^{m\times p}$. Then, we have $M/t(M) \cong D^{1\times p}\, R_{-1} = D^{1\times m}$ and using the fact that

$\operatorname{rank}_D(M/t(M)) = p - q$, we get $m = p - q$. The fact that $D^{1\times q'} Q$ is a free D-module of rank q implies that there exists a full row rank matrix $R' \in D^{q\times p}$ satisfying $D^{1\times q'} Q = D^{1\times q} R'$. Combining this result with the previous exact sequence, we obtain the split exact sequence

$$0 \longrightarrow D^{1\times q} \xrightarrow{.R'} D^{1\times p} \xrightarrow{.R_{-1}} D^{1\times (p-q)} \longrightarrow 0,$$

which shows that $P = R_{-1}$ and $T = S_{-1}$ solve Problem 4.

Let us illustrate this last remark on an example.

Example 5.6 Let us consider again the model of a flexible rod defined in (5.7) and let us compute $T \in D^{1\times 3}$ such that the determinant of the matrix $(R^T \quad T^T)^T$ equals d/dt. In Example 5.3, we proved that we have the exact sequence

$$D^{1\times 3} \xrightarrow{.Q} D^{1\times 3} \xrightarrow{.R_2} D,$$

where $R_2 = \left(-\delta^2 - 1 \quad -2\,\delta \quad \frac{d}{dt}\,\delta^2 - \frac{d}{dt}\right)^T$. R_2 admits a left-inverse T defined by

$$T = \left(1 \quad -\frac{1}{2}\,\delta \quad 0\right),$$

which proves that $M/t(M)$ is a free D-module of rank 1 as we have the isomorphisms

$$M/t(M) = D^{1\times 3}/(D^{1\times 3}\, Q) \cong (D^{1\times 3}\, R_2) \cong D.$$

We finally obtain that the matrix defined by

$$U = \begin{pmatrix} R \\ T \end{pmatrix} = \begin{pmatrix} \frac{d}{dt} & -\frac{d}{dt}\,\delta & -1 \\ 2\,\frac{d}{dt}\,\delta & -\frac{d}{dt}\,\delta^2 - \frac{d}{dt} & 0 \\ 1 & -\frac{1}{2}\,\delta & 0 \end{pmatrix}$$

satisfies $\det U = d/dt$, which solves Problem 4.

To finish, let us show how to handle an example given in [61] by means of Algorithms 5.1 and 5.4.

Example 5.7 Let us consider the commutative polynomial ring $D = \mathbb{Q}[z_1, z_2, z_3]$ and the following matrix defined in [61]:

$$R = \begin{pmatrix} z_1\, z_2^2\, z_3 & 0 & -z_1^2\, z_2^2 - 1 \\ z_1^2\, z_3^2 + z_3 & -z_3 & -z_1^3\, z_3 - z_1 \end{pmatrix} \in D^{2\times 3}.$$

Let us define the D-modules $M = D^{1\times 3}/(D^{1\times 2}\, R)$ and $N = D^{1\times 2}/(D^{1\times 3}\, R^T)$. Computing $\operatorname{ext}^1_D(N, D)$, we then get

$$\begin{cases} t(M) = (D^{1\times 4}\, Q)/(D^{1\times 2}\, R), \\ M/t(M) = D^{1\times 3}/(D^{1\times 4}\, Q), \\ M/t(M) \cong (D^{1\times 3}\, P), \end{cases}$$

with the notations

$$Q = \begin{pmatrix} -z_2^2\,z_3 & z_2^2\,z_3 & z_1\,z_2^2 - z_1\,z_3 \\ -z_3 - z_1^2\,z_3^2 & z_3 & z_1 + z_1^3\,z_3 \\ -z_1^2\,z_3 - 1 & z_1^2\,z_2^2 + 1 & 0 \\ 0 & z_1\,z_2^2\,z_3 & -z_1^2\,z_3 - 1 \end{pmatrix}, \quad P = \begin{pmatrix} z_1^2\,z_2^2 + 1 \\ z_1^2\,z_3 + 1 \\ z_1\,z_2^2\,z_3 \end{pmatrix}. \quad (5.13)$$

Reducing the rows of Q with respect to the rows of R, we obtain that the only torsion element of M is defined by

$$\begin{cases} m = -(z_1^2\,z_3 + 1)\,y_1 + (z_1^2\,z_2^2 + 1)\,y_2, \\ z_3\,m = 0, \end{cases}$$

where y_1, y_2 and y_3 denote the residue classes of the standard basis of $D^{1\times 3}$ in M. We refer the reader to [4] for more details concerning the explicit computations.

We can easily check that P admits the left-inverse $T = (-z_1^2\,z_3 \quad 1 \quad z_1^3)$, a fact showing that $M/t(M)$ is a free D-module of rank 2. Then, the matrix $U = (R^T \quad T^T)^T$ defined by

$$U = \begin{pmatrix} z_1\,z_2^2\,z_3 & 0 & -z_1^2\,z_2^2 - 1 \\ z_1^2\,z_3^2 + z_3 & -z_3 & -z_1^3\,z_3 - z_1 \\ -z_1^2\,z_3 & 1 & z_1^3 \end{pmatrix}$$

satisfies $\det U = z_3$, which solves Problem 4.

Let us solve Problem 3. From the previous result, we know that $\ker_D(.P) = D^{1\times 4}\,Q$ is a free D-module of rank 2. In order to be able to apply a constructive version of the Quillen-Suslin theorem, we first need to compute the first syzygy module of $D^{1\times 4}\,Q$. We obtain that $\ker_D(.Q) = D^{1\times 2}\,Q_2$, where the matrix $Q_2 \in D^{2\times 4}$ is defined by

$$Q_2 = \begin{pmatrix} z_1^2\,z_3 + 1 & z_3 - z_2^2 & -z_3^2 & 0 \\ 0 & 1 & -z_3 & z_1 \end{pmatrix}.$$

Hence, we have $D^{1\times 4}\,Q \cong L = D^{1\times 4}/(D^{1\times 2}\,Q_2)$. Applying a constructive version of the Quillen-Suslin theorem to Q_2, we then obtain $L = \pi_2(D^{1\times 2}\,B)$, where the full row rank matrix B is defined by

$$B = \begin{pmatrix} z_1^4 & 0 & -z_1^2\,z_3 + 1 & 0 \\ 0 & z_1^3\,z_3\,(z_2^2 - z_3) & 0 & 1 \end{pmatrix},$$

and $\pi_2 : D^{1\times 2} \longrightarrow L$ denotes the canonical projection onto L. Hence, we get that the full row rank matrix defined by

$$R' = B\,Q = \begin{pmatrix} R'_{11} & R'_{12} & R'_{13} \\ R'_{21} & R'_{21} & R'_{23} \end{pmatrix} \in D^{2\times 3},$$

where

$$\begin{cases} R'_{11} = -z_1^4\, z_2^2\, z_3 + z_1^4\, z_3^2 - 1, \\ R'_{12} = z_1^2\, z_2^2 - z_1^2\, z_3 + 1, \\ R'_{13} = z_1^5\, (z_2^2 - z_3), \\ R'_{21} = -z_1^3\, z_3^2\, (z_2^2 - z_3)\, (z_1^2\, z_3 + 1), \\ R'_{21} = -z_1^3\, z_3^3 + z_1^3\, z_2^2 + z_1\, z_2^2\, z_3, \\ R'_{23} = -z_1^4\, z_3^2 - z_1^6\, z_3^2 + z_1^4\, z_2^2\, z_3 + z_1^6\, z_2^2\, z_3^2 - z_1^2\, z_3 - 1, \end{cases}$$

satisfies $D^{1\times 4}\, Q = D^{1\times 2}\, R'$ and the two independent rows of R' define a basis of $D^{1\times 4}\, Q$. Finally, we obtain that $R = R''\, R'$, where the matrix R'' is defined by

$$R'' = \begin{pmatrix} -z_1\, z_2^2\, z_3 - z_1^3\, z_2^2\, z_3^2 + z_1^3\, z_3^3 & z_1^2\, z_2^2 - z_1^2\, z_3 + 1 \\ -z_1^2\, z_3^2 - z_3 & z_1 \end{pmatrix}$$

and $\det\, R'' = z_3$, which solves Problem 3.

We note that we can use the fact that P has a full column rank in order to also solve Problem 3. Indeed, we can use a constructive version of the Quillen-Suslin theorem to compute a basis of $\ker_D(.P)$. Indeed, if we transpose the column vector P, we then obtain the row vector defined in Example 3.11. Hence, if we take the last two rows of U^T, where U is the unimodular matrix defined in (3.6), we obtain that the full row rank R'_2 defined by

$$R'_2 = \begin{pmatrix} 1 + z_1^4\, z_2^2\, z_3 + z_1^2\, z_3 & -z_1^2\, z_2^2 - 1 & -z_1^3\, (z_1^2\, z_2^2 + 1) \\ z_1^3\, z_3^2\, z_2^2 & -z_1\, z_2^2\, z_3 & -z_1^4\, z_2^2\, z_3 + 1 \end{pmatrix}, \tag{5.14}$$

satisfies $D^{1\times 4}\, Q = D^{1\times 2}\, R'_2$ and we obtain the factorization $R = R''_2\, R'_2$, where:

$$R''_2 = \begin{pmatrix} z_1\, z_2^2\, z_3 & -z_1^2\, z_2^2 - 1 \\ z_3 & -z_1 \end{pmatrix}, \quad \det\, R''_2 = z_3.$$

6 Computation of (weakly) doubly coprime factorizations of rational transfer matrices

We now turn to another application of the constructive proofs of the Quillen-Suslin theorem in multidimensional systems theory, namely, the problem of finding (weakly) left-/right-/doubly coprime factorizations of rational transfer matrices over the commutative polynomial ring $k[x_1, \ldots, x_n]$ with coefficients in a field k. The general problem of the existence of (weakly) left-/right-/doubly coprime factorizations for general linear systems was recently studied and solved in [48].

Let us recall a few definitions.

Definition 6.1 [48] Let D be a commutative integral domain, its quotient field

$$K = \{n/d \mid 0 \neq d,\, n \in D\},$$

and $P \in K^{q\times r}$ a transfer matrix.

1. A *fractional representation* of P is a representation of P of the form

$$P = D_P\, N_P^{-1} = \widetilde{N}_P\, \widetilde{D}_P^{-1},$$

where

$$\begin{cases} R = (D_P \quad -N_P) \in D^{q\times(q+r)}, \\ \widetilde{R} = \begin{pmatrix} \widetilde{N}_P \\ \widetilde{D}_P \end{pmatrix} \in D^{(q+r)\times r}, \end{cases} \tag{6.1}$$

i. e., the entries of the matrices R and $\widetilde{R}$ belong to the ring D.

2. A fractional representation $P = D_P^{-1}\, N_P$ of P is called a *weakly left-coprime factorization* of P if we have

$$\forall\, \lambda \in K^{1\times q} :\; \lambda R \in D^{1\times(q+r)} \;\Rightarrow\; \lambda \in D^{1\times q}.$$

3. A fractional representation $P = \widetilde{N}_P\, \widetilde{D}_P^{-1}$ is called a *weakly right-coprime factorization* of P if we have

$$\forall\, \lambda \in K^{r} :\; \widetilde{R}\lambda \in D^{(q+r)\times 1} \;\Rightarrow\; \lambda \in D^{r\times 1}.$$

4. A fractional representation $P = D_P^{-1}\, N_P = \widetilde{N}_P\, \widetilde{D}_P^{-1}$ is called a *weakly doubly coprime factorization* of P if $P = D_P^{-1}\, N_P$ is a weakly left-coprime factorization of P and $P = \widetilde{N}_P\, \widetilde{D}_P^{-1}$ is a weakly right-coprime factorization of P.

5. A fractional representation $P = D_P^{-1}\, N_P$ of P is called a *left-coprime factorization* of P if the matrix R admits a right-inverse over D, i. e., if there exists a matrix $S = (X^T \quad Y^T)^T \in D^{(q+r)\times q}$ satisfying

$$R\, S = D_P\, X - N_P\, Y = I_q.$$

6. A fractional representation $P = \widetilde{N}_P\, \widetilde{D}_P^{-1}$ of P is called a *right-coprime factorization* of P if the matrix $\widetilde{R}$ admits a left-inverse over D, namely, if there exists a matrix $\widetilde{S} = (-\widetilde{Y} \quad \widetilde{X}) \in D^{r\times(q+r)}$ satisfying

$$\widetilde{S}\, \widetilde{R} = -\widetilde{Y}\, \widetilde{N}_P + \widetilde{X}\, \widetilde{D}_P = I_r.$$

7. A fractional representation $P = D_P^{-1}\, N_P = \widetilde{N}_P\, \widetilde{D}_P^{-1}$ is called a *doubly coprime factorization* of P if $P = D_P^{-1}\, N_P$ is a left-coprime factorization of P and $P = \widetilde{N}_P\, \widetilde{D}_P^{-1}$ is a right-coprime factorization of P.

In the case of a polynomial ring $D = k[x_1, \ldots, x_n]$, a weakly coprime factorization of a rational transfer matrix is also called a *minor left-coprime factorization*.

The next definition will play an important role in what follows.

Definition 6.2 [48] Let the matrix $R \in D^{q\times p}$ have a full row rank. We call *D-closure* $\overline{D^{1\times q}\, R}$ of the D-submodule $D^{1\times q}\, R$ of $D^{1\times p}$ the D-module defined by

$$\overline{D^{1\times q}\, R} = \{\lambda \in D^{1\times p} \mid \exists\, 0 \neq d \in D:\ d\,\lambda \in D^{1\times q}\, R\}.$$

We have the following characterizations of the closure of a D-submodule of $D^{1\times p}$.

Proposition 6.3 [48] *Let $R \in D^{q\times p}$ be a full row rank matrix and the finitely presented D-module $M = D^{1\times p}/(D^{1\times q}\, R)$. We then have*

1. $\overline{D^{1\times q}\, R} = (K^{1\times q}\, R) \cap D^{1\times p}$, *where K denotes the quotient field of D.*
2. *The following equalities hold:*

$$\begin{cases} t(M) = ((K^{1\times q}\, R) \cap D^{1\times p})/(D^{1\times q}\, R), \\ M/t(M) = D^{1\times p}/((K^{1\times q}\, R) \cap D^{1\times p}). \end{cases}$$

The next theorem gives necessary and sufficient conditions for the existence of a (weakly) left-/right-/doubly coprime factorization of a transfer matrix.

Theorem 6.4 [48] *Let $P \in K^{q\times r}$ and $P = D_P^{-1}\, N_P = \widetilde{N}_P\, \widetilde{D}_P^{-1}$ be a fractional representation of P, where the matrices R and $\widetilde{R}$ are defined by (6.1). Then, we have*

1. *P admits a weakly left-coprime factorization iff the D-module $\overline{D^{1\times q}\, R}$ is free of rank q.*
2. *P admits a weakly right-coprime factorization iff the D-module $\overline{D^{1\times r}\, \widetilde{R}^T}$ is free of rank r.*
3. *P admits a left-coprime factorization iff $\overline{D^{1\times q}\, R}$ is a free D-module of rank q and $D^{1\times (q+r)}/(\overline{D^{1\times q}\, R})$ is a stably free D-module of rank r.*
4. *P admits a right-coprime factorization iff $\overline{D^{1\times r}\, \widetilde{R}^T}$ is a free D-module of rank r and $D^{1\times (q+r)}/(\overline{D^{1\times r}\, \widetilde{R}^T})$ is a stably free D-module of rank q.*
5. *P admits a left-coprime factorization iff $D^{1\times (q+r)}/(\overline{D^{1\times r}\, \widetilde{R}^T})$ is a free D-module of rank q.*
6. *P admits a right-coprime factorization iff $D^{1\times (q+r)}/(\overline{D^{1\times q}\, R})$ is a free D-module of rank r.*

Testing the freeness of modules is a very difficult issue in algebra. Hence, using Theorem 6.4, we deduce that it is generally difficult to check whether or not a transfer matrix $P \in K^{q\times r}$ admits a (weakly) left-/right-/doubly coprime factorization and if so, to compute them. See [48] for results for $D = H_\infty(\mathbb{C}_+)$ or the ring of structural stable multidimensional systems.

However, if we consider the commutative polynomial ring $D = k[x_1, \ldots, x_n]$ over a field k and $K = k(x_1, \ldots, x_n)$ its quotient field, then we can use constructive versions of the Quillen-Suslin theorem in order to effectively compute (weakly) left-/right-/doubly coprime factorizations of a rational transfer matrix. We first note that using Proposition 6.3 and a computation of an extension module, we can explicitly compute the closure $\overline{D^{1\times q}\, R}$ and then test whether the necessary and sufficient conditions given in Theorem 6.4 are fulfilled. The next algorithm gives a constructive way to compute the corresponding factorizations.

Algorithm 6.5
- **Input:** A commutative polynomial ring $D = k[x_1, \ldots, x_n]$ over a computable field k, a fractional representation $P = D_P^{-1}\, N_P$ of a transfer matrix $P \in K^{q\times r}$ which admits a weakly left-coprime factorization over D.
- **Output:** A weakly left-coprime factorization of P.

1. Define the matrix $R = (D_P \quad -N_P) \in D^{q\times(q+r)}$ and the following D-module:
$$M = D^{1\times(q+r)}/(D^{1\times q}\, R).$$
2. Transpose the matrix R and define the finitely presented D-module
$$N = D^{1\times q}/(D^{1\times(q+r)}\, R^T).$$
3. Compute the D-module $\mathrm{ext}^1_D(N, D)$. We obtain a matrix $Q \in D^{q'\times(q+r)}$ such that
$$M/t(M) = D^{1\times(q+r)}/(D^{1\times q'}\, Q).$$
4. Compute a basis of the free D-module $\overline{D^{1\times q}\, R} = D^{1\times q'}\, Q$. We obtain a full row rank matrix $R' \in D^{q\times(q+r)}$ such that $D^{1\times q'}\, Q = D^{1\times q}\, R'$.
5. Write $R' = (D'_P \quad -N'_P)$, where $D'_P \in D^{q\times q}$ and $N'_P \in D^{q\times r}$. If $\det D'_P \neq 0$, then $P = (D'_P)^{-1}\, N'_P$ is a weakly left-coprime factorization of P.

Up to a transposition, weakly right-coprime factorizations can similarly be obtained. Let us illustrate Algorithm 6.5 on an example.

Example 6.6 Let us consider the commutative polynomial ring $D = \mathbb{Q}[z_1, z_2, z_3]$, $K = \mathbb{Q}(z_1, z_2, z_3)$ the quotient field of D and the following rational transfer matrix:

$$P = \begin{pmatrix} \dfrac{z_1^2\, z_2^2 + 1}{z_1\, z_2^2\, z_3} \\[2ex] \dfrac{z_1^2\, z_3 + 1}{z_1\, z_2^2\, z_3} \end{pmatrix} \in K^{2\times 1}. \tag{6.2}$$

Let us check whether or not P admits a weakly left-coprime factorization and if so, let us compute one. We consider the fractional representation $P = D_P^{-1}\, N_P$ of P obtained by cleaning the denominators of P, i. e., D_P and $N_P \in D^{2\times 1}$ are defined by

$$\begin{cases} D_P = \begin{pmatrix} z_1\, z_2^2\, z_3 & 0 \\ 0 & z_1\, z_2^2\, z_3 \end{pmatrix} \in D^{2\times 2}, \\ N_P = \begin{pmatrix} z_1^2\, z_2^2 + 1 \\ z_1^2\, z_3 + 1 \end{pmatrix} \in D^{2\times 1}. \end{cases}$$

We denote by $R = (D_P \quad - N_P) \in D^{2\times 3}$ and define the finitely presented D-modules

$$M = D^{1\times 3}/(D^{1\times 2}\, R), \quad N = D^{1\times 2}/(D^{1\times 3}\, R^T).$$

Computing $\mathrm{ext}^1_D(N, D)$, we then obtain

$$\begin{cases} t(M) = (D^{1\times 4}\, Q)/(D^{1\times 2}\, R), \\ M/t(M) = D^{1\times 3}/(D^{1\times 4}\, Q), \end{cases}$$

where the matrix Q is defined by (5.14) in Example 5.7. Using the results obtained in Example 5.7, we get that the full row rank matrix $R_2' \in D^{2\times 3}$ defined by (5.14) satisfies $D^{1\times 4}\, Q = D^{1\times 2}\, R_2'$. Therefore, if we denote by

$$\begin{cases} D_P' = \begin{pmatrix} 1 + z_1^4\, z_2^2\, z_3 + z_1^2\, z_3 & -z_1^2\, z_2^2 - 1 \\ z_1^3\, z_3^2\, z_2^2 & -z_1\, z_2^2\, z_3 \end{pmatrix}, \\ N_P' = \begin{pmatrix} z_1^3\,(z_1^2\, z_2^2 + 1) \\ z_1^4\, z_2^2\, z_3 - 1 \end{pmatrix}, \end{cases} \tag{6.3}$$

$P = (D_P')^{-1}\, N_P'$ is then a weakly left-coprime factorization of P.

Finally, by construction, the D-module

$$M/t(M) = D^{1\times 3}/(D^{1\times 4}\, Q) = D^{1\times 3}/(D^{1\times 2}\, R_2')$$

is torsion-free, and thus, by Theorem 2.6, we have $\mathrm{ext}^1_D(N', D) = 0$ with the notation $N' = D^{1\times 2}/(D^{1\times 3}\,(R_2')^T)$. Moreover, we can easily check that $\mathrm{ext}^2_D(N', D) = 0$ and $\mathrm{ext}^3_D(N', D) = 0$, which shows that $M/t(M)$ is a projective, and thus, a free D-module by the Quillen-Suslin theorem. Hence, by 3 of Theorem 6.4, we obtain that $P = (D_P')^{-1}\, N_P'$ is a left-coprime factorization of P. We find that the matrix R_2' admits the following right-inverse over D:

$$\begin{pmatrix} 1 & 0 \\ z_1^2\, z_3 & -z_1^3 \\ 0 & 1 \end{pmatrix}.$$

Therefore, we have the Bézout identity $D_P'\, X - N_P'\, Y = I_2$, where

$$X = \begin{pmatrix} 1 & 0 \\ z_1^2\, z_3 & -z_1^3 \end{pmatrix}, \quad Y = (0 \quad 1).$$

The next algorithm gives us a way to compute left-coprime factorizations of a transfer matrix. Up to a transposition, right-coprime factorizations can similarly be obtained.

Algorithm 6.7 • **Input:** A commutative polynomial ring $D = k[x_1, \ldots, x_n]$ over a computable field k, a fractional representation $P = \widetilde{N}_P\, \widetilde{D}_P^{-1}$ of a rational transfer matrix $P \in K^{q \times r}$ which admits a left-coprime factorization over D.

- **Output:** A left-coprime factorization of P.

1. Define the matrix $\widetilde{R} = (\widetilde{N}_P^T \quad \widetilde{D}_P^T)^T \in D^{(q+r) \times r}$ and define the D-module
$$\widetilde{M} = D^{1 \times (q+r)}/(D^{1 \times r}\, \widetilde{R}^T).$$
2. Define the finitely presented D-module
$$\widetilde{N} = D^{1 \times r}/(D^{1 \times (q+r)}\, \widetilde{R}).$$
3. Compute $\mathrm{ext}_D^1(\widetilde{N}, D)$. We obtain a matrix $\widetilde{Q}^T \in D^{r' \times (q+r)}$ such that
$$\widetilde{M}/t(\widetilde{M}) = D^{1 \times (q+r)}/(D^{1 \times r'}\, \widetilde{Q}^T).$$
4. Compute a basis of the free D-module $\widetilde{M}/t(\widetilde{M})$. We obtain a full column rank matrix
$$\widetilde{L}^T = (D'_P \quad -N'_P)^T \in D^{(q+r) \times q},$$
where $D'_P \in D^{q \times q}$ and $N'_P \in D^{q \times r}$, such that we have the following split exact sequence:
$$0 \longleftarrow D^{1 \times q} \xleftarrow{.\widetilde{L}^T} D^{1 \times (q+r)} \xleftarrow{.\widetilde{Q}^T} D^{1 \times r'}.$$
5. Transpose the matrix $\widetilde{L}^T$ to obtain $\widetilde{L} = (D'_P \quad -N'_P) \in D^{q \times (q+r)}$. If $\det D'_P \neq 0$, then $P = (D'_P)^{-1}\, N'_P$ is a left-coprime factorization of P.

Let us illustrate Algorithm 6.7 on an example.

Example 6.8 We consider again Example 6.6 and the rational transfer matrix P defined by (6.2). We have the fractional representation $P = \widetilde{N}_P\, \widetilde{D}_P^{-1}$ of P, where

$$\begin{cases} \widetilde{N}_P = \begin{pmatrix} z_1^2\, z_2^2 + 1 \\ z_1^2\, z_3 + 1 \end{pmatrix} \in D^{2 \times 2}, \\ \widetilde{D}_P = z_1^2\, z_2^2\, z_3 \in D. \end{cases}$$

Let us define the matrix $\widetilde{R} = (\widetilde{N}_P^T \quad \widetilde{D}_P^T)^T$ and the D-modules

$$\widetilde{M} = D^{1 \times (q+r)}/(D^{1 \times r}\, \widetilde{R}^T), \quad \widetilde{N} = D^{1 \times r}/(D^{1 \times (q+r)}\, \widetilde{R}).$$

The row vector $\widetilde{R}^T$ is exactly the one defined in Example 3.11. Hence, using the results obtained in Example 3.11, we obtain that the unimodular matrix U defined by (3.6) satisfies $\widetilde{R}^T U = (1\ 0\ 0)$. Hence, selecting the last two columns of U and transposing the corresponding matrix, we then find again the matrix R_2' defined by (5.14). Hence, using Example 6.6, we obtain that $P = (D_P')^{-1} N_P'$ is a left-coprime factorization of P, where the matrices D_P' and N_P' are defined by (6.3).

7 Decomposition of multidimensional linear systems

It was recently shown in [8] that the computation of bases of free modules plays a central role in the *decomposition problem* of multidimensional linear systems. We shall recall this problem as well as the main important results obtained in [8]. Let us first recall a few definitions and notations.

We shall denote by $\mathrm{end}_D(M)$ the non-commutative ring of D-endomorphisms of the D-module M, i. e., the ring formed by the D-morphisms (namely, the D-linear maps) from M to M. Moreover, we recall that if f is a D-morphism from a D-module M to a D-module N, then $\mathrm{coim} f$ is the D-module defined by $\mathrm{coim}\, f = M/\ker f$, where $\ker f = \{m \in M \mid f(m) = 0\}$ is the kernel of f.

Let M be a finitely presented D-module, i. e., M is of the form

$$M = D^{1\times p}/(D^{1\times q}\, R),$$

where $R \in D^{q\times p}$, and let us denote by $\pi : D^{1\times p} \longrightarrow M$ the canonical projection. We can easily prove that a D-endomorphism f of M is defined by $f(m) = \pi(\lambda\, P)$, where $P \in D^{p\times p}$ is a matrix such that there exists $Q \in D^{q\times q}$ satisfying $R\,P = Q\,R$, and λ is any element of $D^{1\times p}$ satisfying $m = \pi(\lambda)$. See [8] for more details and for constructive algorithms which compute the pairs of matrices (P, Q) satisfying $R\,P = Q\,R$. These algorithms have been implemented in the package MORPHISMS ([9]) of the library OREMODULES ([4]).

We have following results.

Theorem 7.1 [8] *Let $R \in D^{q\times p}$, $M = D^{1\times p}/(D^{1\times q}\, R)$ and $f \in \mathrm{end}_D(M)$ defined by $P \in D^{p\times p}$ and $Q \in D^{q\times q}$, i. e., $R\,P = Q\,R$. If the D-modules*

$$\ker_D(.P), \quad \mathrm{coim}_D(.P), \quad \ker_D(.Q), \quad \mathrm{coim}_D(.Q),$$

are free of rank m, $p-m$, l, $q-l$, then there exist matrices $U_1 \in D^{m\times p}$, $U_2 \in D^{(p-m)\times p}$, $V_1 \in D^{l\times q}$ and $V_2 \in D^{(q-l)\times q}$ such that

$$U = (U_1^T \quad U_2^T)^T \in \mathrm{GL}_p(D), \quad V = (V_1^T \quad V_2^T)^T \in \mathrm{GL}_q(D),$$

and

$$\overline{R} = V\, R\, U^{-1} = \begin{pmatrix} V_1\, R\, W_1 & 0 \\ V_2\, R\, W_1 & V_2\, R\, W_2 \end{pmatrix} \in D^{q\times p},$$

where $U^{-1} = (W_1 \quad W_2)$, $W_1 \in D^{p\times m}$ and $W_2 \in D^{p\times(p-m)}$.

In particular, the full row rank matrix U_1 (resp., U_2, V_1, V_2) defines a basis of the free D-module $\ker_D(.P)$ *(resp.,* $\mathrm{coim}_D(.P)$, $\ker_D(.Q)$, $\mathrm{coim}_D(.Q)$*), i. e., we have*

$$\begin{cases} \ker_D(.P) = D^{1\times m}\, U_1, \\ \mathrm{coim}_D(.P) = \kappa(D^{1\times (p-m)}\, U_2), \\ \ker_D(.Q) = D^{1\times l}\, V_1, \\ \mathrm{coim}_D(.Q) = \rho(D^{1\times (q-l)}\, V_2), \end{cases}$$

where $\kappa : D^{1\times p} \to \mathrm{coim}_D(.P)$ (resp., $\rho : D^{1\times q} \to \mathrm{coim}_D(.Q)$) denotes the canonical projection onto $\mathrm{coim}_D(.P)$ (resp., $\mathrm{coim}_D(.Q)$). An important point in Theorem 7.1 is the computation of bases of the free D-modules $\ker_D(.P), \mathrm{coim}_D(.P), \ker_D(.Q)$ and $\mathrm{coim}_D(.Q)$, which can be solved by means of constructive versions of the Quillen-Suslin theorem and their implementations in computer algebra systems. In order to do that, we use the package QUILLENSUSLIN described in the appendix.

Let us illustrate Theorem 7.1 by means of an explicit example.

Example 7.2 Let us consider the system of partial differential equations defined by

$$\sigma\, \partial_t\, \vec{A} + \frac{1}{\mu}\, \vec{\nabla} \wedge \vec{\nabla}\, \vec{A} - \sigma\, \vec{\nabla}\, V = 0, \tag{7.1}$$

where σ and μ are two constants. The previous system corresponds to the equations satisfied by the electromagnetic quadri-potential $(\vec{A}, V)$ when it is assumed that the term $\partial_t\, \vec{D}$ can be neglected in the Maxwell equations. See [7] for more details. It seems that Maxwell was led to introduce the term $\partial_t\, \vec{D}$ in his famous equations for purely mathematical reasons. See [7] for more details.

Let us consider the ring $D = \mathbb{Q}[\partial_t, \partial_1, \partial_2, \partial_3]$ of differential operators in $\partial_t = \partial/\partial t$ and $\partial_i = \partial/\partial x_i$ with rational constant coefficients, the system matrix of (7.1) defined by

$$R = \begin{pmatrix} \sigma\, \partial_t - \frac{1}{\mu}\, (\partial_2^2 + \partial_3^2) & \frac{1}{\mu}\, \partial_1\, \partial_2 & \frac{1}{\mu}\, \partial_1\, \partial_3 & -\sigma\, \partial_1 \\ \frac{1}{\mu}\, \partial_1\, \partial_2 & \sigma\, \partial_t - \frac{1}{\mu}\, (\partial_1^2 + \partial_3^2) & \frac{1}{\mu}\, \partial_2\, \partial_3 & -\sigma\, \partial_2 \\ \frac{1}{\mu}\, \partial_1\, \partial_3 & \frac{1}{\mu}\, \partial_2\, \partial_3 & \sigma\, \partial_t - \frac{1}{\mu}\, (\partial_1^2 + \partial_2^2) & -\sigma\, \partial_3 \end{pmatrix}$$

and the finitely presented D-module $M = D^{1\times 4}/(D^{1\times 3}\, R)$.

The matrices P and Q defined by

$$P = \begin{pmatrix} 0 & 0 & 0 & 0 \\ 0 & \sigma\, \mu\, \partial_t & 0 & -\sigma\, \mu\, \partial_2 \\ 0 & 0 & \sigma\, \mu\, \partial_t & -\sigma\, \mu\, \partial_3 \\ 0 & \partial_t\, \partial_2 & \partial_t\, \partial_3 & -(\partial_2^2 + \partial_3^2) \end{pmatrix} \in D^{4\times 4},$$

$$Q = \begin{pmatrix} 0 & 0 & 0 \\ -\partial_1\, \partial_2 & \sigma\, \mu\, \partial_t - \partial_2^2 & -\partial_2\, \partial_3 \\ -\partial_1\, \partial_3 & -\partial_2\, \partial_3 & \sigma\, \mu\, \partial_t - \partial_3^2 \end{pmatrix} \in D^{3\times 3},$$

satisfy the relation $R\,P = Q\,R$, and thus, define a D-endomorphism f of M. Moreover, we can check that $\ker_D(.P)$, $\mathrm{coim}_D(.P)$, $\ker_D(.Q)$ and $\mathrm{coim}_D(.Q)$ are free D-modules of rank 2, 2, 1 and 2. Hence, computing bases of these free D-modules by means of a constructive version of the Quillen-Suslin theorem, we obtain

$$\left\{\begin{array}{l} U_1 = \begin{pmatrix} 1 & 0 & 0 & 0 \\ 0 & \partial_2 & \partial_3 & -\sigma\,\mu \end{pmatrix}, \\ U_2 = \dfrac{1}{\sigma\,\mu}\begin{pmatrix} 0 & 1 & 0 & 0 \\ 0 & 0 & 1 & 0 \end{pmatrix}, \end{array}\right. \qquad \left\{\begin{array}{l} V_1 = \begin{pmatrix} 1 & 0 & 0 \end{pmatrix}, \\ V_2 = \begin{pmatrix} 0 & 1 & 0 \\ 0 & 0 & 1 \end{pmatrix}. \end{array}\right.$$

Defining $U = (U_1^T \quad U_2^T)^T \in \mathrm{GL}_4(D)$ and $V = (V_1^T \quad V_2^T)^T \in \mathrm{GL}_3(D)$, we get that $\overline{R} = V\,R\,U^{-1}$ is the block-triangular matrix defined by

$$\begin{pmatrix} \sigma\,\partial_t - \dfrac{1}{\mu}\,(\partial_2^2 + \partial_3^2) & \dfrac{1}{\mu}\,\partial_1 & 0 & 0 \\ \dfrac{1}{\mu}\,\partial_1\,\partial_2 & \dfrac{1}{\mu}\,\partial_2 & \sigma\,(\sigma\,\mu\,\partial_t - (\partial_1^2 + \partial_2^2 + \partial_3^2)) & 0 \\ \dfrac{1}{\mu}\,\partial_1\,\partial_3 & \dfrac{1}{\mu}\,\partial_3 & 0 & \sigma\,(\sigma\,\mu\,\partial_t - (\partial_1^2 + \partial_2^2 + \partial_3^2)) \end{pmatrix}.$$

Now, we recall that an *idempotent* $f \in \mathrm{end}_D(M)$ is a D-endomorphism f of M satisfying $f^2 = f$. We can now state another important result of [8] on the decomposition of D-modules for which the Quillen-Suslin theorem plays a central role.

Theorem 7.3 [8] *Let $R \in D^{q\times p}$, $M = D^{1\times p}/(D^{1\times q}\,R)$ and $f \in \mathrm{end}_D(M)$ be an idempotent defined by two idempotents $P \in D^{p\times p}$ and $Q \in D^{q\times q}$, namely, they satisfy $R\,P = Q\,R$, $P^2 = P$ and $Q^2 = Q$. Then, there exist four matrices $U_1 \in D^{m\times p}$, $U_2 \in D^{(p-m)\times p}$, $V_1 \in D^{l\times q}$ and $V_2 \in D^{(q-l)\times q}$ such that*

$$U = (U_1^T \quad U_2^T)^T \in \mathrm{GL}_p(D), \quad V = (V_1^T \quad V_2^T)^T \in \mathrm{GL}_q(D),$$

and

$$\overline{R} = V\,R\,U^{-1} = \begin{pmatrix} V_1\,R\,W_1 & 0 \\ 0 & V_2\,R\,W_2 \end{pmatrix} \in D^{q\times p},$$

where $U^{-1} = (W_1 \quad W_2)$, $W_1 \in D^{p\times m}$ and $W_2 \in D^{p\times(p-m)}$.

In particular, the full row rank matrix U_1 (resp., U_2, V_1, V_2) defines a basis of the free D-module $\ker_D(.P)$, (resp., $\mathrm{im}_D(.P) = \ker_D(.(I_p - P))$, $\ker_D(.Q)$, $\mathrm{im}_D(.Q) = \ker_D(.(I_q - Q))$) of rank respectively m, $p-m$, l, $q-l$. In other words, we have

$$\left\{\begin{array}{l} \ker_D(.P) = D^{1\times m}\,U_1, \\ \mathrm{im}_D(.P) = D^{1\times(p-m)}\,U_2, \\ \ker_D(.Q) = D^{1\times l}\,V_1, \\ \mathrm{im}_D(.Q) = D^{1\times(q-l)}\,V_2. \end{array}\right.$$

Let us illustrate Theorem 7.3 by means of an example coming from control theory.

Example 7.4 Let us consider the differential time-delay system describing the movement of a vibrating string with an interior mass studied in [31], namely,

$$\begin{cases} \phi_1(t)+\psi_1(t)-\phi_2(t)-\psi_2(t)=0, \\ \dot{\phi}_1(t)+\dot{\psi}_1(t)+\eta_1\,\phi_1(t)-\eta_1\,\psi_1(t)-\eta_2\,\phi_2(t)+\eta_2\,\psi_2(t)=0, \\ \phi_1(t-2\,h_1)+\psi_1(t)-u(t-h_1)=0, \\ \phi_2(t)+\psi_2(t-2\,h_2)-v(t-h_2)=0, \end{cases} \tag{7.2}$$

where h_1 and $h_2 \in \mathbb{R}_+$ are such that $\mathbb{Q}\,h_1+\mathbb{Q}\,h_2$ is a two-dimensional $\mathbb{Q}$-vector space (i. e., there exists no relation of the form $m\,h_1+n\,h_2=0$, where $m,n\in\mathbb{Z}$), η_1 and η_2 are two non-zero constant parameters of the system.

Let us consider the ring $D=\mathbb{Q}(\eta_1,\eta_2)\left[\frac{d}{dt},\sigma_1,\sigma_2\right]$ of differential time-delay operators where $(dy/dt)(t)=\dot{y}(t)$ and $(\sigma_i\,y)(t)=y(t-h_i)$, for $i=1,2$. The condition on h_1 and h_2 implies that the two time-delay operators σ_1 and σ_2 are *incommensurable*, i. e., define two independent variables. Hence, D is a commutative polynomial ring. Let us denote by R the system matrix of (7.2), namely,

$$R=\begin{pmatrix} 1 & 1 & -1 & -1 & 0 & 0 \\ \dfrac{d}{dt}+\eta_1 & \dfrac{d}{dt}-\eta_1 & -\eta_2 & \eta_2 & 0 & 0 \\ \sigma_1^2 & 1 & 0 & 0 & -\sigma_1 & 0 \\ 0 & 0 & 1 & \sigma_2^2 & 0 & -\sigma_2 \end{pmatrix}\in D^{4\times 6},$$

and the finitely presented D-module $M=D^{1\times 6}/(D^{1\times 4}\,R)$.

Computing idempotents of $\mathrm{end}_D(M)$, we obtain an idempotent f defined by the following two idempotent matrices:

$$P=\begin{pmatrix} 1 & 0 & 0 & 0 & 0 & 0 \\ -\sigma_1^2 & 0 & 0 & 0 & \sigma_1 & 0 \\ 0 & 0 & 0 & -\sigma_2^2 & 0 & \sigma_2 \\ 0 & 0 & 0 & 1 & 0 & 0 \\ 0 & 0 & 0 & 0 & 1 & 0 \\ 0 & 0 & 0 & 0 & 0 & 1 \end{pmatrix}, \quad Q=\begin{pmatrix} 1 & 0 & -1 & 1 \\ 0 & 1 & -\dfrac{d}{dt}+\eta_1 & \eta_2 \\ 0 & 0 & 0 & 0 \\ 0 & 0 & 0 & 0 \end{pmatrix}.$$

Moreover, we can check that $\ker_D(.P)$, $\mathrm{im}_D(.P)$, $\ker_D(.P)$ and $\mathrm{im}_D(.P)$ are free D-modules of rank 2, 4, 2 and 2. Computing bases by means of a constructive version of the QuillenSuslin theorem, we then get:

$$\left\{\begin{array}{ll} \ker_D(.P) = D^{1\times 2}\, U_1, & U_1 = \begin{pmatrix} \sigma_1^2 & 1 & 0 & 0 & -\sigma_1 & 0 \\ 0 & 0 & 1 & \sigma_2^2 & 0 & -\sigma_2 \end{pmatrix}, \\ \operatorname{im}_D(.P) = D^{1\times 4}\, U_2, & U_2 = \begin{pmatrix} 1 & 0 & 0 & 0 & 0 & 0 \\ 0 & 0 & 0 & 1 & 0 & 0 \\ 0 & 0 & 0 & 0 & 1 & 0 \\ 0 & 0 & 0 & 0 & 0 & 1 \end{pmatrix}, \\ \ker_D(.Q) = D^{1\times 2}\, V_1, & V_1 = \begin{pmatrix} 0 & 0 & 1 & 0 \\ 0 & 0 & 0 & 1 \end{pmatrix}, \\ \operatorname{im}_D(.Q) = D^{1\times 2}\, V_2, & V_2 = \begin{pmatrix} 1 & 0 & -1 & 1 \\ 0 & -1 & \frac{d}{dt} - \eta_1 & -\eta_2 \end{pmatrix}. \end{array}\right.$$

Forming the matrices $U = (U_1^T \quad U_2^T)^T \in \mathrm{GL}_6(D)$ and $V = (V_1^T \quad V_2^T)^T \in \mathrm{GL}_4(D)$, we obtain that R is then equivalent to the block-diagonal matrix $\overline{R} = V\, R\, U^{-1}$:

$$\begin{pmatrix} 1 & 0 & 0 & 0 & 0 & 0 \\ 0 & 1 & 0 & 0 & 0 & 0 \\ 0 & 0 & 1 - \sigma_1^2 & \sigma_2^2 - 1 & \sigma_1 & -\sigma_2 \\ 0 & 0 & \sigma_1^2 \left(\frac{d}{dt} - \eta_1\right) - \left(\frac{d}{dt} + \eta_1\right) & -\eta_2\,(\sigma_2^2 + 1) & -\sigma_1 \left(\frac{d}{dt} + \eta_1\right) & \eta_2\,\sigma_2 \end{pmatrix}.$$

Now, considering the second diagonal block, namely,

$$S = \begin{pmatrix} 1 - \sigma_1^2 & \sigma_2^2 - 1 & \sigma_1 & -\sigma_2 \\ \sigma_1^2 \left(\frac{d}{dt} - \eta_1\right) - \left(\frac{d}{dt} + \eta_1\right) & -\eta_2\,(\sigma_2^2 + 1) & -\sigma_1 \left(\frac{d}{dt} + \eta_1\right) & \eta_2\,\sigma_2 \end{pmatrix},$$

and the D-module $L = D^{1\times 4}/(D^{1\times 2}\, S)$. Using an algorithm developed in [8], we obtain that an idempotent $g \in \operatorname{end}_D(L)$ is defined by the two idempotent matrices

$$P' = \begin{pmatrix} 1 & 0 & 0 & 0 \\ a & 0 & b & 0 \\ 0 & 0 & 1 & 0 \\ 0 & 0 & 0 & 1 \end{pmatrix}, \quad Q' = \frac{1}{2} \begin{pmatrix} \sigma_2^2 + 1 & \frac{1}{\eta_2}\,(\sigma_2^2 - 1) \\ -\eta_2\,(\sigma_2^2 + 1) & -\sigma_2^2 + 1 \end{pmatrix},$$

with the notations

$$\left\{\begin{array}{l} a = \dfrac{1}{2\,\eta_2} \left(\sigma_1^2 \left(\dfrac{d}{dt} - (\eta_1 + \eta_2)\right) - \dfrac{d}{dt} + (\eta_2 - \eta_1)\right), \\ b = -\dfrac{\sigma_1}{2\,\eta_2} \left(\dfrac{d}{dt} - (\eta_1 + \eta_2)\right). \end{array}\right.$$

We can check that the D-modules $\ker_D(.P')$, $\mathrm{im}_D(.P')$, $\ker_D(.Q')$ and $\mathrm{im}_D(.Q')$ are free and, using a constructive version of the Quillen-Suslin theorem, we obtain that $\ker_D(.P') = D\,U_1'$, $\mathrm{im}_D(.P') = D^{1\times 3}\,U_2'$, $\ker_D(.Q') = D\,V_1'$ and $\mathrm{im}_D(.Q') = D\,V_2'$, where

$$U_1' = 2\eta_2 \begin{pmatrix} a & -1 & b & 0 \end{pmatrix},$$

$$U_2' = \begin{pmatrix} 1 & 0 & 0 & 0 \\ -\sigma_1 & 0 & 1 & 0 \\ \sigma_1^2\,\sigma_2\,(d-\eta_1-\eta_2) - \sigma_2\,(d+\eta_1-\eta_2) & 0 & -\sigma_1\,\sigma_2\,(d-\eta_1-\eta_2) & -2\,\eta_2 \end{pmatrix}$$

and

$$V_1' = (\eta_2 \quad 1), \qquad V_2' = (\eta_2\,(\sigma_2^2+1) \quad \sigma_2^2 - 1).$$

Defining $U' = (U_1'^T \quad U_2'^T)^T \in \mathrm{GL}_4(D)$ and $V' = (V_1'^T \quad V_2'^T)^T \in \mathrm{GL}_2(D)$, we get

$$\overline{S} = V'\,S\,U'^{-1} = \begin{pmatrix} 1 & 0 & 0 & 0 \\ 0 & \dfrac{d}{dt} + \eta_1 + \eta_2 & \sigma_1 \left(\dfrac{d}{dt} + \eta_2 - \eta_1\right) & \sigma_2 \end{pmatrix}.$$

If we denote by $\mathrm{diag}(A, B)$ the diagonal matrix formed by the matrices A and B and define $U'' = \mathrm{diag}(I_2, U') \in \mathrm{GL}_6(D)$ and $V'' = \mathrm{diag}(I_2, V') \in \mathrm{GL}_4(D)$, then we get

$$\overline{\overline{R}} = (V''\,V)\,R\,(U''\,U)^{-1} = \mathrm{diag}(I_2, \overline{S}).$$

The last result proves that the system defined by (7.2) with 6 unknowns and 4 equations is in fact equivalent to the following simple equation:

$$\dot{z}_1(t) + (\eta_1+\eta_2)\,z_1(t) + \dot{z}_2(t-h_1) + (\eta_2-\eta_1)\,z_2(t-h_1) - z_3(t-h_2) = 0. \quad (7.3)$$

Using the results summed up in Figure 2.1, the D-module defined by

$$M' = D^{1\times 3}/\left(D\left(\frac{d}{dt} + \eta_1 + \eta_2 \quad \sigma_1\left(\frac{d}{dt} + \eta_2 - \eta_1\right) \quad \sigma_2\right)\right) \cong M$$

is reflexive but not projective, i. e., not free, as we have

$$J = \mathrm{ann}_D(\mathrm{ext}_D^3(T(M'), D)) = \left(\sigma_1, \sigma_2, \frac{d}{dt} + \eta_1 + \eta_2\right),$$

and $\dim_{\mathbb{C}} V(J) = 0$. As $\sigma_1, \sigma_2 \in J$, we obtain that the $\mathbb{Q}(\eta_1,\eta_2)\left[\frac{d}{dt}, \sigma_1, \sigma_2, \sigma_1^{-1}\right]$-module $\mathbb{Q}(\eta_1,\eta_2)\left[\frac{d}{dt}, \sigma_1, \sigma_2, \sigma_1^{-1}\right] \otimes_D M'$ is free, i. e., (7.3) is σ_1-free ([6, 30]). Computing an injective parametrization of (7.3), we obtain

$$\left\{\begin{array}{l} z_1 = \sigma_1\,\sigma_2\,y_1 + \sigma_1\left(\dfrac{d}{dt} + \eta_2 - \eta_1\right)\,y_2, \\ z_2 = -\sigma_2\,y_1 - \left(\dfrac{d}{dt} + \eta_1 + \eta_2\right)\,y_2, \\ z_3 = -2\,\eta_1\,y_1, \end{array}\right. \quad (7.4)$$

and a basis of $\mathbb{Q}(\eta_1,\eta_2)\left[\frac{d}{dt},\sigma_1,\sigma_2,\sigma_1^{-1}\right]$-module $\mathbb{Q}(\eta_1,\eta_2)\left[\frac{d}{dt},\sigma_1,\sigma_2,\sigma_1^{-1}\right]\otimes_D M'$ is then defined by

$$y_1=-\frac{1}{2\,\eta_1}\,\sigma_1^{-1}\,z_3,\quad y_2=-\frac{1}{2\,\eta_1}\,(\sigma_1^{-1}\,z_1+z_2).$$

Using (7.4) and the transformation $(\phi_1,\psi_1,\phi_2,\psi_2,u,v)^T=(U''\,U)^{-1}\,(z_1,z_2,z_3)^T$, we get an injective parametrization of (7.2) if we also use the advance operator σ_1^{-1}.

Finally, the $\mathbb{Q}(\eta_1,\eta_2)\left[\frac{d}{dt},\sigma_1,\sigma_2,\sigma_2^{-1}\right]$-module $\mathbb{Q}(\eta_1,\eta_2)\left[\frac{d}{dt},\sigma_1,\sigma_2,\sigma_2^{-1}\right]\otimes_D M'$ is free and, from (7.3), we obtain that

$$z_3(t)=\dot z_1(t+h_2)+(\eta_1+\eta_2)\,z_1(t+h_2)+\dot z_2(t-h_1+h_2)+(\eta_2-\eta_1)\,z_2(t-h_1+h_2),$$

showing that the $\mathbb{Q}(\eta_1,\eta_2)\left[\frac{d}{dt},\sigma_1,\sigma_2,\sigma_2^{-1}\right]$-module $\mathbb{Q}(\eta_1,\eta_2)\left[\frac{d}{dt},\sigma_1,\sigma_2,\sigma_2^{-1}\right]\otimes_D M'$ admits the basis $\{z_1,z_2\}$. Using the transformation defined by $(U''\,U)^{-1}$, we get an injective parametrization of (7.2) if we also use the advance operator σ_2^{-1}.

Generalizations of Theorems 7.1 and 7.3 hold for some classes of non-commutative polynomial rings of functional operators. See [8] for more details. However, we need to be able to compute bases of free modules over the corresponding rings. A general algorithm has recently been developed in [49, 51] for the ring of differential operators with polynomial or rational coefficients (the so-called *Weyl algebras*). See [50] for an implementation of this algorithm and a library of examples which illustrates it.

Finally, we refer the reader to [8, 9] for numerous examples of decomposition of classical systems of partial differential equations and of differential time-delay equations appearing in mathematical physics and control theory and for a description of the package MORPHISMS ([9]) as well as a library of examples.

8 Conclusion

In this paper, we have shown new applications of constructive versions of the Quillen-Suslin theorem to mathematical systems theory. In particular, we explained that the construction of bases of a free module over a commutative polynomial ring D gives us a way to obtain flat outputs of a flat multidimensional linear system as well as injective parametrizations of its solutions over a D-module $\mathcal{F}$. We have also shown that a flat multidimensional system was algebraically equivalent to the 1-D controllable linear systems obtained by setting all but one functional operator to particular values in the system matrix. This last result gives an answer to a natural question arising in the study of flat multidimensional linear systems and particularly in the study of differential time-delay systems. Moreover, we gave constructive algorithms for two well-known problems stated by Lin and Bose in the literature of multidimensional systems. These problems are generalizations of Serre's conjecture. We also showed that the computation of (weakly) left-/right-coprime factorizations of rational transfer matrices could constructively be solved by means of the Quillen-Suslin theorem. The need for the computation of bases of free D-modules recently appeared as an important issue in the study of the decomposition problems of multidimensional linear systems. Finally, we have demonstrated the different algorithms by means of the recent implementation of the Quillen-Suslin theorem in the package QUILLENSUSLIN. To our knowledge, this is the first implementation of the Quillen-Suslin theorem in a computer algebra system which is nowadays freely available and dedicated to applications of the Quillen-Suslin theorem and, in particular, to mathematical systems theory and control theory.

New applications of the Quillen-Suslin theorem and of the package QUILLENSUSLIN will be studied in the future (e.g., algebraic geometry, signal processing). Moreover, an interesting but difficult problem is to constructively recognize when a finitely presented $D = A[x]$-module $M = D^{1\times p}/(D^{1\times q}\, R)$, where $R \in D^{q\times p}$ and A is a commutative ring, is *extended*, namely, when there exists $S \in A^{q'\times p'}$ such that $M \cong D \otimes_A P$, where $P = A^{1\times p'}/(A^{1\times q'}\, S)$. See [53] for more details. It is well known that the Quillen-Suslin theorem is a particular case of this problem when M is a projective D-module ([23, 24, 52, 53]). If we can effectively solve this problem for particular classes of D-modules, then, for every D-module $\mathcal{F}$, we obtain $\ker_{\mathcal{F}}(R.) \cong \ker_{\mathcal{F}}(S.)$, which shows that the integration of the system $\ker_{\mathcal{F}}(R.)$ is algebraically equivalent to the integration of the system $\ker_{\mathcal{F}}(S.)$ which contains one functional operator less. Such a result may simplify the explicit integration of these classes of functional systems. Finally, another interesting problem is the computation of a minimal set of generators of a finitely presented $D = A[x]$-module $M = D^{1\times p}/(D^{1\times q}\, R)$, where $R \in D^{q\times p}$. The results recently obtained in [8, 9] were able to explicitly answer these last two questions on particular examples coming from mathematical physics and control theory. However, the general case seems to be far from being solved.

Finally, more heuristic methods need to be developed and implemented in QUILLENSUSLIN in order to avoid as much as we can the use of a general algorithm for solving Problem 2. Different QS-algorithms need also to be implemented in QUILLENQUILLEN and particularly the one recently developed in [27, 58].

9 Appendix: The package QUILLENSUSLIN

9.1 Description of QUILLENSUSLIN: a package for computing bases of free modules over commutative polynomial rings

The package QuillenSuslin is an implementation of a constructive version of the Quillen-Suslin theorem. The main idea of the algorithm was inspired by the article of Logar and Sturmfels [26]. Nevertheless, many important changes were introduced. We have roughly described the implemented algorithm in section 3.4.

The general algorithm proceeds by induction on the number n of the independent variables x_i in the polynomial ring $D = k[x_1, \ldots, x_n]$ and each inductive step, that reduces the problem by one independent variable, consists of the following three main parts:

1. Finding a normalized component in a polynomial vector by means of a change of coordinates (NormalizationStep).

2. Computing a finite number of local solutions (local loop) using Horrocks' theorem (Horrocks).

3. Patching local solutions of Problem 2 together to get a global one (Patch).

This general method is generally quite involved. The package consists of procedures completing a unimodular polynomial row which admits a right-inverse to a square invertible matrix over a given commutative polynomial ring with coefficients in $\mathbb{Q}$ or $\mathbb{Z}$. The implementation was improved by many heuristic methods which are used as soon as it is possible. It allows us to avoid the inductive step and leads to simpler outputs (smaller coefficients and lower degrees).

QUILLENSUSLIN uses the library INVOLUTIVE ([3]) for computing Janet bases over commutative polynomial rings.

```
> with(Involutive):
```

```
> with(QuillenSuslin);
```

[*BasisOfCokernelModule*, *Cofactors*, *CompleteMatrix*, *DenomOf*, *Heuristic*, *Horrocks*, *InjectiveParametrization*, *InvertibleIn*, *IsInS*, *IsMonic*, *IsParkNormalized*, *IsRegular*, *IsUnimod*, *LC*, *LCFactorization*, *LM*, *Laurent2Pol*, *LaurentNormalization*, *LinBose1*, *LinBose2*, *LowestDegree*, *MaxMinors*, *MaximalFF*, *MaximalQQ*, *MaximalZZ*, *NormalizationStep*, *OneLocalSol*, *OneStepEY*, *OneStepQS*, *ParkAlgorithm*, *ParkMatrixNormalization*, *Patch*, *QSAlgorithm*, *ReduceBasisDegree*, *ReduceDeg*, *RightInverse*, *RightInverseFast*, *SHeuristic*, *SetLastVariableA*, *SuslinLemma*, *WLCFactorization*, *WRCFactorization*]

9.1.1 The main functions of the package QUILLENSUSLIN

`QSAlgorithm`	Compute a unimodular matrix U which transforms a row vector admitting a right-inverse into a matrix of the form $(I \quad 0)$
`CompleteMatrix`	Complete a matrix admitting a right-inverse to a unimodular matrix
`HEURISTIC`	Test whether or not a heuristic method can be applied for the given row vector
`BasisOfCokernelModule`	Compute a basis of a free module finitely presented by the given matrix

9.1.2 Important functions of the package QUILLENSUSLIN

`Horrocks`	Implementation of Horrock's theorem which computes a solution of Problem 1 over a given local ring
`IsMonic`	Test whether or not a polynomial row vector has a monic component
`IsRegular`	Test whether or not a polynomial row vector forms a regular sequence
`IsUnimod`	Test whether or not a matrix admits a right-inverse
`MaximalFF`	Find a maximal ideal over a given one in a polynomial ring with coefficient in a finite field
`MaximalQQ`	Find a maximal ideal over a given one in a polynomial ring with rational coefficients
`MaximalZZ`	Find a maximal ideal over a given one in a polynomial ring with integer coefficients
`NormalizationStep`	Compute an invertible transformation and a change of variables such that the last component of the transformed row becomes monic in the last new variable
`OneLocalSol`	Compute a matrix which is unimodular over some localization of the polynomial ring and transforms the given matrix to $(I \quad 0)$
`OneStepEY` `OneStepQS`	One inductive step of the general algorithm: return a unimodular matrix which transforms the given matrix into a matrix where the last variable equals 0
`Patch`	Patching procedure: patch local solutions together
`SuslinLemma`	Implementation of Suslin's lemma which computes a polynomial h in the ideal generated by polynomials p and q such that $\deg(h) = \deg(p) - 1$ and its leading coefficient is a coefficient of the polynomial q

9.1.3 Low level functions of the package QuillenSuslin

`Cofactors`	Compute cofactors of a $(p-1) \times p$-matrix
`DenomOf`	Compute the common denominator of entries of a rational matrix
`LM`	Return the leading monomial of a polynomial with respect to the given variable
`LC`	Return the leading coefficient of a polynomial with respect to the given variable
`MaxMinors`	Return the maximal minors of a given matrix
`ReduceDeg`	Reduce degrees of the components of a polynomial row vector with respect to given variable
`RightInverse`, `RightInverseFast`	Compute a right-inverse of a row vector
`ReduceBasisDegree`	Reduce degrees of the elements of basis of a free module over a commutative polynomial ring

9.1.4 Functions of QuillenSuslin for mathematical systems theory

`InjectiveParametrization`	Compute an injective parametrization of a flat multidimensional linear system
`LCFactorization`	Compute a left-coprime factorization of a rational transfer matrix when it exists
`LinBose1`	Compute a solution of Problem 3 when it exists
`LinBose2`	Compute a solution of Problem 4 when it exists
`RCFactorization`	Compute a right-coprime factorization of a rational transfer matrix when it exists
`SetLastVariableA`	Compute a unimodular matrix which transforms the given matrix into a matrix where the last variable is set to a given constant A
`WLCFactorization`	Compute a weakly left-coprime factorization of a rational transfer matrix when it exists
`WRCFactorization`	Compute a weakly right-coprime factorization of a rational transfer matrix when it exists

9.1.5 Functions of QUILLENSUSLIN for Laurent polynomial rings

`IsParkNormalized`	Test whether or not a Laurent polynomial is normalized, i. e., whether or not all its coefficients are Laurent monomials
`Laurent2Pol`	Compute a transformation which maps a row vector over a Laurent polynomial ring into a row vector over a polynomial ring
`LaurentNormalization`	Return a change of variables which normalizes a Laurent polynomial
`LowestDegree`	Return the lowest degree of a Laurent polynomial with respect to the given variable
`ParkAlgorithm`	Return a unimodular matrix over the Laurent polynomial ring which transforms the given matrix into a matrix of the form $(I \quad 0)$

9.1.6 Functions of QUILLENSUSLIN for localizations

`InvertibleIn`	Find an element in the intersection of an ideal and a multiplicative closed subset of the polynomial ring
`IsInS`	Test whether or not a polynomial belongs to a given multiplicative subset of the polynomial ring
`SHeuristic`	Test whether or not a heuristic method can be used over a localization of the polynomial ring

To our knowledge, the QUILLENSUSLIN package is the only package dedicated to the implementation of the Quillen-Suslin theorem (see [11] for a partial one) and its applications to mathematical physics, control theory and signal processing. An OREMODULES version of QUILLENSUSLIN will soon be available on the OREMODULES web site [4] which will extend [11]. Applications of the Quillen-Suslin theorem to algebraic geometry will be studied in the future.

9.2 Examples

9.2.1 Example taken from [19]

We consider the row vector R over the polynomial ring $D = \mathbb{Z}[x]$ given in [19].

In the QUILLENSUSLIN package, all the computations are performed for a commutative polynomial ring with rational coefficients if the last parameter is set to *true* and with integer coefficients if the last parameter is set to *false*.

We first declare the independent variables x of the polynomial ring by setting

```
> var:=[x];
```

$$var := [x]$$

and then the row vector R:

```
> R:=[13, x^2-1, 2*x-5];
```

$$R := [13,\, x^2 - 1,\, 2\,x - 5]$$

Let us check whether or not R admits a right-inverse over the ring D.

```
> RightInverse(R, var, false);
```

$$[55 - 36\,x + 6\,x^2,\, -6,\, 144 - 36\,x]$$

Applying the `QSAlgorithm` procedure to the row vector R, we then obtain

```
> U:=QSAlgorithm(R, var, false);
```

$$U := [55 - 36\,x + 6\,x^2\,,\, 6481 - 8532\,x + 4175\,x^2 - 900\,x^3 + 72\,x^4\,,$$
$$-(55 - 36\,x + 6\,x^2)\,(2\,x - 5)][-6\,,\, -707 + 468\,x - 72\,x^2\,,\, -30 + 12\,x]$$
$$[144 - 36\,x\,,\, -72\,(x - 4)\,(59 - 39\,x + 6\,x^2)\,,\, 721 - 468\,x + 72\,x^2]$$

The matrix U is unimodular over D and $R\,U = (1\;0\;0)$ as we have

```
> Determinant(U);
```

$$1$$

```
> simplify(Matrix(R).U);
```

$$\begin{bmatrix} 1 & 0 & 0 \end{bmatrix}$$

We note that the `QSAlgorithm` procedure uses a heuristic method as the first two components of the right-inverse of R generate the ring D. Hence, the general algorithm can be avoided in this example:

```
> Heuristic(R, var, false);
```

$$[55 - 36\,x + 6\,x^2\,,\, 6481 - 8532\,x + 4175\,x^2 - 900\,x^3 + 72\,x^4\,,$$
$$-(55 - 36\,x + 6\,x^2)\,(2\,x - 5)]$$
$$[-6\,,\, -707 + 468\,x - 72\,x^2\,,\, -30 + 12\,x]$$
$$[144 - 36\,x\,,\, -72\,(x - 4)\,(59 - 39\,x + 6\,x^2)\,,\, 721 - 468\,x + 72\,x^2]$$

We can check that R is the first row of the inverse U^{-1} of U:

```
> U_inv:=CompleteMatrix(R,var, false);
```

$$U\_inv := \begin{bmatrix} 13 & x^2 - 1 & 2\,x - 5 \\ 6 & 55 - 36\,x + 6\,x^2 & 0 \\ -144 + 36\,x & 1188\,x - 360\,x^2 + 36\,x^3 - 1296 & 1 \end{bmatrix}$$

The residue classes of the last two rows of the matrix U^{-1} define a basis of the D-module $M = D^{1\times 3}/(D\,R)$.

```
> BasisOfCokernelModule(R, var, false);
```

$$\begin{bmatrix} 6 & 55 - 36\,x + 6\,x^2 & 0 \\ -144 + 36\,x & 1188\,x - 360\,x^2 + 36\,x^3 - 1296 & 1 \end{bmatrix}$$

We can reduce the degree of the components of the rows defining the basis:

```
> BasisOfCokernelModule(R, var, false, reduced);
```

$$\begin{bmatrix} 0 & 24 - 6\,x & 1 \\ 72 & 83 & -24 + 12\,x \end{bmatrix}$$

The injective parametrization of the system defined by R is then defined by

```
> InjectiveParametrization(Matrix(R), var, false);
```

$$\begin{bmatrix} 6481 - 8532\,x + 4175\,x^2 - 900\,x^3 + 72\,x^4 & -(55 - 36\,x + 6\,x^2)\,(2\,x - 5) \\ -707 + 468\,x - 72\,x^2 & -30 + 12\,x \\ -72\,(x - 4)\,(59 - 39\,x + 6\,x^2) & 721 - 468\,x + 72\,x^2 \end{bmatrix}$$

9.2.2 Example taken from [22]

We consider the vector vector R with entries in the ring $D = \mathbb{Q}[x, y]$ defined by

```
> var:=[x,y];
```

$$var := [x, y]$$

```
> R := [x^2*y+1, x+y-2, 2*x*y];
```

$$R := [x^2\,y + 1,\, x + y - 2,\, 2\,x\,y]$$

We can check that ideal generated by the entries of R generates D as we have

```
> IsUnimod(R, var);
```

true

Therefore, the row vector R admits a right-inverse over D and then defines a projective D-module $M = D^{1\times 2}/(D\,R)$, i. e., free by the Quillen-Suslin theorem.

As the first and the last components of R generate the ring D, we know that we can use a heuristic method for computing a basis of the D-module M. This last result can be checked as follows once we note that we are working over the field $\mathbb{Q}$ and then need to set the last parameter to *true* in the procedures:

```
> U:=Heuristic(R, var, true);
```

$$U := \begin{bmatrix} 1 & 2 - y - x & -2\,x\,y \\ 0 & 1 & 0 \\ -\frac{x}{2} & \frac{x\,(x + y - 2)}{2} & x^2\,y - 1 \end{bmatrix}$$

We can check that the entries of the inverse U_{inv} of the matrix U belong to D, i. e., $U \in \mathrm{GL}_3(D)$, and its first row is R:

```
> U_inv:=CompleteMatrix(R, var, true);
```

$$U\_inv := \begin{bmatrix} x^2\,y+1 & x+y-2 & 2\,x\,y \\ 0 & 1 & 0 \\ \frac{x}{2} & 0 & 1 \end{bmatrix}$$

The residue classes of the last two rows of U_{inv} in M form a basis of M. This result can directly be obtained as follows:

```
> BasisOfCokernelModule(Matrix(R), var, true);
```

$$\begin{bmatrix} 0 & 1 & 0 \\ \frac{x}{2} & 0 & 1 \end{bmatrix}$$

The injective parametrization of the system defined by R is given by the last two columns of U, a fact that can directly be obtained by

```
> InjectiveParametrization(Matrix(R), var, true);
```

$$\begin{bmatrix} 2-y-x & -2\,x\,y \\ 1 & 0 \\ \frac{x\,(x+y-2)}{2} & x^2\,y+1 \end{bmatrix}$$

9.2.3 Example given by A. van den Essen

The following example was given to us by A. van den Essen (Radboud University Nijmegen). We are grateful to him for letting us using it for illustrating the package QUILLENSUSLIN. We consider the polynomial ring $D = \mathbb{Q}[t, x, y, z]$

```
> var:=[t,x,y,z];
```

$$var := [t,\ x,\ y,\ z]$$

and we consider the row vector R defined by

```
> R:=[2*t*x*z+t*y^2+1, 2*t*x*y+t^2, t*x^2];
```

$$R := [2\,t\,x\,z + t\,y^2 + 1,\ 2\,t\,x\,y + t^2,\ t\,x^2]$$

We check that the ideal of D generated by the entries of R defines the whole ring D:

```
> IsUnimod(R, var, true);
```

$$true$$

Hence, the row vector R admits a right-inverse over D, and thus, the finitely presented D-module $M = D^{1\times 3}/(D\,R)$ is projective, i. e., free by the Quillen-Suslin theorem. Let us solve Problem 2 and compute a basis of the D-module M. In order to do that, we can first check that none of the heuristic methods described in section 3.3 and implemented in QUILLENSUSLIN can be used to solve the problem:

```
> infolevel[QSAlgorithm]:=3;
```

$$infolevel_{QSAlgorithm} := 3$$

```
> U:=QSAlgorithm(R,var, true);

QuillenSuslin/RowQS:    RowQS   [2*t*x*z+t*y^2+1, 2*t*x*y+t^2,
 t*x^2], [t, x, y, z]

QuillenSuslin/RowQS:    Compute RightInverse
```

A right-inverse of the row vector R is defined by

```
QuillenSuslin/RowQS:    RightInverse
[2*t^2*y^2*x*z-2*t*x*z+t^2*y^4-t*y^2+1,
8*x^3*z^2*y^3+8*x^2*z*y^5-4*x^2*z^2*t*y^2-4*t*y^4*x*z+4*x^2*z^2
+2*y^7*x-t*y^6+2*y^2*x*z, -16*y^4*x^2*z^2-16*y^6*x*z-8*y*x*z^2
-4*y^8-4*y^3*z]

QuillenSuslin/RowQS:    RowQS   Test heuristic methods. For more
information set infolevel[Heuristic]:=3

QuillenSuslin/RowQS:    Not easy - no heuristic methods work
```

We obtain that none of the heuristic methods implemented in QUILLENSUSLIN can be applied to R. Hence, we need to use the general algorithm presented in section 3.4. The first step of this algorithm is to compute a transformation which maps R to a row vector with a monic component in the last variable z. We obtain that the permutation of variables $t \mapsto z, x \mapsto t, y \mapsto x, z \mapsto y$ normalizes R:

$$normalization\ over\ QQ$$

```
QuillenSuslin/RowQS:    The row after normalization
[2*z_*t_*y_+z_*x_^2+1, 2*z_*t_*x_+z_^2, z_*t_^2]
```

Let us call the new row vector $\overline{R}$. We can now test whether or not any of heuristic methods can be applied to $\overline{R}$:

```
QuillenSuslin/RowQS:    Test heuristic methods for the
normalized row  [2*z_*t_*y_+z_*x_^2+1, 2*z_*t_*x_+z_^2, z_*t_^2]

QuillenSuslin/RowQS:    No heuristic methods work for the
normalized row
```

No heuristic method can be applied to $\overline{R}$. We can then check if it is possible to reduce the degree of the components of $\overline{R}$ using its monic component $z^2 + 2\,z\,t\,x$:

```
QuillenSuslin/RowQS:    No reduction - the rows is already
reduced  [2*z_*t_*y_+z_*x_^2+1, 2*z_*t_*x_+z_^2, z_*t_^2]
```

No simplification can be done. Now, we enter the general algorithm:

```
QuillenSuslin/RowQS:    OneStep - Enter the inductive procedure
and reduce one variable:
```

```
QuillenSuslin/RowQS:    OneStep - Compute local solutions and
patch them together!

QuillenSuslin/RowQS:    OneStep - For more information set
infolevel[OneStepMore]:=3
```

After one inductive step, we obtain a matrix $U \in \mathrm{GL}_3(D)$ such that $\overline{R}(t,x,y,z)\,U = \overline{R}(t,x,y,0) = (1\;0\;0)$, which directly solves Problem 2:

```
QuillenSuslin/RowQS:    After one step:    [1, 0, 0]

QuillenSuslin/RowQS:    Now repeat the computation for fm
[1, 0, 0]

QuillenSuslin/RowQS:    RowQS    [1, 0, 0], [t_, x_, y_]

QuillenSuslin/RowQS:    Compute RightInverse

QuillenSuslin/RowQS:    RightInverse    [1, 0, 0]

QuillenSuslin/RowQS:    RowQS    Test heuristic methods. For more
information set infolevel[Heuristic]:=3
```

Hence, we obtain that the solution of Problem 2 is given by matrix U defined by

$$U := $$
$$[1 - t\,y^2 + 4\,x^2\,z^2\,t^2 - 2\,t\,x\,z + t^2\,y^4 + 4\,t^2\,y^2\,x\,z + 2\,y^7\,x\,t^2 + 8\,x^3\,z^2\,y^3\,t^2$$
$$+ 8\,x^2\,z\,y^5\,t^2,$$
$$4\,y^6\,t^2\,x^2 + 8\,x^3\,z\,t^2\,y^4 - 2\,t\,x\,y - t^2 + 4\,y\,z\,x^2\,t^2,$$
$$2\,z\,t^2\,x^3 + t^2\,y^2\,x^2 - t\,x^2 + 4\,t^2\,x^4\,z\,y^3 + 2\,y^5\,t^2\,x^3]$$
$$[-8\,z^3\,t\,x^3 - t\,y^6 - 6\,t\,y^4\,x\,z - 12\,x^2\,z^2\,t\,y^2 - 16\,z^3\,x^4\,t\,y^3 - 24\,y^5\,z^2\,x^3\,t$$
$$- 12\,y^7\,x^2\,z\,t - 2\,t\,y^9\,x,$$
$$1 + t\,y^2 - 4\,t\,y^8\,x^2 - 16\,t\,x^4\,z^2\,y^4 - 16\,t\,x^3\,z\,y^6 + 2\,t\,x\,z - 4\,t\,y^3\,z\,x^2 - 8\,x^3\,z^2\,y\,t,$$
$$-4\,z^2\,x^4\,t - 2\,y^7\,x^3\,t - 8\,z^2\,x^5\,t\,y^3 - 8\,z\,x^4\,y^5\,t - 4\,t\,y^2\,z\,x^3 - y^4\,x^2\,t]$$
$$[32\,t\,x^3\,z^3\,y^4 + 48\,t\,z^2\,x^2\,y^6 + 16\,x^2\,z^3\,t\,y + 24\,t\,y^8\,z\,x + 16\,t\,y^3\,z^2\,x + 4\,y^5\,z\,t$$
$$+ 4\,t\,y^{10},$$
$$-4\,t\,y^6 + 16\,x^2\,z^2\,t\,y^2 + 32\,y^5\,z^2\,x^3\,t + 32\,y^7\,x^2\,z\,t - 4\,t\,y\,z + 8\,t\,y^9\,x,$$
$$1 + 4\,t\,y^8\,x^2 + 16\,t\,x^4\,z^2\,y^4 + 16\,t\,x^3\,z\,y^6 + 4\,t\,y^3\,z\,x^2 + 8\,x^3\,z^2\,y\,t]$$

```
> infolevel[QSAlgorithm]:=0;
```

$$infolevel_{QSAlgorithm} := 0$$

We can show that the matrix U is a unimodular matrix satisfying $R\,U = (1\;0\;0)$:

```
> LinearAlgebra[Determinant](U);
```

$$1$$

```
> simplify(Matrix(R).U);
```

$$\begin{bmatrix} 1 & 0 & 0 \end{bmatrix}$$

Hence, the first row of the inverse of U is the row R, a fact which can directly be checked using the command `CompleteMatrix`:

```
> B:=LinearAlgebra[MatrixInverse](U);
```

$B :=$

$[2\,t\,x\,z + t\,y^2 + 1\,,\; 2\,t\,x\,y + t^2\,,\; t\,x^2]$

$[6\,t\,y^4\,x\,z + t\,y^6 + 2\,t\,y^9\,x + 24\,y^5\,z^2\,x^3\,t + 12\,y^7\,x^2\,z\,t + 16\,z^3\,x^4\,t\,y^3 + 12\,x^2\,z^2\,t\,y^2$
$+ 8\,z^3\,t\,x^3,$
$1 + 16\,t\,x^4\,z^2\,y^4 + 16\,t\,x^3\,z\,y^6 + 8\,x^3\,z^2\,y\,t + 4\,t\,y^3\,z\,x^2 + 4\,t^2\,y^2\,x\,z + t^2\,y^4 - 2\,t\,x\,z$
$+ 4\,t\,y^8\,x^2 + 4\,x^2\,z^2\,t^2 + 2\,y^7\,x\,t^2 + 8\,x^3\,z^2\,y^3\,t^2 + 8\,x^2\,z\,y^5\,t^2 - t\,y^2,$
$8\,z^2\,x^5\,t\,y^3 + 4\,t\,y^2\,z\,x^3 + 4\,z^2\,x^4\,t + 2\,y^7\,x^3\,t + 8\,z\,x^4\,y^5\,t + y^4\,x^2\,t]$

$[-4\,t\,y^{10} - 16\,x^2\,z^3\,t\,y - 32\,t\,x^3\,z^3\,y^4 - 48\,t\,z^2\,x^2\,y^6 - 24\,t\,y^8\,z\,x - 16\,t\,y^3\,z^2\,x$
$- 4\,y^5\,z\,t,$
$-16\,t^2\,y^4\,z^2\,x^2 - 16\,y^6\,z\,t^2\,x + 4\,t\,y^6 - 4\,y^3\,z\,t^2 - 8\,y\,x\,z^2\,t^2 + 4\,t\,y\,z$
$- 8\,t\,y^9\,x - 4\,t^2\,y^8 - 32\,y^5\,z^2\,x^3\,t - 32\,y^7\,x^2\,z\,t - 16\,x^2\,z^2\,t\,y^2,$
$1 - 16\,t\,x^4\,z^2\,y^4 - 16\,t\,x^3\,z\,y^6 - 8\,x^3\,z^2\,y\,t - 4\,t\,y^3\,z\,x^2 - 4\,t\,y^8\,x^2]$

The residue classes of the last two rows of the matrix B in M form a basis of the D-module M.

```
> BasisOfCokernelModule(Matrix(R), var, true);
```

normalization over QQ

$[6\,t\,y^4\,x\,z + t\,y^6 + 2\,t\,y^9\,x + 24\,y^5\,z^2\,x^3\,t + 12\,y^7\,x^2\,z\,t + 16\,z^3\,x^4\,t\,y^3 + 12\,x^2\,z^2\,t\,y^2$
$+ 8\,z^3\,t\,x^3, 1 + 16\,t\,x^4\,z^2\,y^4 + 16\,t\,x^3\,z\,y^6 + 8\,x^3\,z^2\,y\,t + 4\,t\,y^3\,z\,x^2 + 4\,t^2\,y^2\,x\,z$
$+ t^2\,y^4 - 2\,t\,x\,z + 4\,t\,y^8\,x^2 + 4\,x^2\,z^2\,t^2 + 2\,y^7\,x\,t^2 + 8\,x^3\,z^2\,y^3\,t^2 + 8\,x^2\,z\,y^5\,t^2 - t\,y^2,$
$8\,z^2\,x^5\,t\,y^3 + 4\,t\,y^2\,z\,x^3 + 4\,z^2\,x^4\,t + 2\,y^7\,x^3\,t + 8\,z\,x^4\,y^5\,t + y^4\,x^2\,t]$

$[-4\,t\,y^{10} - 16\,x^2\,z^3\,t\,y - 32\,t\,x^3\,z^3\,y^4 - 48\,t\,z^2\,x^2\,y^6 - 24\,t\,y^8\,z\,x - 16\,t\,y^3\,z^2\,x$
$- 4\,y^5\,z\,t, -16\,t^2\,y^4\,z^2\,x^2 - 16\,y^6\,z\,t^2\,x + 4\,t\,y^6 - 4\,y^3\,z\,t^2 - 8\,y\,x\,z^2\,t^2 + 4\,t\,y\,z$
$- 8\,t\,y^9\,x - 4\,t^2\,y^8 - 32\,y^5\,z^2\,x^3\,t - 32\,y^7\,x^2\,z\,t - 16\,x^2\,z^2\,t\,y^2,$
$1 - 16\,t\,x^4\,z^2\,y^4 - 16\,t\,x^3\,z\,y^6 - 8\,x^3\,z^2\,y\,t - 4\,t\,y^3\,z\,x^2 - 4\,t\,y^8\,x^2]$

We can try to reduce the degree of the basis elements using the option *reduce*:

```
> BasisOfCokernelModule(Matrix(R), var,true, reduce);
```

normalization over QQ

$[-8\,x^3\,z^2\,y^3 - 8\,x^2\,z\,y^5 - 2\,y^7\,x - y^4 - 4\,y^2\,x\,z - 4\,x^2\,z^2\,,$
$-2\,t\,y^5\,x - 4\,t\,y^3\,z\,x^2 - t\,y^2 - 2\,t\,x\,z + 1\,,\; 0]$

$[16\,y^4\,x^2\,z^2 + 16\,y^6\,x\,z + 4\,y^8 + 4\,y^3\,z + 8\,y\,x\,z^2\,,\; 4\,t\,y^6 + 8\,t\,y^4\,x\,z + 4\,t\,y\,z\,,\; 1]$

We refer the reader to [14] for an explicit description of the corresponding local computations. See also [14] for more examples and, particularly, examples over the Laurent polynomial ring $\mathbb{Q}[x_1, x_1^{-1}, \ldots, x_n, x_n^{-1}]$ using Park's transformation ([34, 37]).

9.3 Equivalences of flat multidimensional linear systems

Examples 4.5, 4.7 and 4.9

We consider the differential time-delay linear system defined by (4.1) ([30]). The matrix R associated with (4.1) is defined by

```
> R:=Matrix([[d-delta+2, 2,-2*delta],
> [d,d,-d*delta-1]]);
```

$$R := \begin{bmatrix} d-\delta+2 & 2 & -2\,\delta \\ d & d & -d\,\delta-1 \end{bmatrix}$$

where d denotes the time-derivative operator and δ the time-delay operator. Hence, we need to consider the commutative polynomial ring $D = \mathbb{Q}[d, \delta]$ and the D-module defined by $M = D^{1\times 3}/(D^{1\times 2}\, R)$.

```
> var:=[d, delta];
```

$$var := [d,\ \delta]$$

Let us check whether or not the matrix R admits a right-inverse over D:

```
> IsUnimod(R, var);
```

$$true$$

As the matrix R admits a right-inverse over D, we then obtain that the D-module M is projective, i. e., free by the Quillen-Suslin theorem. Let us solve Problem 1:

```
> U:=QSAlgorithm(R, var);
```

$$U := \begin{bmatrix} 0 & 0 & -2 \\ \frac{d\,\delta}{2}+\frac{1}{2} & -\delta & d^2\,\delta+d-d\,\delta^2-\delta+2 \\ \frac{d}{2} & -1 & d^2-d\,\delta \end{bmatrix}$$

We can check that U gives a solution of Problem 1 as we have $R\,U = (I_2 \quad 0)$:

```
> simplify(R.U);
```

$$\begin{bmatrix} 1 & 0 & 0 \\ 0 & 1 & 0 \end{bmatrix}$$

and U is a unimodular matrix over D, i. e., $U \in \mathrm{GL}_3(D)$:

```
> LinearAlgebra[MatrixInverse](U);
```

$$\begin{bmatrix} d-\delta+2 & 2 & -2\,\delta \\ d & d & -d\,\delta-1 \\ -1/2 & 0 & 0 \end{bmatrix}$$

The residue class of the last row of the matrix U^{-1} in M defines a basis of M. Moreover, the system defined by R admits the following injective parametrization

```
> Q:=InjectiveParametrization(R, var, true);
```

$$Q := \begin{bmatrix} -2 \\ d^2\,\delta+d-d\,\delta^2-\delta+2 \\ d^2-d\,\delta \end{bmatrix}$$

i.e., for any D-module $\mathcal{F}$ (e.g.., $C^\infty(\mathbb{R})$), every $\mathcal{F}$-solution η of the system $\ker_{\mathcal{F}}(R.)$ has the form $\eta = Q\,\xi$ for a certain $\xi \in \mathcal{F}$. As the system $\ker_{\mathcal{F}}(R.)$ is flat, by Corollary 4.8, we know that $\ker_{\mathcal{F}}(R(d,\delta).)$ is algebraically equivalent to the controllable ordinary differential system without time-delay, i.e., to $\ker_{\mathcal{F}}(R(d,1).)$. We can compute an invertible transformation which maps elements of $\ker_{\mathcal{F}}(R(d,1).)$ to elements of $\ker_{\mathcal{F}}(R(d,\delta).)$:

```
> U[1]:=SetLastVariableA(R, var, 1, true);
```

$$U_1 := \begin{bmatrix} 1 & 0 & 0 \\ \frac{1}{2}\,d\,\delta^2-\frac{1}{2}\,d\,\delta+\frac{1}{2}\,\delta-\frac{1}{2} & 1 & \delta-1 \\ \frac{d\,(\delta-1)}{2} & 0 & 1 \end{bmatrix}$$

We can check that $R(d,\delta)\,U_1 = R(d,1)$:

```
> R[1]:=simplify(R.U[1]);
```

$$R_1 := \begin{bmatrix} d+1 & 2 & -2 \\ d & d & -1-d \end{bmatrix}$$

Then, the inverse transformation, i.e., the transformation sending elements of $\ker_{\mathcal{F}}(R(d,\delta).)$ to elements of $\ker_{\mathcal{F}}(R(d,1).)$, is defined by the matrix U_1^{-1}:

```
> LinearAlgebra[MatrixInverse](U[1]);
```

$$\begin{bmatrix} 1 & 0 & 0 \\ -\frac{1}{2}\,d\,\delta-\frac{1}{2}\,\delta+\frac{1}{2}+\frac{1}{2}\,d & 1 & -\delta+1 \\ -\frac{d\,(\delta-1)}{2} & 0 & 1 \end{bmatrix}$$

As the $E = \mathbb{Q}[d]$-module $N = E^{1\times 3}/(E^{1\times 2}\,R_1)$ is also free, we can find $U_2 \in \mathrm{GL}_3(E)$ such that $R_1\,U_2 = (I_2 \quad 0)$. For instance, we get

```
> U[2]:=QSAlgorithm(R[1], var);
```

$$U_2 := \begin{bmatrix} 0 & 0 & -2 \\ \frac{1}{2} + \frac{d}{2} & -1 & d^2+1 \\ \frac{d}{2} & -1 & d^2-d \end{bmatrix}$$

As the ordinary differential linear system without delay $\ker_{\mathcal{F}}(R(d,1).)$ is flat, it also admits an injective parametrization which can be computed by

```
> Q1:=InjectiveParametrization(R1,var, true);
```

$$Q1 := \begin{bmatrix} -2 \\ d^2+1 \\ d^2-d \end{bmatrix}$$

Similarly, we can prove that the system $\ker_{\mathcal{F}}(R(d,\delta).)$ is algebraically equivalent to the system $\ker_{\mathcal{F}}(R(d,0).)$ by means of the following invertible transformation:

```
> V[1]:=SetLastVariableA(R, var, 0, true);
```

$$V_1 := \begin{bmatrix} 1 & 0 & 0 \\ \frac{1}{2}\, d\,\delta^2 + \frac{1}{2}\,\delta & 1 & \delta \\ \frac{d\,\delta}{2} & 0 & 1 \end{bmatrix}$$

Then, the inverse transformation, i. e., the transformation sending elements of $\ker_{\mathcal{F}}(R(d,\delta).)$ to elements of $\ker_{\mathcal{F}}(R(d,0).)$, is defined by the matrix V_1^{-1}:

```
> LinearAlgebra[MatrixInverse](V[1]);
```

$$\begin{bmatrix} 1 & 0 & 0 \\ -\frac{\delta}{2} & 1 & -\delta \\ -\frac{d\,\delta}{2} & 0 & 1 \end{bmatrix}$$

As the $E = \mathbb{Q}[d]$-module $P = E^{1\times 3}/(E^{1\times 2}\, R_0)$ is also free, we can find $V_2 \in \mathrm{GL}_3(E)$ such that $R_0\, V_2 = (I_2 \quad 0)$. In particular, we have

```
> V[2]:=QSAlgorithm(R[0], var);
```

$$V_2 := \begin{bmatrix} 0 & 0 & -2 \\ \frac{1}{2} & 0 & d+2 \\ \frac{d}{2} & -1 & d^2 \end{bmatrix}$$

See [14] for more examples.

9.4 Pommaret's theorem of the Lin-Bose conjecture

Examples 5.3, 5.5 and 5.6

Let us consider the differential time-delay model of a flexible rod with a force applied on one end defined in Example 5.3 ([30]). The system matrix R with entries in the commutative polynomial ring $D = \mathbb{Q}[d, \delta]$, where d denotes the time-derivative operator and δ the time-delay operator, is defined by

```
> var:=[d, delta];
```

$$var := [d, \delta]$$

```
> R:=Matrix([[d,-d*delta,-1],
> [2*delta*d,-d*delta^2-d,0]]);
```

$$R := \begin{bmatrix} d & -d\,\delta & -1 \\ 2\,d\,\delta & -d\,\delta^2 - d & 0 \end{bmatrix}$$

Let us check whether or not the D-module $M = D^{1\times 3}/(D^{1\times 2}\,R)$ is projective, i. e., free by the Quillen-Suslin theorem:

```
> IsUnimod(R, var);
```

$$false$$

We obtain that R does not admit a right-inverse over D and the D-module M is not free. In particular, there does not exist a matrix $U \in \mathrm{GL}_3(D)$ such that $R\,U = (I_2 \quad 0)$ or, equivalently, R cannot be completed to a unimodular matrix over D. Let us compute the set of all maximal minors of R:

```
> m:=MaxMinors(R);
```

$$m := [d^2\,\delta^2 - d^2,\, 2\,d\,\delta,\, -d\,\delta^2 - d]$$

The ideal I of D defined by the maximal minors is generated by

```
> Involutive[InvolutiveBasis](m, var);
```

$$[d]$$

i. e., $I = (d)$, and thus, d is a greatest common divisor of the maximal minors of R. In particular, using Figure 2.1, we obtain that the torsion D-submodule $t(M)$ of M is not reduced to 0. A solution of Problem 3 can directly be obtained by calling the QUILLENSUSLIN procedure `LinBose1` as follows:

```
> F:=LinBose1(R, var);
```

$$F := [\begin{bmatrix} -1 & 0 \\ 0 & -d \end{bmatrix}, \begin{bmatrix} -d & d\,\delta & 1 \\ -2\,\delta & \delta^2+1 & 0 \end{bmatrix}]$$

The second matrix of the previous output corresponds to the matrix R' solving Problem 3, whereas the first one corresponds to the matrix R'' satisfying $R = R''\,R'$ and $\det R'' = d$, where d denotes the greatest common divisor of the maximal minors of R.

```
> simplify(F[1].F[2]);
```

$$\begin{bmatrix} d & -d\,\delta & -1 \\ 2\,d\,\delta & -d\,\delta^2 - d & 0 \end{bmatrix}$$

```
> LinearAlgebra[Determinant](F[1]);
```

$$d$$

```
> IsUnimod(F[2], var);
```

$$\mathit{true}$$

Let us now solve Problem 4. We can obtain a solution of Problem 4 by using the procedure LinBose2:

```
> C:=LinBose2(R, var);
```

$$C := \begin{bmatrix} d & -d\,\delta & -1 \\ 2\,d\,\delta & -d\,\delta^2 - d & 0 \\ -1 & \frac{\delta}{2} & 0 \end{bmatrix}$$

The determinant of the matrix C equals to

```
> LinearAlgebra[Determinant](C);
```

$$d$$

which solves the problem. See [14] for more examples.

9.5 (Weakly) coprime factorizations of rational transfer matrices

Let us consider the commutative polynomial ring $D = \mathbb{Q}[z_1, z_2, z_3]$:

```
> var:=[z[1],z[2],z[3]];
```

$$\mathit{var} := [z_1,\, z_2,\, z_3]$$

We consider the rational transfer matrix defined in Example 6.6, namely

```
> P:=Matrix(<(z[1]^2*z[2]^2+1)/(z[1]*z[2]^2*z[3]),
> (z[1]^2*z[3]+1)/(z[1]*z[2]^2*z[3]) >);
```

$$P := \begin{bmatrix} \dfrac{z_1^2\, z_2^2 + 1}{z_1\, z_2^2\, z_3} \\ \dfrac{z_1^2\, z_3 + 1}{z_1\, z_2^2\, z_3} \end{bmatrix}$$

Cleaning the denominators of P, we obtain the fractional representation $P = D_P^{-1}\, N_P$ of P, where the matrices $D_P \in D^{2\times 2}$ and $N_P \in D^{2\times 1}$ are defined by

```
> D_P:=LinearAlgebra[ScalarMatrix](DenomOf(P),2,2);
```

$$D\_P := \begin{bmatrix} z_1\, z_2^2\, z_3 & 0 \\ 0 & z_1\, z_2^2\, z_3 \end{bmatrix}$$

```
> N_P:=simplify(D_P.P);
```

$$N\_P := \begin{bmatrix} z_1^2\, z_2^2 + 1 \\ z_1^2\, z_3 + 1 \end{bmatrix}$$

Let us define the matrix $Q = (D_P \quad -N_P) \in D^{2\times 3}$, namely

```
> Q:=Matrix([D_P, -N_P]);
```

$$Q := \begin{bmatrix} z_1\, z_2^2\, z_3 & 0 & -z_1^2\, z_2^2 - 1 \\ 0 & z_1\, z_2^2\, z_3 & -z_1^2\, z_3 - 1 \end{bmatrix}$$

The set of the maximal minors of Q is defined by

```
> m1:=MaxMinors(Q);
```

$$m1 := [z_1^2\, z_2^4\, z_3^2,\ z_1\, z_2^2\, z_3\, (-z_1^2\, z_3 - 1),\ -(-z_1^2\, z_2^2 - 1)\, z_1\, z_2^2\, z_3]$$

The greatest common divisor of the maximal minors of Q is

```
> d:={gcd(m1[1],m1[2]),gcd(m1[1],m1[3]),
> gcd(m1[2],m1[3])};
```

$$d := \{z_1\, z_2^2\, z_3\}$$

Hence, $P = D_P^{-1}\, N_P$ is not a weakly left-coprime factorization of P. Let us check whether or not the rational transfer matrix P admits a weakly left-coprime factorization and, if so, compute one:

```
> WLCF:=WLCFactorization(P,var);
```

$$WLCF := [\begin{bmatrix} -z_1^2\, z_3 - 1 & z_1^2\, z_2^2 + z_1^4\, z_3\, z_2^2 + 1 \\ -z_1\, z_2^2\, z_3 & z_1^3\, z_2^4\, z_3 \end{bmatrix}, \begin{bmatrix} z_1^5\, z_3 + z_1^3 \\ z_1^4\, z_3\, z_2^2 - 1 \end{bmatrix}]$$

We obtain that P admits the weakly left-coprime factorization defined by

$$P = (D'_P)^{-1}\, N'_P,$$

where $D'_P \in D^{2\times 2}$ is the first matrix given in the previous output and $N'_P \in D^{2\times 1}$ is the second one. In particular, we can check that $(D'_P)^{-1}\, N'_P$ is equal to P:

```
> LinearAlgebra[MatrixInverse](WLCF[1]).WLCF[2]);
```

$$\begin{bmatrix} \dfrac{z_1^2\, z_2^2 + 1}{z_1\, z_2^2\, z_3} \\ \dfrac{z_1^2\, z_3 + 1}{z_1\, z_2^2\, z_3} \end{bmatrix}$$

Moreover, if we define the matrix $R = (D'_P \quad -N'_P) \in D^{2\times 3}$, namely

```
> R:=Matrix([WLCF[1],-WLCF[2]]);
```

$$R := \begin{bmatrix} -z_1{}^2\, z_3 - 1 & z_1{}^2\, z_2{}^2 + z_1{}^4\, z_3\, z_2{}^2 + 1 & -z_1{}^5\, z_3 - z_1{}^3 \\ -z_1\, z_2{}^2\, z_3 & z_1{}^3\, z_2{}^4\, z_3 & -z_1{}^4\, z_3\, z_2{}^2 + 1 \end{bmatrix}$$

then, the set of the maximal minors of R is defined by

```
> m2:=MaxMinors(R);
```

$$m2 := [z_1\, z_2{}^2\, z_3,\; -z_1{}^2\, z_3 - 1,\; z_1{}^2\, z_2{}^2 + 1]$$

and the greatest common divisor of the maximal minors of R is then equal to 1 as

```
> {gcd(m2[1],m2[2]),gcd(m2[1],m2[3]),
> gcd(m2[2],m2[3])};
```

$$\{1\}$$

and thus, $P = (D'_P)^{-1}\, N'_P$ is a weakly left-coprime factorization of P. Let us check whether or not the transfer matrix P admits a left-coprime factorization:

```
> LCF:=LCFactorization(P,var);
```

$$LCF := [\begin{bmatrix} z_1{}^2\, z_3 - 1 & -z_1{}^4\, z_3\, z_2{}^2 + z_1{}^2\, z_2{}^2 + 1 \\ -z_1\, z_2{}^2\, z_3 & z_1{}^3\, z_2{}^4\, z_3 \end{bmatrix}, \begin{bmatrix} z_1{}^3\, (z_1{}^2\, z_3 - 1) \\ -z_1{}^4\, z_3\, z_2{}^2 - 1 \end{bmatrix}]$$

$P = (D'_P)^{-1}\, N'_P$ is a left-coprime factorization of P and $R = (D'_P \quad - N'_P)$ admits the following right-inverse over D:

```
> Involutive[PolRightInverse](R,var);
```

$$\begin{bmatrix} z_1{}^2\, z_2{}^2 & z_1{}^3 \\ 1 & 0 \\ 0 & 1 \end{bmatrix}$$

A weakly right-coprime factorization of P can be obtained in a similar way. See [14] for more details.

Acknowledgements. The first author is very grateful to W. Plesken for his support. She would also like to express thanks to F. Castro-Jimenez and J. Gago-Vargas for the opportunity to discuss with them on the Quillen-Suslin theorem and to A. van den Essen for pointing out to her an interesting example that has been included in the paper.

The second author would like to thank Y. Lam, T. Coquand, H. Lombardi and I. Yengui for interesting discussions on the Quillen-Suslin theorem and Z. Lin and H. Park for discussions on applications of the Quillen-Suslin theorem to mathematical systems theory, control theory and signal processing.

Bibliography

[1] M. Amidou and I. Yengui, An algorithm for unimodular completion over Laurent polynomial rings, preprint 2006, private communication.

[2] J. E. Bjork, *Rings of Differential Operators*, North Holland, 1979.

[3] Y. A. Blinkov, C. F. Cid, V. P. Gerdt, W. Plesken, and D. Robertz, *The MAPLE Package "Janet": I. Polynomial Systems*. In the Proceedings of *Computer Algebra in Scientific Computing CASC 2003*, edited by V. G. Ganzha, E. W. Mayr, E. V. Vorozhtsov, pp. 31–40. Also available with the package from `http://wwwb.math.rwth-aachen.de/Janet`.

[4] F. Chyzak, A. Quadrat, and D. Robertz, OREMODULES project, directly available from `http://wwwb.math.rwth-aachen.de/OreModules`.

[5] F. Chyzak, A. Quadrat, and D. Robertz, Effective algorithms for parametrizing linear control systems over Ore algebras, *Appl. Algebra Engrg. Comm. Comput.*, 16 (2005), pp. 319–376.

[6] F. Chyzak, A. Quadrat, and D. Robertz, OREMODULES: A symbolic package for the study of multidimensional linear systems", *Applications of Time-Delay Systems*, J. Chiasson and J.-J. Loiseau (eds.), LNCIS 352, Springer, 2007, pp. 233–264.

[7] K. Eriksson, D. Estep, P. Hansbo, and C. Johnson, *Computational Differential Equations*, Cambridge University Press, 1996.

[8] T. Cluzeau and A. Quadrat, Factoring and decomposing a class of linear functional systems, to appear in *Linear Algebra Appl.*

[9] T. Cluzeau and A. Quadrat, MORPHISMS: A homological algebra package for factoring and decomposing linear functional systems, 2007, in preparation.

[10] D. Eisenbud, *Commutative Algebra with a View Toward Algebraic Geometry*, Graduate Texts in Mathematics 150, Springer-Verlag, 1994.

[11] J. Evers, Implementation of the Quillen-Suslin theorem in OREMODULES, Internship at INRIA Sophia Antipolis (06-08/2005) under the supervision of A. Quadrat.

[12] A. Fabiańska, QUILLENSUSLIN project: A package for computing bases of free modules over commutative polynomial rings, `http://wwwb.math.rwth-aachen.de/QuillenSuslin`.

[13] A. Fabiańska and A. Quadrat, Flat multidimensional linear systems with constant coefficients are equivalent to controllable 1-D linear systems, Proceedings of *MTNS 2006*, Kyoto (Japan), 24-28/07/06.

[14] A. Fabiańska and A. Quadrat, Applications of the Quillen-Suslin theorem to multidimensional systems theory, INRIA report 6126, February 2007, `https://hal.inria.fr/inria-00131035`.

[15] N. Fitchas and A. Galligo, Nullstellensatz effectif et conjecture de Serre (Théorème de Quillen-Suslin) pour le calcul formel, *Math. Nachr.* 149 (1990), pp. 231–253.

[16] M. Fliess, J. Lévine, P. Martin, and P. Rouchon, Flatness and defect of nonlinear systems: introductory theory and examples, *Internat. J. Control* 61 (1995), pp. 1327–1361.

[17] M. Fliess, J. Lévine, P. Martin, and P. Rouchon, A Lie-Bäcklund approach to equivalence and flatness of nonlinear systems, *IEEE Trans. Automatic Control* 40 (1999), pp. 922–937.

[18] M. Fliess and H. Mounier, Controllability and observability of linear delay systems: an algebraic approach, *ESAIM Control Optim. Calc. Var.* 3 (1998), pp. 301–314.

[19] J. Gago-Vargas, Constructions in $R[x_1, \ldots, x_n]$: applications to K-theory, *J. Pure Appl. Algebra* 171 (2002), pp. 185–196.

[20] G.-M. Greuel and G. Pfister, *A Singular Introduction to Commutative Algebra*, Springer, 2002.

[21] M. Kashiwara, *Algebraic Study of Systems of Partial Differential Equations*, Master's Thesis, Tokyo University 1970, English translation, Société Mathématique de France, Mémoire 63, 1995.

[22] R. C. Laubenbacher and C. J. Woodburn, A new algorithm for the Quillen-Suslin Theorem, *Beiträge Algebra Geom.* 41 (2000), pp. 23–31.

[23] T. Y. Lam, *Serre's Conjecture*, Lecture Notes in Mathematics 635, Springer Verlag, 1978.

[24] T. Y. Lam, *Serre's Problem on projective Modules*, Springer Monograph in Mathematics, Springer Verlag, 2006.

[25] Z. Lin and N. K. Bose, A generalization of Serre's conjecture and related issues, *Linear Algebra Appl.* 338 (2001), pp. 125–138.

[26] A. Logar and B. Sturmfels, Algorithms for the Quillen-Suslin theorem, *J. Algebra* 145 (1992), pp. 231–239.

[27] H. Lombardi and I. Yengui, Suslin's algorithms for reduction of unimodular rows, *J. Symbolic Comput.* 39 (2005), pp. 707–717.

[28] B. Malgrange, Systèmes à coefficients constants, *Séminaire Bourbaki* 1962/63, pp. 1–11.

[29] G. Monge, Supplément où l'on fait voir que les équations aux différences ordinaires pour lesquelles les conditions d'intégrabilité ne sont pas satisfaites sont susceptibles d'une véritable intégration, *Mémoire Académie des Sciences*, Paris, 1784, pp. 502–576.

[30] H. Mounier, *Propriétés structurelles des systèmes linéaires à retards: aspects théoriques et pratiques*, PhD thesis, University Paris XI, France, 1995.

[31] H. Mounier, J. Rudolph, M. Fliess, and P. Rouchon, Tracking control of a vibrating string with an interior mass viewed as delay system, *ESAIM Control Optim. Calc. Var.* 3 (1998), pp. 315–321.

[32] U. Oberst, Multidimensional constant linear systems, *Acta Appl. Math.* 20 (1990), pp. 1–175.

[33] V. P. Palamodov, *Linear Differential Operators with Constant Coefficients*, Springer, 1970.

[34] H. Park, *A Computational Theory of Laurent Polynomial Rings and Multidimensional FIR Systems*, University of Berkeley, 1995.

[35] H. Park and C. Woodburn, An algorithmic proof of Suslin's stability theorem for polynomial rings, *J. Algebra*, 178 (1995), pp. 277–298.

[36] H. Park, Prime factorization of n-D polynomial matrices, *Proceedings of IEEE International Symposium on Circuits and Systems* vol. 3 (2003), pp. 666–669.

[37] H. Park, Symbolic computation and signal processing, *J. Symbolic Comput.* 37 (2004), pp. 209–226.

[38] N. Petit and P. Rouchon, Dynamics and solutions to some control problems for water-tank systems, *IEEE Trans. Automat. Control* 47 (2002), pp. 595–609.

[39] H. K. Pillai and S. Shankar, A behavioural approach to control of distributed systems, *SIAM J. Control Optim.* 37 (1999), pp. 388–408.

[40] J. W. Polderman and J. C. Willems, *Introduction to Mathematical Systems Theory. A Behavioral Approach*, Texts in Applied Mathematics 26, Springer, 1998.

[41] J.-F. Pommaret, Solving Bose conjecture on linear multidimensional systems, *Proceedings of European Control Conference (ECC)*, Porto (Portugal), 04-07/09/01.

[42] J.-F. Pommaret and A. Quadrat, Generalized Bezout Identity, *Appl. Algebra Engrg. Comm. Comput.* 9 (1998), pp. 91–116.

[43] J.-F. Pommaret and A. Quadrat, Localization and parametrization of linear multidimensional control systems, *Systems Control Lett.* 37 (1999), pp. 247–260.

[44] J.-F. Pommaret and A. Quadrat, Algebraic analysis of linear multidimensional control systems, *IMA J. Math. Control Inform.* 16 (1999), pp. 275–297.

[45] J.-F. Pommaret and A. Quadrat, Equivalences of linear control systems, Proceedings of the 14th International Symposium on Mathematical Theory of Networks and Systems (MTNS2000), Perpignan (France), 19-23/06/00.

[46] J.-F. Pommaret and A. Quadrat, A differential operator approach to multidimensional optimal control, *Internat. J. Control* 77 (2004), pp. 821–836.

[47] A. Quadrat, Extended Bézout identities, Proceedings of the *European Control Conference (ECC)*, Porto (Portugal), 04-07/09/01.

[48] A. Quadrat, The fractional representation approach to synthesis problems: an algebraic analysis viewpoint. Part I: (weakly) doubly coprime factorizations, Part II: internal stabilization, *SIAM J. Control Optim.* 42 (2003), pp. 266–299, pp. 300–320.

[49] A. Quadrat and D. Robertz, Computation of bases of free modules over the Weyl algebras, to appear in *J. Symbolic Comput.*

[50] A. Quadrat and D. Robertz, The Stafford project, directly available from `wwwb.math.rwth-aachen.de/OreModules`.

[51] A. Quadrat and D. Robertz, Constructive computation of flat outputs of multidimensional linear systems, Proceedings of *MTNS06*, Kyoto (Japan), 20-24/07/06.

[52] D. Quillen, Projective modules over polynomial rings, *Invent. Math.* 36 (1976), pp. 167–171.

[53] J. J. Rotman, *An Introduction to Homological Algebra*, Academic Press, 1979.

[54] V. Srinivas, A generalized Serre problem, *J. Algebra* 278 (2004), pp. 621–627.

[55] A. A. Suslin, Projective modules over polynomial rings are free, *Dokl. Akad. Nauk. S.S.S.R* 229 (1976), (*Soviet Math. Dokl.* 17 (1976)), pp. 1160–1164.

[56] L. N. Vaseršteín and A. A. Suslin, Serre's Problem on projective modules over polynomial rings and algebraic K-theory, *Math. USSR Izviestija* 10 (1976), no. 5, pp. 937–1001.

[57] M. Vidyasagar, *Control System Synthesis. A Factorization Approach*, MIT Press, 1985.

[58] I. Yengui, Suslin's lemma for elimination, Proceedings of MEGA 2007, Stobl (Austria), June 25–29, 2007.

[59] D. C. Youla and P. F. Pickel, The Quillen-Suslin theorem and the structure of n-dimensional elementary polynomial matrices, *Trans. Circuits and Systems* 31 (1984), pp. 513–517.

[60] M. Wang and D. Feng, On Lin-Bose problem, *Linear Algebra Appl.* 390 (2004), pp. 279–285.

[61] M. Wang and C. P. Kwong, On multivariate polynomial matrix factorizations problems, *Math. Control Signals Systems* 17 (2005), pp. 297–311.

[62] J. Wood, Modules and behaviours in nD systems theory, *Multidimens. Syst. Signal Process.* 11 (2000), pp. 11–48.

[63] E. Zerz, *Topics in Multidimensional Linear Systems Theory*, Lecture Notes in Control and Information Sciences 256, Springer, 2000.

[64] P. Zervos, *Le problème de Monge*, Mémorial des sciences mathématiques, fasicule LIII, Gauthier-Villars, 1932.

Author information

Anna Fabiańska, Lehrstuhl B für Mathematik,
RWTH - Aachen,
Templergraben 64, 52056 Aachen, Germany.
Email: `fabianska@momo.math.rwth-aachen.de`

Alban Quadrat, INRIA Sophia Antipolis,
APICS project,
2004 Route des Lucioles BP 93, 06902 Sophia Antipolis Cedex, France.
Email: `Alban.Quadrat@sophia.inria.fr`

Radon Series Comp. Appl. Math **3**, 107–125

Normal forms in statistical signal processing

Jerome Lebrun

Key words. Generalized normal forms, matrix of traces, parametric rational representations, randomized algorithms.

AMS classification. 68W30, 68W20, 94A14

1 Introduction

In these last years, some tools from algebraic geometry found their way into many applications in signal processing and systems theory. By using these new algorithms, many real world problems in filter bank design [26, 9, 40, 38, 18, 39, 20, 31, 30], multidimensional systems [4, 5, 25, 8] or image processing [13, 46] can now be solved in a way that is very competitive with numerical methods. The success of algebraic geometry in these applications is closely linked to new efficient ways to deal with one of the central application of computer algebra: solving systems of multivariate polynomial equations [6, 28, 41].

Among these most promising approaches to solve systems of polynomial equations, Gröbner bases, homotopic continuation, or resultants methods show however some strong limitations [12, 20] (typically very high computational cost in time and memory, difficulties to deal with parametric equations and/or limitations to rational parameters). This hinders seriously their usability in settings like statistical signal processing with only limited computational power (e. g. the CPU of a mobile phone) and stringent time-constraints (systems with parameters evolving real-time).

In the continuation of our recent paper [19], we detail in this contribution an alternative algebraic approach to solve some problems arising in statistical signal processing for communications that can be described as or reduced to systems of multivariate quadratic polynomial equations.

Based on methods from computational algebraic geometry, this ad-hoc approach derived from the works in [11, 23, 34, 42] achieves a full description of the solution space and thus avoids the local minima issues of the usual gradient-descent based adaptive algorithms of statistical signal processing.

Furthermore, unlike most classical algebraic geometry based methods, the computational cost is kept low by a subtle split of the problem into two stages.

First, using generalized normal forms [42], the symbolic pre-computation of a generic multiplication matrix $\mathbf{M}_u$ and a very convenient, but quite expensive, parametric representation of the problem in the form of a matrix of traces $\mathrm{T_R}\mathbf{M}$ is done *offline* once for all sets of parameters.

Now, *online*, the solutions of the problem are easily obtained for any set of parameters from these two parametric matrices. From the computation of the rank of the instantiated matrix of traces $\mathrm{T_R}\mathbf{M}$ and the minimal polynomial $\tilde{\chi}_u(t)$ of a generic separating polynomial u, we get a geometric representation of the solution set $\mathcal{Z}_{\mathbb{C}}(I)$ of the type

$$\mathcal{Z}_{\mathbb{C}}(I) = \left\{ \left[\frac{g_1(\zeta)}{g_0(\zeta)}, \frac{g_2(\zeta)}{g_0(\zeta)}, \dots, \frac{g_N(\zeta)}{g_0(\zeta)} \right] \middle| \, \tilde{\chi}_u(\zeta) = 0 \right\}$$

for some univariate polynomials $\tilde{\chi}_u, g_0, g_1, \dots, g_N$.

In our previous paper [19], the computation of a generic separating polynomial was obtained through an exhaustive search within a set of polynomials. Here, we generate at random a polynomial that can be later certified as separating (Las-Vegas algorithm).

The computation of the minimal polynomial associated with u is also of the probabilistic type: by generating a Wiedemann [44] sequence of scalars from the matrix of traces, we compute a tentative minimal polynomial using the Berlekamp-Massey algorithm. The validity of the minimal polynomial is again checked later through certificates.

As a consequence, the *online* computational cost has been lowered from our previous implementation [19] and lies now mostly in the computation of the rank and getting the roots of the univariate minimal polynomial $\tilde{\chi}_u(t)$.

The approach described is also general enough that it may easily be applied to a wide variety of other problems of statistical signal processing.

2 Statistical signal processing

In this paper, we will mostly deal with two families of parametric systems of polynomial equations. These systems of equations arise from problems of identification in statistical signal processing, more precisely digital communications, especially cellular.

The basic principle and problem of digital communications is the transmission of an input signal $\mathbf{x}$ trough a propagation channel $\mathbf{H}$ (typically a copper/fiber wire or using antennas) that corrupts the signal by multiple reflections and fading [29]. The signal received $\mathbf{y}$ is thus quite different from the signal sent $\mathbf{x}$. The central problem is then to mitigate the effects of the propagation channel so as to recover the unknown signal $\mathbf{x}$ from the received signal $\mathbf{y}$. This is the role of the so-called equalizer step.

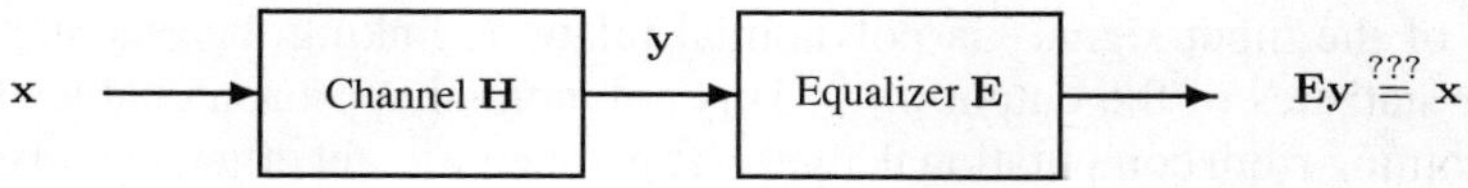

Many reliable equalizing methods have been developed [29], but usually need prior knowledge of the channel. A good estimate of the channel, usually referred to as channel identification, is thus necessary and quite critical. We will consider here the case of a linear and time-invariant (LTI) scalar single-input single-output (SISO) communication channel. Such a channel can be described as the convolutive filtering of an input signal $x[n]$ by a filter $h[n]$. We will assume furthermore here that $h[n]$ has a finite number of coefficients and so finite impulse response (FIR).

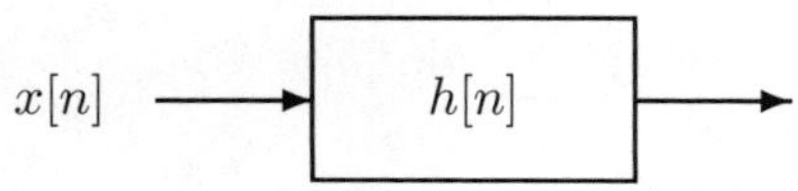

$$y[n] = \sum_{k=0}^{N-1} h[k]x[n-k]$$

Most identification algorithms rely then on the knowledge of the output $y_\star[n]$ of the channel for a special given test input $x_\star[n]$. So-called pilot sequences $\mathbf{x}_\star$ are usually transmitted, either in the middle of each data block as in the GSM telecom standard, or as background signal transmitted in a parallel channel as in the UMTS third generation standard. Most algorithms achieve the construction of an equalizer $\mathbf{E}_\star$ by minimizing the error on pilot sequences

$$\mathbf{E}_\star = \underset{\mathbf{x}_\star \text{ pilot}}{\mathrm{ArgMin}} ||\mathbf{E}(\mathbf{H}\mathbf{x}_\star) - \mathbf{x}_\star||$$

for a well-chosen norm $||\cdot||$ by using a gradient-descent approach [29].

Unlike this, our concern lies here in the *blind* identification of the channel taps $h[n]$, that is, identification without any exact knowledge of any input symbols $x[n]$. Advantages of such an approach include in particular the possibility to reduce or remove the pilot sequence, which permits an increase in the throughput but especially also the stealth interception of digital communications having no or encrypted pilot sequences.

Blind identification or equalization is not a new subject, for it has been addressed as early as around the eighties. However, most of the algorithms are *adaptive* again, of the gradient-descent type, that is, recursive in time, and converge quite slowly, and sometimes even to local minima. Improvements made since early algorithms include the use of the diversity induced by space, time, or excess bandwidth, to modify the model into a Single Input Multiple Output problem and block calculations (i. e. removal of time recursions). Our present contribution introduces a block blind identification algorithm in a framework where diversity cannot be exploited. With this respect, our approach is similar to [43], where inputs were assumed to belong to the unit circle, and to [45, 33] where they were assumed to belong to a finite alphabet. This underlying idea makes sense in digital communications for the emitted signal $x[n]$ normally comes from well-defined modulation schemes (typically BPSK, MSK, QPSK, $\pi/4$-DQPSK, 8-PSK or

$3\pi/8$-D8PSK, or even some type of QAM). Our algorithm is based on the algebraic properties of the input signal via polynomial relations linking the channel taps with high order statistics of the output $y[n]$. This polynomial approach enables the use of methods coming from computational algebraic geometry to get efficient and exhaustive estimates of $h[0], \ldots, h[N-1]$ from the sole observations $\{y[n]\}$.

The input signals considered in our case are of PSK-type i. e. the symbols are roots of unity [19]. Then, by introducing the *non-circular* statistics on $y[n]$, we get the following polynomial equations in $h[n]$ with no $x[n]$ in the equations since $\mathbb{E}(x[n]^{2^M}) = 1$. For far more details, the reader is referred to [19].

BPSK, QPSK, 8-PSK and 2^M-PSK:

For BPSK, $x[n]$ is iid discrete-uniform $\{-1, 1\}$. We get then for $p = 0, \ldots, N-1$,

$$\gamma_p := \gamma_y[p] = \mathbb{E}(y[n]y[n-p]) = \sum_{m=p}^{N-1} h[m]h[m-p]. \tag{2.1}$$

For QPSK, $x[n]$ is iid discrete-uniform $\{1, i, -1, -i\}$, which gives

$$\mathbb{E}(y[n]y[n-p_1]y[n-p_2]y[n-p_3]) = \sum_{m=\max(p_1,p_2,p_3)}^{N-1} h[m]h[m-p_1]h[m-p_2]h[m-p_3].$$

This case is easily reduced back to the BPSK case by taking $p_1 = 0, p_3 = p_2 = p$ and $g[n] := h[n]^2$. Namely,

$$\gamma_p := \gamma_{y^2}[p] = \mathbb{E}(y[n]^2 y[n-p]^2) = \sum_{m=p}^{N-1} h[m]^2 h[m-p]^2.$$

Alike, the 8-PSK and in general all 2^M-PSK modulations can be reduced back to the BPSK equations (2.1) by the change of variable $g[n] := h[n]^{2^{M-1}}$.

MSK, $\frac{\pi}{4}$-DQPSK, $\frac{3\pi}{8}$-D8PSK and D2^M-PSK:

For MSK, we have $x[n] = i^n b[n]x[0]$ with $b[n]$ BPSK. So, for $p = 0, \ldots, N-1$,

$$\gamma_y[p, n] := \mathbb{E}(y[n]y[n-p]|x[0]) = x[0]^2 \sum_{m=p}^{N-1} (-1)^{n-m} h[m]h[m-p]. \tag{2.2}$$

As above, the $\frac{\pi}{4}$-DQPSK, $\frac{3\pi}{8}$-D8PSK and D2^M-PSK cases can all be reduced back to the MSK equations (2.2).

The systems of polynomials equations are then typically of the form

$$\gamma_p - \sum_{m=p}^{N-1} g[m]g[m-p] \qquad \text{for} \quad p = 0, \ldots, N-1 \tag{2.3}$$

or

$$\gamma_p - \sum_{m=p}^{N-1} (-1)^{m-p} g[m]g[m-p] \qquad \text{for} \quad p = 0, \ldots, N-1 \tag{2.4}$$

where the bona-fide unknowns, the $h[m]$, are linked to the $g[m]$ by $g[m] := (h[m])^{2^{M-1}}$ for some fixed positive integer M and $[\gamma_0, \ldots, \gamma_{N-1}] \in \mathbb{C}^N$ are some parameters evaluated real-time from the statistics of the signal received $y[n]$. In [19], we worked out the systems of equations of type (2.4) by using a purely deterministic approach. Here, we will deal with the other family (2.3) and show what can be gained by switching to probabilistic algorithms with certification (*Las-Vegas* algorithms).

E. g. for an 8-PSK modulation and an estimated length of channel $N = 3$, we get the following system of polynomial equations, where $\gamma_0, \gamma_1, \gamma_2$ are the parameters obtained by evaluating the second order non-circular statistics on $y^4[n]$:

$$\begin{cases} \gamma_0 - h[0]^8 - h[1]^8 - h[2]^8 = 0 \\ \gamma_1 - h[0]^4 h[1]^4 - h[1]^4 h[2]^4 = 0 \\ \gamma_2 - h[0]^4 h[2]^4 = 0. \end{cases} \tag{2.5}$$

From Bézout's theorem [6], this system has either infinitely many solutions, either exactly 512 (with multiplicities), or no solution.

To illustrate our approach, we will focus on this example. Everything is easily generalizable to any N (practically up to $N = 9$) and the two afore-mentioned families of modulations (BPSK, QPSK, 8-PSK and MSK, $\frac{\pi}{4}$-DQPSK, $\frac{3\pi}{8}$-D8PSK).

3 Parametric rational representation

3.1 Generalized normal forms

By the following generic change of variables, $x_1 := h[0]^4$, $x_2 := h[1]^4 - x_1$ and $x_3 := h[2]^4 - (x_1 + x_2)$, system (2.5) can be rewritten as

$$\begin{cases} \gamma_0 - 3x_1^2 - 4x_1x_2 - 2x_2^2 - 2x_1x_3 - 2x_2x_3 - x_3^2 \\ \gamma_1 - 2x_1^2 - 3x_1x_2 - x_1x_3 - x_2^2 - x_2x_3 \\ \gamma_2 - x_1^2 - x_1x_2 - x_1x_3. \end{cases} \tag{P}$$

Now by solving for x_1^2, x_2^2, x_3^2, we get

$$\begin{cases} x_1^2 = \gamma_2 - x_1x_2 - x_1x_3 \\ x_2^2 = \gamma_1 - 2\gamma_2 - x_1x_2 + x_1x_3 - x_2x_3 \\ x_3^2 = \gamma_0 - 2\gamma_1 + \gamma_2 + x_1x_2 - x_1x_3. \end{cases} \tag{3.1}$$

Therefore, the monomials x_1^2, x_2^2, x_3^2 can be expressed as linear combinations of monomials in the basis $\mathscr{B} = \{\omega_1, \ldots, \omega_d\}$ given by

$$\mathscr{B} := \{1, x_1, x_2, x_3, x_1x_2, x_1x_3, x_2x_3, x_1x_2x_3\}.$$

Now, by solving again for

$$\{x_1^2x_2, x_1^2x_3, x_2^2x_1, x_2^2x_3, x_3^2x_1, x_3^2x_2\}$$

and then again for

$$\{x_1^2x_2x_3, x_1x_2^2x_3, x_1x_2x_3^3\},$$

we can express all the monomials of

$$x_1\mathscr{B} = \{x_1, x_1^2, x_1x_2, x_1x_3, x_1^2x_2, x_1^2x_3, x_1x_2x_3, x_1^2x_2x_3\}$$

but also of $x_2\mathscr{B}$, and $x_3\mathscr{B}$ as linear combinations of monomials in $\mathscr{B}$.

Using Bézout's theorem [6], it is then easily proved that $\mathscr{B}$ is indeed a bona-fide linear base of the d-dimensional quotient algebra $\mathcal{A} := \mathbb{C}[x_1, \ldots, x_N]/I$ where $I = \langle P \rangle$ denotes the ideal generated by (P). However, we shall notice that this monomial basis is not compatible with multiplication for any monomial order in the sense of Gröbner bases. Nevertheless, this basis is stable by division and compatible with reducing graduations as defined in [24]. Consequently, all the multiplication applications defined on $\mathcal{A}$ will also commute, thus providing a nice "Gröbner-like" framework for $\mathcal{A}$.

By working in this setting, solving system (P) is now just a problem of linear algebra (for more details on the use of linear algebra in algebraic geometry, one can read [23] or [41]). Intuitively, starting from a list (P) of polynomials such that the generated ideal $I = \langle P \rangle$ is zero-dimensional (finite number of solutions), the quotient space $\mathcal{A} := \mathbb{C}[x_1, \ldots, x_N]/I$ inherits a structure of finite-dimensional algebra. Constructing the multiplication table $[\omega_k\omega_l]_{k,l}$ of $\mathcal{A}$ gives then a full description of the linear algebraic framework associated to system (P). The central problem is then to choose (or construct) a convenient linear basis $\mathscr{B} := \{\omega_1, \ldots, \omega_d\}$ called the *monomial basis* of $\mathcal{A}$ and to describe the associated normal form $\bar{p} = \mathrm{NF}(p, I) \in \mathcal{A}$ for any $p \in \mathbb{C}[x_1, \ldots, x_N]$. A usual way to get a convenient monomial basis and its associated normal form is through the computation of a reduced Gröbner basis and the so-called monomials under the staircase scheme [11, 34]. This is however usually very expensive from a computational point of view, very tricky when you have parameters in the equations but also quite useless here, since the monomial basis given by

$$\mathscr{B} := \{1, x_1, x_2, \ldots, x_N, x_1x_2, x_1x_3, \ldots, x_{N-1}x_N, \ldots, x_1x_2 \ldots x_N\}$$

is in our case always a bona-fide linear basis of $\mathcal{A}$ as easily proved by induction [7].

Now, any element $\bar{p} \in \mathcal{A}$ can be thus expressed as a vector $[p]$ in $\mathscr{B}$ as $\bar{p} = \sum_{k=1}^{d} [p]_k \omega_k$. For example, using (3.1) we get

$$[x_1^2] = \begin{bmatrix} \gamma_2 & 0 & 0 & 0 & -1 & -1 & 0 & 0 \end{bmatrix}^\top.$$

A complete description of the normal form is thus easily obtained by multiplying and solving system (3.1) and its iterations for the monomials in $\mathscr{B}$ [12, 24]. Now, we introduce the linear operator $\mathbf{M}_u$ on $\mathcal{A}$,

$$\begin{aligned} \mathbf{M}_u : \mathcal{A} &\to \mathcal{A} \\ \bar{p} &\mapsto \mathbf{M}_u\bar{p} := \overline{up}, \end{aligned}$$

and link it to its $\mathbb{C}^{d\times d}$ matrix representation in the monomial basis of $\mathcal{A}$. This matrix is computed by expressing $u\omega_k$ in the monomial basis $\mathscr{B}$ which gives the kth column of $\mathbf{M}_u$. For $[u] \in \mathcal{A}$, we thus get the multiplication matrix $\mathbf{M}_u[v] := [uv]$ on $\mathbb{C}^d$. E. g. for $u = x_1$, since

$$x_1\mathscr{B} = \{x_1, x_1^2, x_1x_2, x_1x_3, x_1^2x_2, x_1^2x_3, x_1x_2x_3, x_1^2x_2x_3\},$$

we get from (3.1) and the full computation of the normal form,

$$\mathbf{M}_{x_1} = \begin{bmatrix} 0 & \gamma_2 & 0 & 0 & 0 & 0 & 0 & \gamma_2(\gamma_1 - 2\gamma_2) \\ 1 & 0 & 0 & 0 & (2\gamma_1 - \gamma_2 - \gamma_0) & (-\gamma_1 + 2\gamma_2) & 0 & 0 \\ 0 & 0 & 0 & 0 & 0 & \gamma_2 & 0 & 0 \\ 0 & 0 & 0 & 0 & \gamma_2 & 0 & 0 & 0 \\ 0 & -1 & 1 & 0 & 0 & 0 & 0 & -\gamma_1 + \gamma_2 \\ 0 & -1 & 0 & 1 & 0 & 0 & 0 & \gamma_2 \\ 0 & 0 & 0 & 0 & 0 & 0 & 0 & -\gamma_2 \\ 0 & 0 & 0 & 0 & -1 & 0 & 1 & 0 \end{bmatrix}$$

and similarly $\mathbf{M}_{x_2}, \ldots, \mathbf{M}_{x_N}$.

The computation of $\mathbf{M}_u$ gives some important information on the set of solutions of the system $\mathcal{Z}_{\mathbb{C}}(I) := \{\boldsymbol{\alpha} \in \mathbb{C}^N \mid \forall p \in P,\ p(\boldsymbol{\alpha}) = 0\}$ and the system in general.

Namely, denoting by $\mu(\boldsymbol{\alpha})$ the multiplicity of a solution $\boldsymbol{\alpha}$ of (P), Stickelberger's theorem [22, 41] states that the multiplication matrix $\mathbf{M}_u$ has eigenvalues $u(\boldsymbol{\alpha})$ each with multiplicity $\sum_{\boldsymbol{\beta}\in\mathcal{Z}(I), u(\boldsymbol{\beta})=u(\boldsymbol{\alpha})} \mu(\boldsymbol{\beta})$. As a consequence, we have the following properties

$$\begin{aligned} \det(\mathbf{M}_u) &= \prod_{\boldsymbol{\alpha}\in\mathcal{Z}_{\mathbb{C}}(I)} u(\boldsymbol{\alpha})^{\mu(\boldsymbol{\alpha})}, \\ \operatorname{trace}(\mathbf{M}_u) &= \sum_{\boldsymbol{\alpha}\in\mathcal{Z}_{\mathbb{C}}(I)} \mu(\boldsymbol{\alpha})u(\boldsymbol{\alpha}), \\ \chi_u(t) := \det(t\mathbf{I} - \mathbf{M}_u) &= \prod_{\boldsymbol{\alpha}\in\mathcal{Z}_{\mathbb{C}}(I)} (t - u(\boldsymbol{\alpha}))^{\mu(\boldsymbol{\alpha})}. \end{aligned}$$

Now, factorizing $\chi_u(t)$, the characteristic polynomial of $\mathbf{M}_u$, would give $u(\boldsymbol{\alpha})$ for any solution $\boldsymbol{\alpha}$ of the system. We would then get the solutions of the system by taking $u = x_1, \ldots, x_N$ so as to obtain the coordinates of $\boldsymbol{\alpha}$.

3.2 Rational representation

Unfortunately, computing directly the characteristic polynomial (incl. the determinant) of a matrix like $\mathbf{M}_u$ and factorizing it, is very time and memory consuming (and in fact usually intractable). We detail here an alternative method using a geometric representation of the roots of polynomials systems called rational representation. This approach dates back to Kronecker [17]. It has been known also under the name of effective primitive element theorem [2] and studied extensively [32, 11, 10, 37, 16]. The approach detailed here relies on the computation of the traces of multiplication matrices [27, 1, 34, 14] and takes advantage of the special structure of the matrices $\mathbf{M}_u$.

Introducing the minimal polynomial $\tilde{\chi}_u(t)$,

$$\tilde{\chi}_u(t) := \prod_{\boldsymbol{\alpha}\in\mathcal{Z}_{\mathbb{C}}(I)} (t - u(\boldsymbol{\alpha})) = \frac{\chi_u(t)}{\gcd(\chi_u(t), \chi_u'(t))}, \tag{3.2}$$

and assuming that u is separating $\mathcal{Z}_{\mathbb{C}}(I)$, i. e. on $\mathcal{Z}_{\mathbb{C}}(I), \boldsymbol{\alpha} \neq \boldsymbol{\beta} \Rightarrow u(\boldsymbol{\alpha}) \neq u(\boldsymbol{\beta})$ (it implies that $\mathbf{M}_u$ has eigenvalues $u(\boldsymbol{\alpha})$ with multiplicity exactly $\mu(\boldsymbol{\alpha})$), we introduce the polynomial mapping

$$\begin{aligned} g_u : \mathcal{A} &\to \mathbb{C}[x_1, \ldots, x_N] \\ \bar{v} &\mapsto g_u(v,t) := \textstyle\sum_{\boldsymbol{\alpha}\in\mathcal{Z}_{\mathbb{C}}(I)} \mu(\boldsymbol{\alpha}) v(\boldsymbol{\alpha}) \frac{\tilde{\chi}_u(t)}{t-u(\boldsymbol{\alpha})}. \end{aligned}$$

From (3.2), this can be rewritten as

$$g_u(v,t) = \sum_{\boldsymbol{\alpha}\in\mathcal{Z}_{\mathbb{C}}(I)} \mu(\boldsymbol{\alpha}) v(\boldsymbol{\alpha}) \prod_{\boldsymbol{\beta}\in\mathcal{Z}_{\mathbb{C}}(I)\setminus\{\boldsymbol{\alpha}\}} (t - u(\boldsymbol{\beta})). \tag{3.3}$$

So, for $\boldsymbol{\alpha} \in \mathcal{Z}_{\mathbb{C}}(I)$ and $t = u(\boldsymbol{\alpha})$, we have

$$g_u(v, u(\boldsymbol{\alpha})) = \mu(\boldsymbol{\alpha}) v(\boldsymbol{\alpha}) \prod_{\boldsymbol{\beta}\in\mathcal{Z}_{\mathbb{C}}(I)\setminus\{\boldsymbol{\alpha}\}} (u(\boldsymbol{\alpha}) - u(\boldsymbol{\beta})).$$

Hence, we get a geometric representation [36] of the solutions of (P) since

$$\frac{g_u(v, u(\boldsymbol{\alpha}))}{g_u(1, u(\boldsymbol{\alpha}))} = v(\boldsymbol{\alpha}),$$

by taking $v = x_1, \ldots, x_N$, we have

$$\boldsymbol{\alpha} = \left[\frac{g_u(x_1, u(\boldsymbol{\alpha}))}{g_u(1, u(\boldsymbol{\alpha}))}, \frac{g_u(x_2, u(\boldsymbol{\alpha}))}{g_u(1, u(\boldsymbol{\alpha}))}, \ldots, \frac{g_u(x_N, u(\boldsymbol{\alpha}))}{g_u(1, u(\boldsymbol{\alpha}))}\right].$$

Consequently, we have the following theorem giving a one-to-one mapping of the solutions of the multivariate system (P) onto the roots of the univariate polynomial $\chi_u(t)$.

Theorem 1 *If α is a solution of the system, then $u(\boldsymbol{\alpha})$ is a root of characteristic polynomial $\chi_u(t)$ with the same multiplicity and conversely, if ζ is a root of $\chi_u(t)$, then*

$$\left[\frac{g_u(x_1, \zeta)}{g_u(1, \zeta)}, \frac{g_u(x_2, \zeta)}{g_u(1, \zeta)}, \ldots, \frac{g_u(x_N, \zeta)}{g_u(1, \zeta)}\right]$$

is a solution of the system with the same multiplicity.

Proof. For a complete proof, the reader is referred to [34]. □

And if, as in our case, we have no special interest in the multiplicities of the solutions of (P), we have the following simpler theorem:

Theorem 2 *If α is a solution of the system, then $u(\alpha)$ is a root of the minimal polynomial $\tilde{\chi}_u(t)$ and conversely, if ζ is a root of $\tilde{\chi}_u(t)$, then*

$$\left[\frac{g_u(x_1,\zeta)}{g_u(1,\zeta)}, \frac{g_u(x_2,\zeta)}{g_u(1,\zeta)}, \ldots, \frac{g_u(x_N,\zeta)}{g_u(1,\zeta)}\right]$$

is a solution of the system.

And $\tilde{\chi}_u(t)$ is usually much easier to factorize since all its roots are simple.

3.3 Algorithmic aspects

Yet, this approach would be quite useless without a practical way to compute $\chi_u(t)$, $\tilde{\chi}_u(t)$ and $g_u(v,t)$. We will detail here an approach based on the computation of the traces of some multiplication matrices.

3.3.1 Computation of $\chi_u(t)$

Writing $\chi_u(t) = \sum_{k=0}^{d} b_k t^{d-k}$ and let $\chi'_u(t)$ be its derivative, then

$$\frac{\chi'_u(t)}{\chi_u(t)} = \sum_{\alpha\in\mathcal{Z}_{\mathbb{C}}(I)} \frac{\mu(\alpha)}{t-u(\alpha)} = \sum_{\alpha\in\mathcal{Z}_{\mathbb{C}}(I)} \frac{1}{t}\frac{\mu(\alpha)}{1-\frac{u(\alpha)}{t}}$$

$$= \sum_{\alpha\in\mathcal{Z}_{\mathbb{C}}(I)} \sum_{k\geq 0} \frac{1}{t}\mu(\alpha)u^k(\alpha)t^{-k} = \sum_{k\geq 0} \mathrm{trace}(\mathbf{M}_{u^k})t^{-(k+1)}.$$

So, we have

$$\chi'_u(t) = \chi_u(t)\sum_{k\geq 0} \mathrm{trace}(\mathbf{M}_{u^k})t^{-(k+1)}$$

and since

$$\chi'_u(t) = \sum_{k=0}^{d-1}(d-k)b_k t^{d-1-k}$$

this yields, using the definition of $\chi_u(t)$,

$$(d-k)b_k = \sum_{l=0}^{k} \mathrm{trace}(\mathbf{M}_{u^l})b_{k-l}.$$

We then get the values of b_k starting from $b_0 = 1$ (by definition of $\chi_u(t)$) and iterating for $k = 1, \ldots, d$,

$$b_k = -\tfrac{1}{k}\sum_{l=1}^{k} \mathrm{trace}(\mathbf{M}_{u^l})b_{k-l}.$$

We then get b_k for $k = 0, \dots, d$ and thus $\chi_u(t)$. Therefore, the computation of $\chi_u(t)$ relies solely on the computation of the scalars $\mathrm{trace}(\mathbf{M}_{u^k})$ for $k = 0, \dots, d$.

3.3.2 Computation of $\tilde{\chi}_u(t)$

The minimal polynomial $\tilde{\chi}_u(t) = \sum_{k=0}^{r} a_{r-k} t^k$ of an element u of the algebra $\mathcal{A}$ can be computed by building a linear dependency sequence between the powers of u: $1, u, u^2, \dots$.

Also, from its definition we have $\tilde{\chi}_u(\mathbf{M}_u) = 0$, we then get

$$\tilde{\chi}_u(\mathbf{M}_u) = \sum_{k=0}^{r} a_{r-k} \mathbf{M}_u^k = \sum_{k=0}^{r} a_{r-k} \mathbf{M}_{u^k} = 0.$$

So for a random polynomial $z \in \mathcal{A}$, we get

$$\mathbf{M}_z \sum_{k=0}^{r} a_{r-k} \mathbf{M}_{u^k} = \sum_{k=0}^{r} a_{r-k} \mathbf{M}_{u^k z} = 0.$$

By taking the trace, we then get

$$\sum_{k=0}^{r} a_{r-k} \mathrm{trace}(\mathbf{M}_{u^k z}) = 0.$$

We can then apply the Wiedemann algorithm [44] on the sequence $\mathrm{trace}(\mathbf{M}_{u^k z})$ for $k = 0, \dots, 2r-1$ and z taken at random in $\mathcal{A}$. The Berlekamp-Massey algorithm gives then a potential $\tilde{\chi}_u(t)$ after processing the first $2r$ elements of this sequence. A certificate for the computation is obtained [35] by checking that $a_1/a_0 = -\mathrm{trace}(\mathbf{M}_u) = 0$.

And so, the minimal polynomial $\tilde{\chi}_u(t)$ may again be easily computed from the scalars $\mathrm{trace}(\mathbf{M}_{u^k z})$ for $k = 0, \dots, r$ and a random polynomial $z \in \mathcal{A}$.

3.3.3 Computation of $g_u(v, t)$

Concerning the $g_u(v,t)$, following ideas of Hermite-Sylvester [3, 1] and in a similar way to what was done to compute $\chi_u(t)$, we have

$$\begin{aligned} \frac{g_u(v,t)}{\tilde{\chi}_u(t)} &= \sum_{\boldsymbol{\alpha} \in \mathcal{Z}_{\mathbb{C}}(I)} \frac{\mu(\boldsymbol{\alpha}) v(\boldsymbol{\alpha})}{t - u(\boldsymbol{\alpha})} = \sum_{\boldsymbol{\alpha} \in \mathcal{Z}_{\mathbb{C}}(I)} \frac{1}{t} \frac{\mu(\boldsymbol{\alpha}) v(\boldsymbol{\alpha}}{1 - \frac{u(\boldsymbol{\alpha})}{t}} \\ &= \sum_{\boldsymbol{\alpha} \in \mathcal{Z}_{\mathbb{C}}(I)} \sum_{k \geq 0} \frac{1}{t} \mu(\boldsymbol{\alpha}) u^k(\boldsymbol{\alpha}) v(\boldsymbol{\alpha}) t^{-k} = \sum_{k \geq 0} \mathrm{trace}(\mathbf{M}_{u^k v}) t^{-(k+1)}. \end{aligned}$$

So,

$$g_u(v,t) = \tilde{\chi}_u(t) \sum_{k \geq 0} \mathrm{trace}(\mathbf{M}_{u^k v}) t^{-(k+1)}.$$

Now, writing $\tilde{\chi}_u(t) = \sum_{l=0}^{r} a_l t^{r-l}$ where $r = \#\mathcal{Z}_{\mathbb{C}}(I)$ and since from (3.3), $g_u(v,t)$ has degree at most $r-1$ in t, we have

$$
\begin{aligned}
g_u(v,t) &= \sum_{l=0}^{r} \sum_{k \geq 0} \operatorname{trace}(\mathbf{M}_{u^k v}) a_l t^{r-1-k-l} \\
&= \sum_{l=0}^{r-1} \sum_{k=0}^{r-1-l} \operatorname{trace}(\mathbf{M}_{u^k v}) a_l t^{r-1-k-l} \\
&= \sum_{k=0}^{r-1} \operatorname{trace}(\mathbf{M}_{u^k v}) \sum_{l=0}^{r-1-l} a_l t^{r-1-k-l}.
\end{aligned}
$$

Introducing the Horner sequence of polynomials $H_k(\tilde{\chi}_u)(t) := \sum_{l=0}^{k} a_l t^{k-l}$, we get

$$
g_u(v,t) = \sum_{k=0}^{r-1} \operatorname{trace}(\mathbf{M}_{u^k v}) H_{r-1-k}(\tilde{\chi}_u)(t).
$$

So, the $g_u(v,t)$ are easily computed from $\tilde{\chi}_u(t)$ and the scalars $\operatorname{trace}(\mathbf{M}_{u^k v})$ for $k = 0,\ldots,r$ and $v = x_1,\ldots,x_N$.

3.3.4 Computation of traces

We showed above that the geometric representation of (P) relies solely on the computation of the scalars $\operatorname{trace}(\mathbf{M}_{u^k v})$ for $k = 0,\ldots,r$ and $v = 1, x_1,\ldots,x_n$ or random. There is now an easy way to compute these traces by noticing that $\operatorname{trace}(\mathbf{M}_{fg}) = \mathrm{Tr}(f)[g]$ where

$$
\mathrm{Tr}(f) := \left[\operatorname{trace}(\mathbf{M}_{f\omega_1}),\ldots,\operatorname{trace}(\mathbf{M}_{f\omega_d})\right].
$$

Also, since $\mathrm{Tr}(u^{k+1}) = \mathrm{Tr}(u^k)\mathbf{M}_u$, by iterating we get $\operatorname{trace}(\mathbf{M}_{u^{k+1}}) = \mathrm{Tr}(u^k)[u]$ and $\operatorname{trace}(\mathbf{M}_{u^k v}) = \mathrm{Tr}(u^k)[v]$.

As a result, all the scalars $\operatorname{trace}(\mathbf{M}_{u^k v})$ with $k = 0,\ldots,r$ and $v = 1, x_1,\ldots,x_N$ are easily derived from the trace matrix $\mathrm{Tr}\mathbf{M}$ defined by $[\mathrm{Tr}\mathbf{M}]_{k,l} := \operatorname{trace}(\mathbf{M}_{\omega_k \omega_l})$, i. e.

$$
\mathrm{Tr}\mathbf{M} := \begin{bmatrix} \operatorname{trace}(\mathbf{M}_{\omega_1\omega_1}) & \ldots & \operatorname{trace}(\mathbf{M}_{\omega_1\omega_d}) \\ \vdots & & \vdots \\ \operatorname{trace}(\mathbf{M}_{\omega_d\omega_1}) & \ldots & \operatorname{trace}(\mathbf{M}_{\omega_d\omega_d}) \end{bmatrix}.
$$

Namely, since

$$
\operatorname{trace}(\mathbf{M}_{u^k v}) = [v]^\top \mathrm{Tr}\mathbf{M}[u^k],
$$

let $\mathbf{v}^\top = [1]^\top \mathrm{Tr}\mathbf{M}$ and $\mathbf{w}_0 = [1]$, then by iterating $\mathbf{w}_{k+1} = \mathbf{M}_u \mathbf{w}_k$ we get

$$
\operatorname{trace}(\mathbf{M}_{u^k}) = \mathbf{v}^\top \mathbf{w}_k.
$$

In the same way, we have for $k = 0,\ldots,r$ and $l = 1,\ldots,N$

$$
\operatorname{trace}(\mathbf{M}_{u^k x_l}) = ([x_l]^\top \mathrm{Tr}\mathbf{M})\, \mathbf{w}_k,
$$

and again for a random $z \in \mathcal{A}$, we get

$$\mathrm{trace}(\mathbf{M}_{u^k z}) = ([z]^\top \mathrm{T_R}\mathbf{M})\, \mathbf{w}_k.$$

So, we only have to compute $\mathbf{w}_k$ for $k = 0, \ldots, d$ (and even only up to r when $r < d$ since we don't need to compute $\chi_u(t)$ in that case).

4 Implementation

We have now almost all the tools to describe our algorithm. We will detail in this part the practical implementation of the methods described above and apply it to our example of 8-PSK modulation. We then describe our two-parts-algorithm.

4.1 Offline part

1. Compute the normal form $[\cdot]$ iterating from (P) on $\mathscr{B}$.
2. Compute $\mathbf{M}_{x_1}, \ldots, \mathbf{M}_{x_N}$.
3. Compute $\mathbf{M}_u$ with $u = \theta_1 x_1 + \ldots + \theta_N x_N$ and **store it**
4. Compute the matrix of traces $\mathrm{T_R}\mathbf{M} = [\mathrm{trace}(\mathbf{M}_{\omega_k \omega_l})]_{k,l}$ and **store it**.

For example, for 8-PSK and $N = 3$, we get

$$\mathbf{M}_u = \left[\begin{array}{ccccc}
0 & \theta_1\gamma_2 & \theta_2(\gamma_1-2\gamma_2) & \theta_3(-2\gamma_1+\gamma_0+\gamma_2) & 0 \\
\theta_1 & 0 & 0 & 0 & \theta_1(2\gamma_1-\gamma_2-\gamma_0)+\theta_2(-2\gamma_1+\gamma_0+\gamma_2) \\
\theta_2 & 0 & 0 & 0 & \theta_2\gamma_2 \\
\theta_3 & 0 & 0 & 0 & \theta_1\gamma_2-\theta_2\gamma_2 \\
0 & -\theta_1+\theta_2 & \theta_1-\theta_2 & \theta_3 & 0 \\
0 & -\theta_1+\theta_3 & \theta_2 & \theta_1-\theta_3 & 0 \\
0 & 0 & -\theta_2+\theta_3 & \theta_2 & 0 \\
0 & 0 & 0 & 0 & -\theta_1+\theta_3
\end{array}\right|$$

$$\left|\begin{array}{cc}
0 & 0 \\
\theta_1(-\gamma_1+2\gamma_2)+\theta_3(\gamma_1-2\gamma_2) & \theta_2(-3\gamma_2+3\gamma_1-\gamma_0)+\theta_3(-2\gamma_1+\gamma_0+\gamma_2) \\
\theta_1\gamma_2-\theta_3\gamma_2 & \theta_2(-3\gamma_2+2\gamma_1-\gamma_0)+\theta_3(-2\gamma_1+2\gamma_2+\gamma_0) \\
\theta_3\gamma_2 & \theta_2\gamma_1-\theta_3\gamma_2 \\
0 & 0 \\
0 & 0 \\
0 & 0 \\
\theta_2-\theta_3 & \theta_1-\theta_2-\theta_3
\end{array}\right|$$

$$\left|\begin{array}{c}
\theta_1\gamma_2(\gamma_1-2\gamma_2)-\theta_2(-2\gamma_1+\gamma_0+\gamma_2)\gamma_2+\theta_3(3\gamma_2-3\gamma_1+\gamma_0)\gamma_2 \\
0 \\
0 \\
0 \\
\theta_1(-\gamma_1+\gamma_2)-\theta_2\gamma_2+\theta_3\gamma_1 \\
\theta_1\gamma_2+\theta_2(-2\gamma_1+2\gamma_2+\gamma_0)+\theta_3(-3\gamma_2+2\gamma_1-\gamma_0) \\
-\theta_1\gamma_2+\theta_2\gamma_2+\theta_3\gamma_2 \\
0
\end{array}\right]$$

and

$$
\mathrm{T_R M} = \left[\begin{array}{cccccccc}
8 & 0 & 0 & 0 & -2\gamma_0+4\gamma_1+4\gamma_2 & -4\gamma_1+8\gamma_2 & -4\gamma_0+8\gamma_1-16\gamma_2 & 0 \\
0 & 2\gamma_0-4\gamma_2 & -2\gamma_0+4\gamma_1+4\gamma_2 & -4\gamma_1+8\gamma_2 & 0 & 0 & 0 & 2\gamma_0\gamma_1+2\gamma_0\gamma_2-4\gamma_1^2-4\gamma_1\gamma_2+12\gamma_2^2 \\
0 & -2\gamma_0+4\gamma_1+4\gamma_2 & 6\gamma_0-8\gamma_1+4\gamma_2 & -4\gamma_0+8\gamma_1-16\gamma_2 & 0 & 0 & 0 & -4\gamma_0\gamma_1-2\gamma_0\gamma_2+8\gamma_1^2-4\gamma_1\gamma_2-12\gamma_2^2 \\
0 & -4\gamma_1+8\gamma_2 & -4\gamma_0+8\gamma_1-16\gamma_2 & 6\gamma_0-8\gamma_1+4\gamma_2 & 0 & 0 & 0 & 2\gamma_0\gamma_1-4\gamma_0\gamma_2-4\gamma_1^2+16\gamma_1\gamma_2-16\gamma_2^2 \\
-2\gamma_0+4\gamma_1+4\gamma_2 & 0 & 0 & 0 & 2\gamma_0^2-4\gamma_0\gamma_1-4\gamma_0\gamma_2+16\gamma_1\gamma_2-8\gamma_2^2 & 2\gamma_0\gamma_1+2\gamma_0\gamma_2-4\gamma_1^2-4\gamma_1\gamma_2+12\gamma_2^2 & 2\gamma_0\gamma_1-4\gamma_0\gamma_2-4\gamma_1^2+16\gamma_1\gamma_2-16\gamma_2^2 & 0 \\
-4\gamma_1+8\gamma_2 & 0 & 0 & 0 & 2\gamma_0\gamma_1+2\gamma_0\gamma_2-4\gamma_1^2-4\gamma_1\gamma_2+12\gamma_2^2 & -4\gamma_0\gamma_2+4\gamma_1^2-8\gamma_1\gamma_2 & 2\gamma_0\gamma_1-4\gamma_0\gamma_2-4\gamma_1^2+16\gamma_1\gamma_2-16\gamma_2^2 & 0 \\
-4\gamma_0+8\gamma_1-16\gamma_2 & 0 & 0 & 0 & -4\gamma_0\gamma_1-2\gamma_0\gamma_2+8\gamma_1^2-4\gamma_1\gamma_2-12\gamma_2^2 & 2\gamma_0\gamma_1-4\gamma_0\gamma_2-4\gamma_1^2+16\gamma_1\gamma_2-16\gamma_2^2 & 4\gamma_0^2-8\gamma_0\gamma_1+24\gamma_0\gamma_2-32\gamma_1\gamma_2+40\gamma_2^2 & 0 \\
0 & 2\gamma_0\gamma_1+2\gamma_0\gamma_2-4\gamma_1^2-4\gamma_1\gamma_2+12\gamma_2^2 & -4\text{-}_0\gamma_1-2\gamma_0\gamma_2+8\gamma_1^2-4\gamma_1\gamma_2-12\gamma_2^2 & 2\gamma_0\gamma_1-4\gamma_0\gamma_2-4\gamma_1^2+16\gamma_1\gamma_2-16\gamma_2^2 & 0 & 0 & 0 & -4\gamma_0^2\gamma_2+4\gamma_0\gamma_1^2+4\gamma_0\gamma_1\gamma_2-22\gamma_0\gamma_2^2-8\gamma_1^3+16\gamma_1^2\gamma_2+24\gamma_1\gamma_2^2-36\gamma_2^3
\end{array}\right].
$$

In this approach, most of the computational cost lies in getting the generic multiplication matrix M_u and the parametric matrix of traces $\mathrm{T_R M}(\gamma_0, \ldots, \gamma_{N-1})$ of the system. This expensive symbolic computation is however done (using Maple) once for all, i. e. for any value of the parameters $\gamma_0, \ldots, \gamma_{N-1}$ and this holds for any type of the modulations afore-mentioned. The rest of the algorithm (the online part) will rely solely on these two matrices M_u and $\mathrm{T_R M}$.

4.2 Online part

Two remaining points still need to be clarified: how to get a separating polynomial?, and its subsequent question: how to efficiently compute $\mathrm{rank}(\mathrm{T_R M})$?

In [34], it was proved that a polynomial u is separating $\mathcal{Z}_{\mathbb{C}}(I)$ if and only if its minimal polynomial satisfies $\deg(\tilde{\chi}_u) = \#\mathcal{Z}_{\mathbb{C}}(I) =: r$. Since furthermore $\mathrm{rank}(\mathrm{T\scriptstyle{R}}\mathbf{M}) = \#\mathcal{Z}_{\mathbb{C}}(I)$, a good certificate to check whether u is separating is to compute firsthand $\mathrm{rank}(\mathrm{T\scriptstyle{R}}\mathbf{M})$ and compare $\deg(\tilde{\chi}_u)$ with it. This is easily done by first noticing that $\mathrm{T\scriptstyle{R}}\mathbf{M}$ is sparse block and that its columns of odd degree are orthogonal to the columns of even degree, i. e. $\mathrm{T\scriptstyle{R}}(\omega_k) := [\mathrm{trace}(\mathbf{M}_{\omega_k\omega_1}), \ldots, \mathrm{trace}(\mathbf{M}_{\omega_k\omega_d})]$ is orthogonal to $\mathrm{T\scriptstyle{R}}(\omega_l)$ if $\deg(\omega_k) \neq \deg(\omega_l) \mod 2$. For example, in the 8-PSK case, computing the rank of $\mathrm{T\scriptstyle{R}}\mathbf{M}$ which is 8×8 reduces to computing the rank of two 4×4 matrices.

In our previous paper [19], the separating polynomial u was obtained by an exhaustive search, since the set of linear polynomials $\mathcal{S} := \{x_1 + kx_2 + \ldots + k^{N-1}x_N \,|\, 0 \leq k \leq (N-1)\binom{r}{2}\}$ contains at least one separating polynomial [34].

However, in order to avoid this enumeration, it can be proved, using the Schwartz-Zippel lemma [21], that one can construct a separating polynomial with probability $1-\epsilon$ by taking at random $u = \theta_1 x_1 + \ldots + \theta_N x_N$ where $\theta_1, \ldots, \theta_N$ are iid uniform randomly chosen in a finite subset $\mathcal{S}$ of $\mathbb{C}$ of size larger than $\binom{r}{2}/\epsilon$.

We have now all the elements to describe the online part of the algorithm:

1. Evaluate the statistics $\gamma_0, \ldots, \gamma_{N-1}$ and instantiate $\mathrm{T\scriptstyle{R}}\mathbf{M}$.
2. Compute $r = \mathrm{rank}(\mathrm{T\scriptstyle{R}}\mathbf{M}) = \#\mathcal{Z}_{\mathbb{C}}(\langle P\rangle)$ (noticing that $\mathrm{T\scriptstyle{R}}\mathbf{M}$ is sparse block).
3. Generate at random $\theta_1, \ldots, \theta_N$ and using the statistics $\gamma_0, \ldots, \gamma_{N-1}$, instantiate $\mathbf{M}_u$.
4. Compute the minimal polynomial $\tilde{\chi}_u(t)$:
   - If $r = d$, $\chi_u(t)$ is already minimal, so compute $\chi_u(t)$ from the scalars $\mathrm{trace}(\mathbf{M}_{u^k})$ for $k = 0, \ldots, d$.
   - Otherwise (if $r < N$), compute $\tilde{\chi}_u(t)$ from the scalars $\mathrm{trace}(\mathbf{M}_{u^k z})$ using the Wiedemann approach and Berlekamp-Massey algorithm. If FAIL choose another u.
   - Isolate r different roots ζ of $\tilde{\chi}_u(t)$. If FAIL go back to 3. and choose another u.
5. Compute $g_u(1,t), g_u(x_1,t), \ldots, g_u(x_n,t)$ from the scalars $\mathrm{trace}(\mathbf{M}_{u^k v})$ where $k = 0, \ldots, r$ and $v = 1, x_1, \ldots, x_N$.
6. Compute the solutions $\left[\dfrac{g_u(x_1,\zeta)}{g_u(1,\zeta)}, \dfrac{g_u(x_2,\zeta)}{g_u(1,\zeta)}, \ldots, \dfrac{g_u(x_N,\zeta)}{g_u(1,\zeta)}\right]$ of the system.

4.3 8-PSK example

We now illustrate the online part of the algorithm for our running case ($N = 3$ and 8-PSK) by plugging in a set of parameters obtained by the non-circular statistics from a block of received signal $y[n]$. E. g. for system (P), by taking arbitrarily $\gamma_0 = 2, \gamma_1 = 0$

and $\gamma_2 = 1$, we get

$$\mathrm{T_R}\mathbf{M}(1,0,1) = \begin{bmatrix} 8 & 0 & 0 & 0 & 2 & 8 & -20 & 0 \\ 0 & -2 & 2 & 8 & 0 & 0 & 0 & 14 \\ 0 & 2 & 10 & -20 & 0 & 0 & 0 & -14 \\ 0 & 8 & -20 & 10 & 0 & 0 & 0 & -20 \\ 2 & 0 & 0 & 0 & -10 & 14 & -14 & 0 \\ 8 & 0 & 0 & 0 & 14 & -4 & -20 & 0 \\ -20 & 0 & 0 & 0 & -14 & -20 & 68 & 0 \\ 0 & 14 & -14 & -20 & 0 & 0 & 0 & -62 \end{bmatrix}.$$

Since rank $\mathrm{T_R}\mathbf{M}(2,0,1) = 8$, we test that $u := x_1 - x_2 + x_3$ is separating, and get the following representation for (P):

$$\begin{aligned} \tilde{\chi}_u(t) &= t^8 - 35t^6 + 504t^4 - 2275t^2 + 8281 \\ \text{and} \quad g_u(1,t) &= 70t^6 - 2016t^4 + 19650t^2 - 66248, \\ g_u(x_1,t) &= 4t^7 - 246t^5 + 3372t^3 - 14378t, \\ g_u(x_2,t) &= -28t^7 + 870t^5 - 8604t^3 + 29666t, \\ g_u(x_3,t) &= 38t^7 - 900t^5 + 7674t^3 - 22204t. \end{aligned}$$

Now, the roots of $\tilde{\chi}_u(t)$ can be easily isolated using the trace matrix [22]. In this special case, an exact factorization can even be computed (this was lucky and pure show-off),

$$\begin{aligned} \tilde{\chi}_u(t) = &\left(t - \tfrac{1}{2}\sqrt{\tfrac{1}{2} + 6i\sqrt{3}}\right)\left(t + \tfrac{1}{2}\sqrt{\tfrac{1}{2} + 6i\sqrt{3}}\right)\left(t - \tfrac{1}{2}\sqrt{\tfrac{1}{2} - 6i\sqrt{3}}\right) \\ &\left(t + \tfrac{1}{2}\sqrt{\tfrac{1}{2} - 6i\sqrt{3}}\right)\left(t - \sqrt{11 + 4i\sqrt{3}}\right)\left(t + \sqrt{11 + 4i\sqrt{3}}\right) \\ &\left(t - \sqrt{11 - 4i\sqrt{3}}\right)\left(t + \sqrt{11 - 4i\sqrt{3}}\right), \end{aligned}$$

and so we get the following eight solutions for $[h[0]^4, h[1]^4, h[2]^4]$:

$$\begin{aligned} \Big\{ &\left[\tfrac{1}{2}\sqrt{3} + \tfrac{1}{2}i, 0, \tfrac{1}{2}\sqrt{3} - \tfrac{1}{2}i\right], \left[-\tfrac{1}{2}\sqrt{3} - \tfrac{1}{2}i, 0, -\tfrac{1}{2}\sqrt{3} + \tfrac{1}{2}i\right], \\ &\left[\tfrac{1}{2}\sqrt{3} - \tfrac{1}{2}i, 0, \tfrac{1}{2}\sqrt{3} + \tfrac{1}{2}i\right], \left[-\tfrac{1}{2}\sqrt{3} + \tfrac{1}{2}i, 0, -\tfrac{1}{2}\sqrt{3} - \tfrac{1}{2}i\right], \\ &\left[i, -\sqrt{3}, -i\right], \left[-i, \sqrt{3}, i\right], \left[-i, -\sqrt{3}, i\right], \left[i, \sqrt{3}, -i\right] \Big\}. \end{aligned}$$

By solving now for $[h[0], h[1], h[2]]$, we thus get 512 possible solutions for system (2.5).

This second (on-line) stage of the algorithm does not require any symbolic computation (except in this special case for the show-off on the exact factorization of $\chi_u(t)$).

The full computation of the geometric representation of the system is done in floating point arithmetic in a Matlab-like environment (Scilab) from the instantiated matrix of traces and a generic multiplication matrix.

4.4 Identification and tracking

The finite set of solutions can now be pruned by using other statistics on the received signal $y[n]$. For example, we can identify the best solution using the circular statistics of $y[n]$, as done in [12, 19],

$$c_p := \mathbb{E}(y[n]y^*[n-p]) = \sum_{m=p}^{N-1} h[m]h^*[m-p].$$

Alternatively, it may also be wiser to use higher-order statistics on $y[n]$ to parse directly through the tree of solutions in $[h[0]^4, h[1]^4, h[2]^4]$. This would reduce greatly the size of the set of valid solutions.

Furthermore, this approach by matrix of traces is quite robust to added noise on the estimates of the non-circular statistics $\gamma_0, \ldots, \gamma_{N-1}$ and this is especially true by using $\tilde{\chi}_u(t)$ instead of $\chi_u(t)$. However, in some quite degenerated cases, a precise estimate of the rank of TrM can be quite difficult to obtain especially when the solutions come in clusters. A recent paper [14] details the computation of approximated matrices of traces to alleviate this issue.

Also, geometric representations suggest the possibility to track on the space of solutions when the channel evolves in time. The solution space depends then on the roots of the univariate polynomial $\tilde{\chi}_u(t)$ which can now be seen as a root process [15] whose coefficients are random variables evolving in time (following channel evolution).

5 Conclusion

In the continuation of our previous work described in [19], we detailed here an algebraic alternative to some problems arising in statistical signal processing. Based on the offline computation of matrix of traces and an online randomized algorithm of Wiedemann-type, the approach is radical in many sense. First unlike the usual gradient-based algorithms of statistical signal processing, we are able to deal with an exhaustive description of the solution space. Moreover, the classically high computational cost of algebraic geometry based algorithms has been greatly alleviated by a subtle *offline/online* split of the approach. The proposed algorithm has a small on-line computational cost since the expensive symbolic computation of the parametric matrix of traces has been done once for all offline. Hence in our case, the solutions of the problem are easily obtained through the randomized computation of a univariate minimal polynomial.

Acknowledgements. We would like to thank the reviewers for suggesting useful remarks.

Bibliography

[1] E. Becker and T. Woermann, *On the trace formula for quadratic forms and some applications*, Recent Advances in Real Algebraic Geometry and Quadratic Forms, Contemporary Mathematics, vol. 155, AMS, 1994, pp. 271–291.

[2] J. F. Canny, *Some algebraic and geometric computations in PSPACE*, Proc. ACM Symp. on Theory of Computing, 1988, pp. 460–467.

[3] ———, *Generalized characteristic polynomials*, J. Symb. Comput. 9 (1990), pp. 241–250.

[4] C. Charoenlarpnopparut and N. K. Bose, *Multidimensional FIR filter bank design using Gröbner bases*, IEEE Trans. on CAS II 46 (1999), pp. 1475–1486.

[5] P. Ciblat and A. Quadrat, *New proof of a blind equalization result: a module theory approach*, Proc. First IFAC Symp. on Syst. Struct. and Cont. (Praha), 2001.

[6] D. Cox, J. Little, and D. O'Shea, *Using Algebraic Geometry*, GTM, vol. 185, Springer-Verlag, 1998.

[7] I. Emiris and A. Rege, *Monomial bases and polynomial system solving*, Proc. ACM Intern. Symp. on Symbolic and Algebraic Computation, 1994, pp 114–122.

[8] A. Fabiańska and A. Quadrat, *Applications of the Quillen-Suslin theorem to multidimensional systems theory*, Gröbner Bases in Control Theory and Signal Processing (H. Park and G. Regensburger, eds.), Radon Series Comp. Appl. Math. 3, de Gruyter, 2007, pp. 23–106 (this volume).

[9] J.-C. Faugère, F. Moreau de Saint-Martin, and F. Rouillier, *Design of regular nonseparable bidimensional wavelets using Gröbner basis techniques*, IEEE Trans. Signal Process. 46 (1998), pp. 845–857.

[10] M. Giusti, G. Lecerf, and B. Salvy, *A Gröbner free alternative for polynomial systems solving*, J. of Complexity 17 (2001), pp. 154–211.

[11] L. Gonzalez-Vega, F. Rouillier, and M.-F. Roy, *Some Tapas of Computer Algebra*, ch. Symbolic recipes for polynomial system solving, Springer-Verlag, 1999.

[12] O. Grellier, P. Comon, B. Mourrain, and P. Trebuchet, *Analytical blind channel identification*, IEEE Trans. Signal Process. 50 (2002), pp. 2196–2207.

[13] R. J. Holt, T. S. Huang, and A. N. Netravali, *Algebraic methods for image processing and computer vision*, IEEE Trans. Image Process. 5 (1996), pp. 976–986.

[14] I. Janovitz-Freireich, L. Rónyai, and A. Szántó, *Approximate radical of ideals with clusters of roots*, Proc. ACM Intern. Symp. on Symbolic and Algebraic Computation, 2006.

[15] W. S. Kendall, *Zeros of Brownian Polynomials*, Tech. Report 353, University of Warwick, 2000.

[16] J. Keyser, K. Ouchi, and J. M. Rojas, *The Exact Rational Univariate Representation for Detecting Degeneracies*, Geometric and Algorithmic Aspects of Computer-Aided Design and Manufacturing, DIMACS, AMS, 2005, pp. 299–328.

[17] L. Kronecker, *Grundzüge einer arithmetischen Theorie der algebraischen Größen*, J. reine angew. Math. (1882).

[18] J. Lebrun, *Balancing MultiWavelets*, Ph.D. thesis, Swiss Federal Institute of Technology (EPFL), Lausanne, 2000.

[19] J. Lebrun and P. Comon, *Blind identification of communication channels - Symbolic solution algorithms*, Appl. Algebra Eng. Commun. Comput. 17 (2006), pp. 471–485.

[20] J. Lebrun and I. Selesnick, *Gröbner bases and wavelet design*, J. Symb. Comput. 35 (2004), pp. 227–259.

[21] M. Moreno Maza, E. Schost, and W. Zhou, *Primary decomposition of zero-dimensional ideals: Putting Monico's algorithm into practice*, Transgressive Computing (Granada, Spain), 2006.

[22] B. Mourrain, *Computing isolated polynomial roots by matrix methods*, J. Symb. Comp. 26 (1998), pp. 715–738.

[23] ______, *An introduction to linear algebra methods for solving polynomial equations*, Proc. HERCMA'9 (E.A. Lipitakis, ed.), 1999, pp. 179–200.

[24] B. Mourrain and Ph. Trébuchet, *Generalized Normal Forms and Polynomial System Solving*, Proc. ACM Intern. Symp. on Symbolic and Algebraic Computation, 2005, pp. 253–260.

[25] H. Park, *Symbolic computation and signal processing*, J. Symb. Comput. 35 (2004), pp. 209–226.

[26] H. Park, T. Kalker, and M. Vetterli, *Gröbner Bases and Multidimensional FIR Multirate Systems*, Multidimensional Syst. Signal Process. 8 (1997), pp. 11–30.

[27] P. Pedersen, M.-F. Roy, and A. Szpirglas, *Counting real zeros in the multivariate case*, Proc. Mega'92 Int. Symp. on Effective Methods in Algebraic Geometry (Birkhauser, ed.), Progress in Mathematics, 1992, pp. 203–224.

[28] G. Pistone, E. Riccomagno, and H. P. Wynn, *Algebraic Statistics: Computational Commutative Algebra in Statistics*, Chapman & Hall, CRC Press, 2000.

[29] J. G. Proakis, *Digital Communications*, McGraw-Hill, 1995, 3rd edition.

[30] G. Regensburger, *Parametrizing compactly supported orthonormal wavelets by discrete moments*, Appl. Algebra Eng. Commun. Comput. (2007), to appear.

[31] G. Regensburger and O. Scherzer, *Symbolic Computation for Moments and Filter Coefficients of Scaling Functions*, Ann. Comb. 9 (2005), pp. 223–243.

[32] J. Renegar, *On the Computational Complexity and Geometry of the First-Order Theory of the Reals, Part I, II & III*, J. Symb. Comput. 13 (1992), pp. 255–352.

[33] L. Rota and P. Comon, *Blind Equalizers based on Polynomial Criteria*, Proc. ICASSP'04 (Montreal), May 17–21 2004.

[34] F. Rouillier, *Solving zero-dimensional systems through the rational univariate representation*, Appl. Algebra Eng. Commun. Comput. 9 (1999), pp. 433–461.

[35] B. D. Saunders, A. Storjohann, and G. Villard, *Matrix Rank Certification*, Elec. J. Lin. Algebra 11 (2004), pp. 16–23.

[36] E. Schost, *Computing Parametric Geometric Resolutions*, Appl. Algebra Eng. Commun. Comput. 13 (2001), pp. 349–393.

[37] ______, *Computing Parametric Geometric Resolutions*, Appl. Algebra Eng. Commun. Comput. 13 (2003), pp. 349–393.

[38] I. W. Selesnick, *Interpolating multiwavelet bases and the sampling theorem*, IEEE Trans. Signal Process. 47 (1999), pp. 1615–1621.

[39] ______, *Balanced Multiwavelet Bases based on Symmetric FIR filters*, IEEE Trans. Signal Process. 48 (2000), pp. 184–191.

[40] I. W. Selesnick and C. S. Burrus, *Maximally Flat Low-pass FIR Filters with Reduced Delay*, IEEE Trans. on CAS II 45 (1998), pp. 53–68.

[41] B. Sturmfels, *Solving Systems of Polynomial Equations*, CBMS series, no. 97, AMS, 2002.

[42] P. Trébuchet, *Vers une résolution stable et rapide des équations algébriques*, Ph.D. thesis, INRIA - Sophia-Antipolis, 2002.

[43] A. J. van der Veen and A. Paulraj, *An Analytical Constant Modulus Algorithm*, IEEE Trans. Signal Process. 44 (1996), pp. 1136–1155.

[44] D. H. Wiedemann, *Solving sparse linear equations over finite fields*, IEEE Trans. Inform. Theory 32 (1986), pp. 54–62.

[45] D. Yellin and B. Porat, *Blind identification of FIR systems excited by discrete-alphabet inputs*, IEEE Trans. Signal Process. 41 (1993), pp. 1331–1339.

[46] J. Zhou and M. N. Do, *Multidimensional multichannel FIR deconvolution using Gröbner bases*, IEEE Trans. Image Process. 15 (2006), pp. 2998–3007.

Author information

Jerome Lebrun, CNRS UMR6070,
F-06903 Sophia-Antipolis, France.
Email: `lebrun@i3s.unice.fr`

Radon Series Comp. Appl. Math **3**, 127–149

Obstructions to genericity in study of parametric problems in control theory

Viktor Levandovskyy and Eva Zerz

Key words. Generic properties, systems with parameters, Gröbner bases, parametric Gröbner bases, non-commutative Gröbner bases.

AMS classification. 13P10, 93B25

1 Introduction

Motivation. Problems from control theory often involve a set of physical parameters, for instance, masses, spring constants, and damping coefficients with mechanical systems, or resistances, capacitances, and inductances with electrical circuits. The structural properties of the control system may depend crucially on the specific choice of concrete parameter values. In many relevant examples, a system is generically controllable (i. e., controllable for almost all possible parameter values), but becomes uncontrollable when certain relations between the parameters are fulfilled.

It has been shown for many system classes of practical interest that controllability amounts to the torsion-freeness of a module associated to the system. For example, if $A = \mathbb{K}[\partial_1, \dots, \partial_n]$ for a field $\mathbb{K}$ and $\mathcal{F} = \mathcal{C}^\infty(\mathbb{R}^n, \mathbb{K})$, the system given by the linear constant-coefficient partial differential equations $R(\partial_1, \dots, \partial_n)w = 0$, where $R \in A^{p\times q}$ and $w \in \mathcal{F}^q$, is controllable if and only if $M = A^{1\times q}/A^{1\times p}R$ is torsion-free. Then the **parametric controllability** problem can be formulated as follows: Given $R \in A^{p\times q}$, where $A = \mathbb{K}(p_1, \dots, p_t)[\partial_1, \dots, \partial_n]$ for some parameters $p_1, \dots, p_t$, find out whether M is generically torsion-free, and moreover determine the relations among the p_i that will cause torsion elements in M. In general, we pose this question for a left module M over a (non-commutative) algebra A. The antipode notion of controllability

First author: Partially supported by SFB project F1301 of the Austrian FWF.

is autonomy, which happens for systems being torsion. For a corresponding system module M, this means that M is annihilated by a non-zero ideal in A.

To a system module M over a system algebra A one associates the **transposed module** $N = N(M)$, defined as follows. Let the left module M be presented by a matrix $R \in A^{p\times q}$, then $N(M)$ is a right module with the presentation matrix $R^T \in A^{q\times p}$.

Then, there is an alternative description of torsion-freeness (controllability) and torsion (autonomy) in the language of homological algebra. Namely,

1. M is torsion-free if and only if $\mathrm{Ext}^1_A(N(M), A) = 0$, and
2. M is a torsion module if and only if $\mathrm{Ext}^0_A(M, A) = \mathrm{Hom}_A(M, A) = 0$.

For a survey of the correspondence between control systems and their system modules, see [9] and the references therein, in particular, the works of Oberst, Pommaret and Quadrat. A general approach to parametric modules including the case when the parameters are non-constant was introduced in [14, 15]. The authors showed that so-called trees of integrability conditions, depending on parameters of the system, determine the control-theoretic properties of the system. These trees result into systems of partial differential equations and nonlinear differential conditions.

The situation described above was our original motivation for studying parameter-dependent questions of homological algebra such as the specific problem outlined above. It turned out that apart from the concrete application area, it is a challenging task for computer algebra to investigate parametric modules, and in particular, to get a grip on the special values of parameters that cause a qualitative change of structural module properties. These questions reach far beyond the limited set of algebras that typically arise in control theory. Roughly speaking, the problem can be tackled from the computational point of view for virtually every algebra that is accessible to Gröbner basis techniques. The main idea is simple but effective: it consists in a careful monitoring of denominators of cofactors that appear during Gröbner basis computation. Thus in this article we continue with the investigations, started in the articles [9, 10].

Outline of the paper. In this paper, we give an algorithm for answering the following

Question from Control Theory. Given a linear system S, depending on a finite number of parameters. Determine the control-theoretic properties (such as the decomposition into a controllable and an autonomous part) of S for all the values of involved parameters.

Since there is, for certain system classes [2, 10, 13], a one-to-one correspondence between control-theoretic properties of a system S and the homological properties of an associated module $M(S)$, we can reformulate the question as follows:

Question for Computer Algebra. Given a finite presentation of a parametric module M over a (non-commutative) algebra A, determine the properties (e. g. homological) of M for all the values of involved parameters.

We present detailed solutions for the bipendulum equations (Example 3.1), and for the "two pendula mounted on a cart" problem, for both negligible friction (section 4.1) and essential friction (section 4.2). The latter problem, to the best of our knowledge, has not been yet solved completely in an explicit way. We also present and comment on several curious examples. Namely, we show the existence of a non-generic controllability in the generically autonomous system (Example 3.2), and present a system, where both controllability and autonomy properties appear only in the non-generic situation (Example 3.3). In the treatment of the case 3 of 4.2, we illustrate the ability of our method to treat nested obstructions, that is, investigating sub-obstructions of a given obstruction to genericity.

Preliminaries. Algebraically speaking, a **parameter** is a non-zero (and thus invertible) element of the ground field. In this article we deal with parameters which mutually commute with the elements of the algebra. In other words, the action of operators of the algebra on parameters is just the commutative multiplication.

In this article, we use the following definition of a property being generic.

Definition 1.1 Let $\mathcal{P}(p_1, \ldots, p_n)$ be a polynomial expression in p_i over some domain $\mathcal{D}$, on which a measure μ exists. The identity $\mathcal{P} = 0$ holds **generically** in $\mathcal{D}$ if $\mathcal{P}(\xi_1, \ldots, \xi_n) = 0$, for almost all $(\xi_1, \ldots, \xi_n) \in \mathcal{D}^n$. In other words, $\mathcal{P} = 0$ holds in $\mathcal{D}^n \setminus E$, where $E \subset \mathcal{D}^n$ and $\mu(E) = 0$.

For instance, if $\mathcal{D} = \mathbb{C}$ and $\mathcal{P}$ vanishes on the complement of a nonzero algebraic set $E \subset \mathcal{D}^n$, then $\mathcal{P} = 0$ holds generically in $\mathcal{D}^n$. The Krull dimension $\mathrm{Kr.dim}$ of the coordinate ring $\mathbb{K}[V]$ of a variety V can be used for defining a measure on closed subsets $E \subset \mathcal{D}^n$ by assigning $\mu(E) = 0$ if $\mathrm{Kr.dim}\,\mathbb{K}[E] < n$, and $\mu(E) = 1$ otherwise.

Notations. For a matrix M, M^T denotes its transposed matrix. By ${}_A\langle F \rangle$ we denote a left A-submodule, generated by the finite set F. The subscript A is dropped when A is commutative. For an ideal I in a commutative ring $\mathbb{K}[x_1, \ldots, x_n]$, we denote by $V(I) \subseteq \mathbb{K}^n$ the set of common zeros of polynomials in I.

2 Genericity of Gröbner bases of parametric modules

Since the major role in computations (of e. g. homological properties) is played by Gröbner bases, we investigate their behaviour in the case when a ground field involves parameters.

Let $\mathbb{K}$ be a field. Let A be a (non-commutative) algebra over $\mathbb{K}(p_1, \ldots, p_t)$. Suppose that in this algebra the notion of algorithmic left Gröbner basis exists (e. g. A can be a ring of solvable type [6] or, more restrictively, an Ore algebra [2]).

Let us recall the definition of a ring of solvable type.

Definition 2.1 Let $K \supseteq \mathbb{K}$ be a skew field and let $R' := K[x_1, \ldots, x_n]$ be a commutative ring over K. Suppose that $\prec$ is a fixed term ordering on R'. Let R be a ring generated by $\{x_1, \ldots, x_n\}$ subject to the new multiplication $*$. If the properties 1 and 2 below hold and $(R, *)$ is an associative ring, R is called a **ring of solvable type**.

1. $\forall 1 \le i < j \le n$, $x_i * x_j = x_i x_j$ and $x_j * x_i = c_{ij} x_i x_j + p_{ij}$, where $0 \neq c_{ij} \in K$ and $p_{ij} \in R'$, such that $\mathrm{lm}(p_{ij}) \prec x_i x_j$,
2. $\forall 1 \le i \le n$, $\forall a \in K$, $a * x_i = a x_i$ and $x_i * a = c_{ai} a x_i + p_{ai}$, where $0 \neq c_{ai} \in K$ and $p_{ai} \in K$.

Good examples of rings of solvable type are the rings of (partial) differential-difference operators.

The elements of a free module A^m are represented as the vectors $\bar{t} = \sum_{k=1}^{m} t_i e_i$, where $t_i \in A$, and e_i is the i-th canonical basis vector. By $\bar{0}$ we denote the zero vector $(0, \ldots, 0)^T \in A^m$. The set of vectors $\bar{t}_1, \ldots, \bar{t}_l$, for instance the set of generators of a submodule of a free module A^m, will be often identified with the matrix $T \subset \mathrm{Mat}(m \times l, A)$. A single vector $\bar{t}_i$ corresponds to the i-th column of T and vice versa.

Given a monomial well-ordering $\prec$ on A, there are several ways to extend it to a monomial module ordering $\prec_M$ on A^m, that is, an ordering consisting of two components $(\prec, \prec_C)$, where $\prec_C$ is an ordering on the components e_i. In the following, we need a so-called **term over position ordering**, that is, $m_1 e_i \prec_M m_2 e_j$ if and only if $m_1 \prec m_2$ or, if $m_1 = m_2$, then $e_i \prec_C e_j$ for monomials $m_i \in A$.

Recall, that a **left syzygy** of a finite set of elements $\{f_1, \ldots, f_m\}$, $f_i \in A$, is a tuple $(b_1, \ldots, b_m)^T \in A^m$, such that $b_1 f_1 + \ldots + b_m f_m = 0$. The set of all left syzygies of a given set of m elements is a left submodule of A^m. It is often denoted as $\mathrm{Syz}(\{f_1, \ldots, f_m\})$.

In this article, we work with left submodules, left syzygies etc. It is clearly possible to do the same also from the right. However, two-sided (bimodule) problems deserve, except for the commutative case, a fairly distinct treatment. Most (if not all) problems, originating from applications of e. g. control theory, are formulated in terms of left modules.

2.1 LIFT and LEFTINVERSE algorithms

Proposition 2.2 *Suppose that a left submodule L of a free module A^m is generated by the set of column vectors $F = \{\bar{f}_1, \ldots, \bar{f}_l\} \subset A^m$. Consider the set $\widetilde{F} := \{\bar{f}_1 + e_{m+1}, \ldots, \bar{f}_l + e_{m+l}\}$ and assume, that the fixed ordering $\prec$ on A^m, naturally extended to the ordering $\prec_l$ on A^{m+l}, satisfies $x^\alpha e_{m+i} \prec_l x^\beta e_j$, for all $1 \le i \le l$, $1 \le j \le m$ and for all α, β. Suppose that the left Gröbner basis $\widetilde{G}$ of $\widetilde{F}$ is finite. Then we reorder the columns of $\widetilde{G}$ in such a way, that the elements, whose first m components are zero, are moved to the left. This process is schematically presented in the following picture:*

$$\widetilde{F} = \begin{pmatrix} \bar{f}_1 & \cdots & \bar{f}_l \\ 1 & & 0 \\ & \ddots & \\ 0 & & 1 \end{pmatrix} \xrightarrow{\text{LEFTGB}} \left(\begin{array}{ccc|ccc} \bar{0} & \ldots & \bar{0} & \bar{h}_1 & \ldots & \bar{h}_t \\ \hline & & & & & \\ & \mathbf{S} & & & \mathbf{T} & \\ & & & & & \end{array}\right) = \widetilde{G}.$$

Let $H = \{\bar{h}_1, \dots, \bar{h}_t\}$ be a left Gröbner basis of F. Recall that we identify F with the matrix $(\bar{f}_1, \dots, \bar{f}_l) \in A^{m \times l}$, T with an $l \times t$ matrix over A, and H with the matrix $(\bar{h}_1, \dots, \bar{h}_t) \in A^{m \times t}$. Then

- $\mathbf{T}$ *is a left transformation matrix between two generating sets of F, i.e. $H^T = \mathbf{T}^T F^T$,*
- *the columns of* $\mathbf{S}$ *form a left Gröbner basis of* $\operatorname{Syz}(\{\bar{f}_1, \dots, \bar{f}_l\})$.

Proof. Since $\bar{h}_i = \sum_{k=1}^{l} a_{ik} \bar{f}_k$, we have

$$\sum_{k=1}^{l} a_{ik}(\bar{f}_k + e_{m+k}) = \bar{h}_i + \sum_{k=1}^{l} a_{ik} e_{m+k}.$$

Hence, the i-th column of $\mathbf{T}$ is $(a_{i1}, \dots, a_{il})^T$, and $H^T = (\bar{h}_1, \dots, \bar{h}_t)^T = \mathbf{T}^T \cdot F^T$.

Let $\mathbf{S} = \{\bar{s}_1, \dots, \bar{s}_r\}$ and $\widetilde{S} := \{(\bar{0}, \bar{s}_1)^T, \dots, (\bar{0}, \bar{s}_r)^T\}$. Since $\widetilde{G}$ is a left Gröbner basis of $\widetilde{F}$, for any f in ${}_A\langle \widetilde{F} \rangle \cap (\{\bar{0}\}^m \times A^l) = {}_A\langle \widetilde{F} \rangle \cap \oplus_{k=m+1}^{m+l} A e_k$ there exists $g \in \widetilde{G}$, such that $\operatorname{lm}(g)$ divides $\operatorname{lm}(f)$. Then $\operatorname{lm}(g) \in \{\bar{0}\}^m \times A^l$, hence, by the property of the ordering, $g \in \widetilde{G} \cap (\{0\} \times A^l) = \widetilde{S}$. Thus $\widetilde{S}$ is a left Gröbner basis of $\widetilde{F} \cap (\{\bar{0}\}^m \times A^l)$ and, in particular, $\widetilde{S}$ generates the latter. Since

$$\sum_{k=1}^{l} b_k(\bar{f}_k + e_{m+k}) = \sum_{k=1}^{l} b_k e_{m+k} \quad \text{holds if and only if} \quad \sum_{k=1}^{l} b_k \bar{f}_k = \bar{0},$$

$\mathbf{S}$ consists of columns $(b_1, \dots, b_l)^T$, which are the syzygies of the set $\{\bar{f}_1, \dots, \bar{f}_l\}$. □

Remark 2.3 Clearly, for a Noetherian algebra A the algorithm terminates.

More generally, if the left Gröbner basis of F is finite, we get the transformation matrix in finitely many steps. Namely, in the generalized Buchberger's algorithm for computing left Gröbner basis, we do not consider S-polynomials between elements whose leading monomials include components greater than m.

If the algebra A is commutative, the transformation matrix property translates into $H = F \cdot \mathbf{T}$.

We call the algorithm computing the transformation matrix as above LIFT(F, H). Note that with this algorithm we are able to trace any computation which uses Gröbner bases. It is worth mentioning that Proposition 2.2 shows, that with basically one Gröbner basis computation we can get three important objects, namely a Gröbner basis of a module, a Gröbner basis of the first syzygy module and a transformation matrix. These three applications are sometimes called **Gröbner trinity** and play a fundamental role in computer algebra.

Many problems in control theory involve parameters, which are known to be non-zero, or even strictly positive, for physical reasons. However, it might happen that the vanishing of certain algebraic expressions in the parameters has a direct impact on the control-theoretic properties. Very often we observe generically controllable parametric systems which, for some values of parameters, become uncontrollable.

As a further application of the algorithm LIFT, we compute a left inverse of a given polynomial matrix in the case it exists. Below, the algorithm RMLEFTGROEBNERBASIS(M) computes the monic reduced minimal left Gröbner basis of a submodule M, which is unique for a fixed ordering ([6, 7]).

Algorithm 2.4 LEFTINVERSE(matrix M)
Input: $M \in \mathrm{Mat}_{m\times n}(A)$
Output: $L \in \mathrm{Mat}_{n\times m}(A)$, such that $L \cdot M = Id_{n\times n}$
or $0 \in \mathrm{Mat}_{1\times 1}(A)$, if no left inverse exists

module $G :=$ RMLEFTGROEBNERBASIS(M)
if $G \neq Id_{n\times n}$ **then**
report "No left inverse exist"
return 0
endif

module $N :=$ TRANSPOSE(M)
module $K :=$ LIFT$(N, Id_{n\times n})$
return TRANSPOSE(K)

Proof. The algorithm LEFTINVERSE terminates as soon as LIFT does. Note that $LM = Id_{n\times n}$ can happen only in the case when the monic reduced minimal left Gröbner basis of a free submodule generated by the columns $\{N_j\}$ of $N = M^T$ is equal to $Id_{n\times n}$.

In the setup of the LIFT algorithm, we use $H = Id_{n\times n}$. Denote by K the result of LIFT$(N, Id_{n\times n})$. Then, by the Proposition 2.2, $Id_{n\times n} = Id_{n\times n}^T = K^T M^T = K^T N$. Hence, for $L = K^T$ we have $LM = Id_{n\times n}$. □

The existence of a left inverse (or, more generally, a generalized inverse G, such that $G \cdot M \cdot G = G$) often gives us the information on genericity of parameters. Namely, one analyzes the possible vanishing of denominators of a generalized inverse, as it is done in e. g. [2]. In the special case where $A = \mathbb{K}[\partial]$ is a principal ideal domain, consider the module $M = A^{1\times q}/A^{1\times p}R$. Without loss of generality, we can assume R has full row rank. Then M is torsion-free if and only if there exists a right inverse to R.

As we have shown, computing the inverse is a special case of computing the transformation matrix with the algorithm LIFT. In comparison with LEFTINVERSE, LIFT allows us to deal effectively with more general problems.

We call the polynomials in parameters, whose vanishing implies the failure of generic properties, **obstructions to genericity**. We can compute them as described above using the LIFT algorithm.

There is a need for complete information on the parametric module. It consists of the list of properties, computed for the generic and all the non-generic cases. In the context of generically controllable problems, we are interested in computing e. g. an annihilator of a torsion submodule for the each non-generic case. Thus, we need to stratify the set of obstructions.

2.2 Stratification of obstructions to genericity

Let $\mathbb{K}$ be a field of characteristic 0. Recall that a set is called **locally closed**, if it is a difference of two closed sets. A finite union of locally closed sets is called a **constructible** set.

Suppose we are given a set of polynomials $P = \{p_1, \ldots, p_n\} \subset \mathbb{K}[a_1, \ldots, a_m]$, which are irreducible over $\mathbb{K}$.

We associate to P a set $C(P) := \{\bar{\xi} = (\xi_1, \ldots, \xi_m) \in \mathbb{K}^m \mid \prod_{i=1}^n p_i(\bar{\xi}) = 0\} \subset \mathbb{K}^m$.

Lemma 2.5 *The set $C(P)$ is constructible.*

Proof. Let $\Omega := \{(\Lambda', \Lambda'') \mid \Lambda' \cup \Lambda'' = \{1, \ldots, n\}, \Lambda' \cap \Lambda'' = \emptyset\}$ be the set of all divisions of $\{1, \ldots, n\}$ into two disjoint complementary subsets. Let, furthermore, $\Sigma := \Omega \setminus (\emptyset, \{1, \ldots, n\})$. Then,

$$\begin{aligned} C(P) &= \bigcup_{(j,k)\in\Sigma} \{\bar{\xi} \mid \forall j \in \Lambda' \; p_j(\bar{\xi}) = 0, \quad \forall k \in \Lambda'', \; p_k(\bar{\xi}) \neq 0\} \\ &= \bigcup_{(j,k)\in\Sigma} V(\langle\{p_j \mid j \in \Lambda'\}\rangle) \setminus V(\langle\{p_k \mid k \in \Lambda''\}\rangle) \\ &= \bigcup_{(j,k)\in\Sigma} \left(\bigcap_{j\in\Lambda'} V(\langle p_j \rangle) \setminus \bigcap_{k\in\Lambda''} V(\langle p_k \rangle) \right), \end{aligned}$$

and, indeed, we see that $C(P)$ is a disjoint union of locally closed sets. Note that in $C(P)$ there is a closed subset $\cap_i V(p_i)$; the rest of subsets are locally closed. □

It is convenient to represent $C(P)$ as a binary tree, where the vertices are the decision points, associated to polynomials p_i, and the edges represent the logical conditions $(p_i = 0)$ and $(p_i \neq 0)$, respectively. In such a way it is easy to see, that starting from n elements in the set P, we will have $2^n - 1$ algebraic systems describing the locally closed components of $C(P)$.

Given two ideals $I, J \in \mathbb{K}[a_1, \ldots, a_m]$, an algebraic data describing a locally closed set $V(I) \setminus V(J) = V(I) \setminus (V(I) \cap V(J))$ can be computed with a factorizing Gröbner basis algorithm (e. g. [4]). Such an algorithm takes I, J as input and returns a list of ideals, where the zero set of the intersection of the output ideals is contained in the $V(I)$ and contains the complement of the $V(J)$ in $V(I)$. We refer to this algorithm as to FACTGB(I, J).

Example 2.6 Let $P = \{p_1, p_2\}$, then the binary tree for $C(P)$ consists of the following 3 systems of equations and inequations: $\{p_1 = 0, p_2 = 0\}$, $\{p_1 \neq 0, p_2 = 0\}$ and $\{p_1 = 0, p_2 \neq 0\}$.

Denote $V_i := V(\langle p_i \rangle)$ for $i = 1, 2$, and $V_{12} := V(\langle p_1, p_2 \rangle) = V_1 \cap V_2$. Then, the decomposition of $C(P)$ can be written as

$$V_{12} \uplus (V_1 \setminus V_2) \uplus (V_2 \setminus V_1) = V_{12} \uplus (V_1 \setminus V_{12}) \uplus (V_2 \setminus V_{12}),$$

where $\uplus$ denotes the disjoint union.

Computationally, we need to compute the Gröbner basis of an ideal $I_{12} := \langle p_1, p_2 \rangle$, and two lists $L_i := \text{FACTGB}(\{p_i\}, I_{12})$, obtained with the factorizing Gröbner basis algorithm, which describe $V_i \setminus V_{12}$.

Given a set of polynomials $\{f_1, \ldots, f_s\} \subset \mathbb{K}[a_1, \ldots, a_m]$, we factorize them and form a set of pairwise different irreducible factors $P := \{p_1, \ldots, p_n\}$. We sort p_i by using a positively graded degree ordering, starting with the smaller elements. With such an ordering, it is easier to compute with locally closed sets. Namely, the bigger elements will often reduce to simpler polynomials with respect to the smaller elements. Thus, also the detection of empty components (that is, systems with no solutions) can be achieved faster.

Lemma 2.5 is constructive indeed. Together with the presentation of locally closed sets using the algorithm FACTGB above, we call the whole procedure STRATIFYLC(list L). It takes a finite list of irreducible polynomials on the input and returns a list of systems of equations and inequations, corresponding to $C(P)$.

2.3 The GENERICITY algorithm

Let A be a $\mathbb{K}$-algebra and suppose that the coefficients of a given system S involve parameters $p_1, \ldots, p_t$. We interpret the parameters as generators of the transcendental field extension of $\mathbb{K}$ and we use the natural $\mathbb{K}(p_1, \ldots, p_t)$-algebra structure on A.

Algorithm 2.7 GENERICITY(matrix M)
Assume, that a monomial module ordering on the algebra A is fixed.
Input: $M \in \text{Mat}_{m \times n}(A)$
Output: $\{h_1, \ldots, h_s\} \subset \mathbb{K}[p_1, \ldots, p_t]$, such that if a specialization of the parameters implies $h_i(p_1, \ldots, p_t) = 0$, then a left Gröbner basis of M is different from the generic one

```
module G := RMLEFTGROEBNERBASIS(M)      \\ G = {ḡ_1, ..., ḡ_ℓ} ∈ Mat_{m×ℓ}(A)
matrix T := LIFT(M, G)                  \\ T ∈ Mat_{n×ℓ}(A)
list S, H;  int i, j
for j = 1 to ℓ
    i := the leading component of ḡ_j
    for k = 1 to n
        if (M_ik ≠ 0) then
            if (T_kj ≠ 0) then   S := S∪ DENOMINATOR(T_kj)
            end if
        end if
    end for
end for
if (S ≠ empty list) then
```

$H := \text{FACTORIZE}(S)$

$H := \text{SIMPLIFY}(H)$

else $H :=$ empty list

return H

Proof. The algorithm FACTORIZE(list L) returns a list of monic factors of every polynomial of the list L. The algorithm SIMPLIFY(list L) refines a list L by removing doubled appearances of same elements. We may assume it also sorts L by an ordering, putting with the smaller elements in the beginning of the output.

The algorithm GENERICITY terminates as soon as LIFT terminates. Now, we prove the correctness. Suppose that the leading term of $\bar{g}_j$ lies in the i-th module component. From the property $G^T = T^T M^T$ it follows, that there is a presentation of the element $G_{ij} \in A$ as the sum

$$G_{ij} = \sum_{k=1}^{n} T_{kj} M_{ik} = \sum_{k=1}^{n} T^T_{jk} M^T_{ki}.$$

Hence, it suffices to collect only the denominators of $T_{kj} \neq 0$ with $M_{ik} \neq 0$, since only such elements contribute to the leading coefficient of $\bar{g}_j$.

If some leading coefficient of the unique generic Gröbner basis vanishes for some specialization of parameters, then the Gröbner basis under such a specialization is different from the generic one. □

Note that with the algorithm we obtain the expressions in the parameters which lead to non-generic Gröbner bases. In order to obtain Gröbner bases under specialization, provided by h_i, one cannot use the generic Gröbner basis. Instead, one has to compute the specialized Gröbner basis from scratch.

Suppose that the output of GENERICITY is the list of irreducible polynomials H. In practice, we exclude from H the polynomials, which do not satisfy the problem-specific constraints for e. g. physical admissibility like non-negativity. Then, we apply the algorithm STRATIFYLC(H) and obtain a complete stratification of a given system with respect to its parameters.

2.4 Comparison with other methods

2.4.1 Comprehensive Gröbner bases

Comprehensive Gröbner bases (see e. g. [18]) were introduced by Weispfenning and generalized to rings of solvable type by Kredel [6].

A comprehensive Gröbner basis, by definition, is a finite subset G of a parametric polynomial ideal I such that $\sigma(G)$ constitutes a Gröbner basis of the ideal generated by $\sigma(I)$ under all specializations σ of the parameters in arbitrary fields ([18]).

The construction of a comprehensive Gröbner basis follows the lines of Buchberger's algorithm. However, the result will be a union of trees of ideal bases (called Gröbner systems), where each basis is accompanied with a set of conditions of parameters. Being a powerful theoretical instrument, comprehensive Gröbner bases are quite complicated to compute. To the best of our knowledge, there is no implementation yet, which is able to treat serious examples.

In our approach we separate two processes, which are unified in the comprehensive Gröbner basis method. Namely, we compute the tree of sets of conditions of parameters *after* the Gröbner basis and transformation matrix computations. In such a way we avoid repeated computations in trees of ideals and sets of conditions, which might occur during the computation of a comprehensive Gröbner basis.

2.4.2 The Leykin-Walther method

The method has been formalized by Leykin for the case of ideals [11] and has been generalized to modules by Walther [17]. The idea behind the method has been used before, however Leykin and Walther formulated and proved the whole framework in a complete way. In the following, we reformulate the Lemma 2.3 from [17].

Let $\mathbb{K}$ be a field of characteristic 0. Given a $\mathbb{K}$-algebra A, we consider parameters as new commutative variables and perform further computations in the $\mathbb{K}$-algebra $\widetilde{A} := A \otimes_{\mathbb{K}} \mathbb{K}[p_1, \dots, p_m]$. We use in $\widetilde{A}$ an elimination ordering $\prec_A$ for the variables of A. Such an ordering is characterized by the property $p_1^{\alpha_1} \dots p_m^{\alpha_m} \prec_A t$, for any monomial $t \in A$ and any $\alpha \in \mathbb{N}^m$.

Let $G = \{g_1, \dots, g_\ell\}$ be a reduced Gröbner basis for the left submodule $N \subset \widetilde{A}^s$ with respect the position over term ordering, induced by $\prec_A$ on $\widetilde{A}^s$. Let, moreover, $Q_N \subset \mathbb{K}[p_1, \dots, p_t]$ be the ideal $\{p \in \mathbb{K}[p_1, \dots, p_t] \mid p\widetilde{A}^s \subseteq N\}$. For $g_i \neq Q_N \widetilde{A}^s$, multiply all the leading coefficients with respect to $\mathbb{K}[p_1, \dots, p_t]$ of such g_i and denote the result by h. Let $\sigma : \mathbb{K}[p_1, \dots, p_t] \to \mathbb{K}$ be a specialization, then if $\sigma(h) \neq 0$, then $\sigma(G) = \{\sigma(g_1), \dots, \sigma(g_t)\}$ is a Gröbner basis.

This method has some drawbacks in practice. Suppose that the number of parameters is big and there are many obstructions, which appear in several components as, say, leading coefficients by a monomial 1. This situation is typical for generically controllable systems. Then, using the method of Leykin-Walther, we are forced to compute Gröbner basis of a submodule of elements as described above, whereas a better solution would be just to collect the leading coefficients in parameters. Secondly, in a similar situation we get many elements in Gröbner basis and the analysis of the impact of obstructions, e. g. the computation of the stratification, becomes very involved.

On the other hand, this method allows us to handle the cases, when the parameters satisfy algebraic identities between themselves or when there are more general identities, involving both variables and parameters. We believe, that this method will be enhanced in order to overcome the described difficulties.

3 Implementation of algorithms

The described method for detecting the obstructions to genericity of parametric modules is implemented in the procedure `genericity` of control theory toolbox CONTROL.LIB [1], which is realized as a library in the computer algebra system SINGULAR.

SINGULAR [5] is the specialized computer algebra system for polynomial computations, well-known for its high performance (especially in Gröbner bases-related computations) and rich functionality. It uses intuitive C-like programming language, in which the libraries are written. It is important to mention, that SINGULAR is distributed under GPL license, that is, it is free for academic purposes.

The current implementation of the procedure `genericity` works in a little different way, compared with the Algorithm 2.7. Namely, it takes as input a matrix T, which is assumed to be the result of the LIFT algorithm. This minor modification allows us to compute the data, which are independent from the choice of a monomial module ordering. The output of the procedure `genericity` is a list of strings and thus it is ring-independent. In the first item of the list the names of parameters, by which we have divided in the algorithm, are collected. Every further item of the list contains a single non-trivial polynomial in the parameters.

There are several algorithms in SINGULAR, which compute (left) Gröbner bases of modules over commutative polynomial algebras and non-commutative GR-algebras [7, 8]. It is recommended to use the heuristic routine `groebner`, which often provides the best match for a concrete example. For more details on SINGULAR, consult with the book [4] and with the website of the system [5], which contains among other the online documentation. The algorithm FACTGB is implemented in SINGULAR and is accessible via the function `facstd`.

In the library CONTROL.LIB, we have implemented several functions for supporting the research in systems and control theory. Among others, there are the procedures `LeftInverse` and `LeftKernel`, their counterparts `RightInverse` and `RightKernel`, as well as `canonize` and `iostruct`.

The main purpose of the library is to provide maximal relevant information based on a simple input. This principle led us to the development of heuristic procedures `control` and `autonom`, which use homological computations. Respectively, for systems with a full row rank presentation matrix, there are dimension-guided procedures `controlDim` and `autonomDim`.

Given a system algebra and a system module over it, both procedures compute relevant properties of a given module from the point of view of controllability (with the procedure `control` or `controlDim`) or autonomy analysis (with the procedure `autonom` or `autonomDim`). The procedure `canonize` takes the output of either `control` or `autonom` procedure and computes reduced and tail-reduced Gröbner bases of the objects, thus simplifying and *canonizing* the output.

We illustrate the functionality of the library and the flexibility of SINGULAR with the following example.

Example 3.1 Consider a bipendulum, that is, a system, describing a bar with two fixed pendula of length ℓ_1 and ℓ_2 respectively (e. g. [13, 3]). The system algebra is a commutative algebra in variable ∂ over a field of rationals with parameters g, ℓ_1, ℓ_2, that is, $\mathbb{Q}(g, \ell_1, \ell_2)[\partial]$. A system module is presented via the matrix

$$\begin{pmatrix} \partial^2 + \frac{g}{\ell_1} & 0 & -\frac{g}{\ell_1} \\ 0 & \partial^2 + \frac{g}{\ell_2} & -\frac{g}{\ell_2} \end{pmatrix}.$$

We run the following code in a SINGULAR session.

```
LIB "control.lib";
option(redSB); option(redTail);
```

With the `LIB` command we load the library. The `option` commands tell SINGULAR to compute reduced bases (`option(redSB)`), and also reduce not only leading terms, but any terms in the occurring polynomials (`option(redTail)`).

It is important to mention, that any polynomial computation in SINGULAR requires the definition of a ground ring.

```
ring r1 = (0,g,l1,l2),(d),(c,dp);
module RR = [d^2+g/l1, 0, -g/l1], [0, d^2+g/l2, -g/l2];
```

The ring we set bears the name `r1`, it has $\mathbb{Q}(g, \ell_1, \ell_2)$ as the ground field (0 stands for the characteristic of a field, `g, l1, l2` is a list of names for parameters), and the only variable `d`. The last comma-separated block describes the monomial module ordering on `r1`. In this case `(c,dp)` means the following. We use the descending ordering `c` on the module components and the degree reverse lexicographical ordering `dp` on the monomials in the same component.

```
module R = transpose(RR);
list L = canonize(control(R));
L;
```

We have to transpose the module R, because SINGULAR takes the columns of a given matrix presentation as the generators of a module. Here is the output of SINGULAR:

```
[1]:
   number of first nonzero Ext:
[2]:
   -1
[3]:
   strongly controllable(flat), image representation:
[4]:
   _[1]=[(-g*l2)*d^2+(-g^2),(-g*l1)*d^2+(-g^2),
            (-l1*l2)*d^4+(-g*l1-g*l2)*d^2+(-g^2)]
[5]:
   left inverse to image representation:
[6]:
   _[1,1]=(-l1)/(g^2*l1-g^2*l2)
   _[1,2]=(l2)/(g^2*l1-g^2*l2)
   _[1,3]=0
[7]:
   dimension of the system:
[8]:
   1
[9]:
   Parameter constellations which might lead to a non-controllable
   system:
[10]:
   [1]:
      g
   [2]:
      l1-l2
```

As one can see, in the output of the procedure we provide both textual comments on the properties of a system and the corresponding data. The heuristics says that the modules $\mathrm{Ext}^i_{r1}(R, r1)$ of a transposed module indeed vanish for $i \geq 1$ (-1 is returned in this situation). Hence, the system is generically controllable (the notion of strong controllability from above coincides with classical controllability for systems of ordinary differential equations). Moreover, the procedure computes the image representation, left inverse to the image representation and the dimension of the system. The 10-th item is the output of the procedure `genericity`, that is, a list of strings. The polynomial obstruction to genericity in this example is $\ell_1 - \ell_2$. The monomial obstruction g is not physically admissible.

Let us analyze the properties of the system in the non-generic case $\ell_1 = \ell_2 = \ell$. We do this with the help of the following code in the same SINGULAR session:

```
ring r2 = (0,g,l),(d),(c,dp);
module RR = [d^2+g/l, 0, -g/l], [0, d^2+g/l, -g/l];
module R = transpose(RR);
list L = canonize(control(R));
L;
```

We get the following output:

```
[1]:
   number of first nonzero Ext:
[2]:
   1
[3]:
   not controllable , image representation for controllable part:
[4]:
   _[1]=[(g),(g),(l)*d2+(g)]
[5]:
   kernel representation for controllable part:
[6]:
   _[1]=[0,1]
   _[2]=[1]
[7]:
   obstruction to controllability
[8]:
   _[1]=[0,1]
   _[2]=[(-l)*d2+(-g)]
[9]:
   annihilator of torsion module (of obstruction to
   controllability)
[10]:
   _[1]=(-l)*d2+(-g)
[11]:
   dimension of the system:
[12]:
   1
```

We see that the system is not controllable, since it contains a torsion submodule annihilated by $\langle \ell\partial^2 + g \rangle$. However, we give both image and kernel representations for the controllable part of the system and describe the obstruction to controllability explicitly. Now, we are interested in the autonomy analysis of this non-controllable system, what can be achieved with the following code:

```
list A = canonize(autonom(R));
A;
```

This gives us the following output:

```
[1]:
   number of first nonzero Ext:
[2]:
   0
[3]:
   not autonomous
[4]:
   kernel representation for controllable part
[5]:
   _[1]=[0,1]
   _[2]=[(-1)*d2+(-g),-1]
   _[3]=[(g)]
[6]:
   column rank of the matrix
[7]:
   2
[8]:
   dimension of the system:
[9]:
   1
```

Since the 0-th Ext module of the system module RR (in other words, $\mathrm{Hom}_{r2}(RR, r2)$) does not vanish, the system is not autonomous. In addition, we compute a kernel representation for the controllable part, the column rank of the presentation matrix and the dimension of the system.

Parametric systems quite often are generically controllable and contain an autonomous subsystem for some special values of parameters. In the following example, we show that also a generically autonomous system might be controllable in a non-generic case.

Example 3.2 Let $R = \mathbb{K}(a, b)[\partial]$ be a ring. A module $N = R/\langle a\partial + b\rangle$ is generically autonomous. However,

1. if $a = 0, b \neq 0$, then $M = 0$ and thus M is autonomous,
2. if $a = 0, b = 0$, then M is free of rank 1 and hence M is controllable.

A general system might specialize to controllable and autonomous system in non-generic cases, as the next example shows.

Example 3.3 Let $R = \mathbb{K}(a, b)[\partial]$ be a ring. Consider a module

$$M = R^2 / \begin{pmatrix} 0 & 0 \\ 0 & a\partial + b \end{pmatrix}.$$

Generically, it is neither controllable nor autonomous, the annihilator of a torsion submodule is $\langle a\partial + b\rangle$.

The stratification of M with respect to parameters looks as follows:

1. if $a \neq 0, b = 0$, a torsion submodule of M is annihilated by $\langle \partial \rangle$,
2. if $a = 0, b \neq 0$, then M is free of rank 1,
3. if $a = 0, b = 0$, then M is free of rank 2.

Assume, that $a, b \in \mathcal{D} \supseteq \mathbb{K}$. Then the space of parameters $\mathcal{D}^2$ decomposes into a direct sum of subspaces $G \uplus E_1 \uplus E_2 \uplus E_3$, where $G = \{(a,b) \mid a \neq 0, b \neq 0\}$, $E_1 = \{(a,b) \mid a = 0, b \neq 0\}$, $E_2 = \{(a,b) \mid a \neq 0, b = 0\}$ and $E_3 = \{(a,b) \mid a = 0, b = 0\}$. Denote by $\bar{E}$ the closure of E, then $\dim \bar{E}_1 = \dim \bar{E}_2 = 1$, $\dim \bar{E}_3 = 0$ in $\mathcal{D}^2$. Hence, all $\bar{E}_i$ have measure 0 and $\bar{G} = \mathcal{D}^2$ has measure 1.

Remark 3.4 There are packages like D-modules for MACAULAY2, [16], and ORE-MODULES for MAPLE, [3], which have a functionality to treat some of the problems above. The latter package provides the possibility to reveal dangerous parametric denominators via the computation of generalized inverse.

4 Example: Two pendula, mounted on a cart

Consider the Example 5.2.28 from [12] (see also the examples and solutions to them in [3]) describing two pendula, mounted on a cart.

In this example, m_i is the mass and L_i is the length of the i-th pendula. Respectively, k_i and d_i are the coefficients, characterizing the friction at the joints of pendula. M_0 denotes the mass of the cart and g is a gravitational constant. All these parameters can take only non-negative values.

Let us denote $z_i := k_i - m_i L_i g$ for $i = 1, 2$. Then the presentation matrix for a system module is constituted by the raws of the following matrix

$$\begin{pmatrix} m_1 L_1 \partial^2 & m_2 L_2 \partial^2 & (m_1 + m_2 + M_0)\partial^2 & -1 \\ m_1 L_1^2 \partial^2 + d_1 \partial + z_1 & 0 & m_1 L_1 \partial^2 & 0 \\ 0 & m_2 L_2^2 \partial^2 + d_2 \partial + z_2 & m_2 L_2 \partial^2 & 0 \end{pmatrix}.$$

We take the transposed module of the matrix. It is convenient to consider the columns of the matrix above as the generators of submodule of a free module. Since the last generator then is just $(-1, 0, 0)^T$, we perform reduction and simplification of first components with respect to this generator. In such a way we obtain much easier presentation matrix.

4.1 Negligible friction

Let us assume, that the friction is negligible (that is, $d_i = 0$ and $k_i = 0$). We get the simplified presentation matrix of the transposed module as follows:

$$\begin{pmatrix} L_1\partial^2 - g & 0 & m_1L_1\partial^2 \\ 0 & L_2\partial^2 - g & m_2L_2\partial^2 \end{pmatrix}.$$

The generic reduced minimal Gröbner basis is the 2×2 identity matrix. With the LIFT algorithm we obtain the transformation matrix

$$\begin{pmatrix} \frac{L_1L_2}{g^2L_1 - g^2L_2}\partial^2 - \frac{1}{g} & -\frac{m_1L_1L_2}{g^2m_2L_1 - g^2m_2L_2}\partial^2 \\ \frac{m_2L_1L_2}{g^2m_1L_1 - g^2m_1L_2}\partial^2 & -\frac{L_1L_2}{g^2L_1 - g^2L_2}\partial^2 - \frac{1}{g} \\ -\frac{L_1L_2}{g^2m_1(L_1 - L_2)}\partial^2 + \frac{L_1}{gm_1(L_1 - L_2)} & \frac{L_1L_2}{g^2m_2(L_1 - L_2)}\partial^2 - \frac{L_2}{gm_2(L_1 - L_2)} \end{pmatrix}.$$

Collecting the denominators, we can see that their lcm is $m_1m_2g^2(L_1 - L_2)$. Since m_i and g are strictly positive, the only obstruction to genericity appears when $L_1 - L_2 = 0$.

Indeed, in the case $L_1 = L_2 = L$ the generic Gröbner basis is

$$\begin{pmatrix} 0 & 1 \\ L\partial^2 - g & \frac{m_2}{m_1} \end{pmatrix},$$

hence the system is not controllable. The torsion submodule is annihilated by the ideal $\langle L\partial^2 - g \rangle$, but the system is not completely autonomous.

4.2 Essential friction

Now, all the parameters are strictly positive. The simplified presentation matrix of the transposed module is the following

$$\begin{pmatrix} L_1\partial^2 + d_1'\partial + z_1' & 0 & m_1L_1\partial^2 \\ 0 & L_2\partial^2 + d_2'\partial + z_2' & m_2L_2\partial^2 \end{pmatrix}$$

where $z_i' := \frac{z_i}{m_iL_i} = \frac{k_i}{m_iL_i} - g$ and $d_i' := \frac{d_i}{m_iL_i}$ for $i = 1, 2$.

The generic reduced minimal Gröbner basis is the 2×2 identity matrix. The output of the Algorithm 2.7 delivers the list of three polynomials $\{z_1', z_2', P\}$, where

$$P = L_1^2 {z'}_2^2 - 2L_1L_2z_1'z_2' - L_1d_1'd_2'z_2' + L_1{d'}_2^2 z_1' + L_2^2{z'}_1^2 + L_2{d'}_1^2 z_2' - L_2d_1'd_2'z_1'.$$

$z_i' = 0$ means, that $k_i = m_iL_ig$. This is physically admissible situation. Let us analyze P for the admissibility. Indeed, P is irreducible but it has a special form, namely

$$P = (L_2z_1' - L_1z_2')^2 + (L_2d_1' - L_1d_2') \cdot (d_1'z_2' - d_2'z_1'). \tag{4.1}$$

In particular, P vanishes if both z_1' and z_2' do, so P is admissible.

The stratification consists of 6 cases, namely

1. $k_1 = m_1 L_1 g, k_2 = m_2 L_2 g, P = 0$
2. $k_1 = m_1 L_1 g, k_2 \neq m_2 L_2 g, P = 0$
3. $k_1 = m_1 L_1 g, k_2 \neq m_2 L_2 g, P \neq 0$
4. $k_1 \neq m_1 L_1 g, k_2 = m_2 L_2 g, P = 0$
5. $k_1 \neq m_1 L_1 g, k_2 = m_2 L_2 g, P \neq 0$
6. $k_1 \neq m_1 L_1 g, k_2 \neq m_2 L_2 g, P = 0$

The setup for SINGULAR treatment of the cases is the following:

```
LIB "control.lib";
ring T = (0,g),(m1,m2,L1,L2,d1,d2,k1,k2),dp;
poly P = k1^2*L2^4*m2^2-2*k1*k2*L1^2*L2^2*m1*m2-k1*d1*d2*L2^2*m2+
k1*d2^2*L1^2*m1+2*k1*g*L1^2*L2^3*m1*m2^2-2*k1*g*L1*L2^4*m1*m2^2+
k2^2*L1^4*m1^2+k2*d1^2*L2^2*m2-k2*d1*d2*L1^2*m1-
2*k2*g*L1^4*L2*m1^2*m2+2*k2*g*L1^3*L2^2*m1^2*m2-d1^2*g*L2^3*m2^2+
d1*d2*g*L1^2*L2*m1*m2+d1*d2*g*L1*L2^2*m1*m2-d2^2*g*L1^3*m1^2+
g^2*L1^4*L2^2*m1^2*m2^2-2*g^2*L1^3*L2^3*m1^2*m2^2+
g^2*L1^2*L2^4*m1^2*m2^2;
poly z1 = k1 - m1*L1*g;
poly z2 = k2 - m2*L2*g;
```

In particular, we can see the expression for P in terms of original variables. The name of a ring, where the interesting parameters live as polynomials, is T. In SINGULAR, we can switch between different rings and also map objects.

Case 1) $k_1 = m_1 L_1 g, k_2 = m_2 L_2 g, P = 0$

Note that these three equations describe an algebraic variety, that is a closed set. The Gröbner basis of the ideal $k_1 - m_1 L_1 g, k_2 - m_2 L_2 g, P$ is $k_1 - m_1 L_1 g, k_2 - m_2 L_2 g$, since P vanishes, when both $k_1 = m_1 L_1 g$ and $k_2 = m_2 L_2 g$. Hence, it suffices to plug the values for k_i in the corresponding system. For this, we run the following code:

```
ring r1 = (0,g,m1,m2,L1,L2,d1,d2,k1,k2),(d),(c,dp);
poly z1 = 0; poly z2 = 0;
module RR =
          [m1*L1^2*d^2+d1*d+z1, 0, m1*L1*d^2],
          [0, m2*L2^2*d^2+d2*d+z2, m2*L2*d^2];
module R = transpose(RR);
list LC = canonize(control(R));
list LA = canonize(autonom(R));
```

From the output of `control` and `autonom` procedures, we conclude, that this system is neither controllable nor autonomous. In particular, the torsion submodule is annihilated by $\langle \partial \rangle$.

Case 2) $k_1 = m_1 L_1 g, k_2 \neq m_2 L_2 g, P = 0$

Here we deal with the locally closed set $V(\langle k_1 - m_1 L_1 g, P \rangle) \setminus V(\langle k_2 - m_2 L_2 g \rangle)$. Using the following code, we get its better description. We employ a technical trick by modifying a ground ring in such a way, that k_i have priority over the rest of polynomials. In such a way during the computations the relation $k_1 = m_1 L_1 g$ will be used as replacing k_1 with $m_1 L_1 g$. This is achieved by using a different ordering like e. g. the **elimination** ordering (see e. g. [4]) for k_1, k_2.

```
ring T2 = (0,g),(k1,k2,m1,m2,L1,L2,d1,d2),(a(1,1),dp);
poly z1 = ...; poly z2 = ...; poly P = ...; // from above
ideal I2 = P,z1;
I2 = groebner(I2);
facstd(I2,z2);
```

The output of `facstd` command gives us the only component

```
[1]:
   _[1]=k1+(-g)*m1*L1
   _[2]=k2*m1^2*L1^4+(-g)*m1^2*m2*L1^4*L2+m2*L2^2*d1^2
        -m1*L1^2*d1*d2
```

We are able to extract e. g. k_2 from the last equation explicitly:

$$k_2 = m_2 L_2 g + \frac{m_1 L_1^2 d_2 - m_2 L_2^2 d_1}{m_1^2 L_1^4} d_1.$$

Alternatively, we can express d_2 in terms of variables m_i, L_i, k_2, d_1.

Computing with substitutions, we see that this system is neither controllable nor autonomous. The torsion submodule is annihilated by $\langle m_1 L_1^2 \partial^2 + d_1 \partial \rangle$.

Case 3) $k_1 = m_1 L_1 g, k_2 \neq m_2 L_2 g, P \neq 0$

We use the computations of the case 2 and describe a locally closed set via the following system of equations and inequations:

$$k_1 = m_1 L_1 g, \quad k_2 - m_2 L_2 g \neq 0, \quad k_2 - m_2 L_2 g \neq \frac{m_1 L_1^2 d_2 - m_2 L_2^2 d_1}{m_1^2 L_1^4} d_1.$$

In order to treat both inequations involving $k_2 - m_2 L_2 g$, we introduce a new parameter u (thus, u is mutually non-zero in the ground field) and plug in the transposed system module the fake equation $k_2 - m_2 L_2 g = u$.

Also this system is generically neither controllable nor autonomous. The torsion submodule is annihilated by $\langle \partial \rangle$. Compare with the annihilator for the case 2, which is $\langle m_1 L_1^2 \partial^2 + d_1 \partial \rangle$. Let us investigate, for which u the properties change.

```
LIB "control.lib";
ring r3 = (0,g,m1,m2,L1,L2,d1,d2,k1,k2,u),(d),(c,dp);
poly z1 = 0; poly z2 = u;
module RR =
                [m1*L1^2*d^2+d1*d+z1, 0, m1*L1*d^2],
                [0, m2*L2^2*d^2+d2*d+z2, m2*L2*d^2];
module R = transpose(RR);
module S = groebner(R);
matrix T = lift(R,S);
genericity(T);
```

The output of `genericity` delivers

```
[1]:
   u,m2,L2,d1
[2]:
   m1^2*L1^4*u-m1*L1^2*d1*d2+m2*L2^2*d1^2
```

That is, the generic annihilator of a torsion submodule of the system subject to constraints $k_1 - m_1L_1g = 0, k_2 - m_2L_2g = u \neq 0$ is indeed $\langle\partial\rangle$. However, if $u = k_2 - m_2L_2g = \frac{m_1L_1^2d_2 - m_2L_2^2d_1}{m_1^2L_1^4}d_1$, the non-generic annihilator equals $\langle m_1L_1^2\partial^2 + d_1\partial\rangle$. This illustrates the difference between two components, corresponding to cases 2 and 3.

Case 4) $k_1 \neq m_1L_1g, k_2 = m_2L_2g, P = 0$

and

Case 5) $k_1 \neq m_1L_1g, k_2 = m_2L_2g, P \neq 0$

The simplified presentation matrix for the transposed module is symmetric, that is, exchanging $m_1 \leftrightarrow m_2$, $L_1 \leftrightarrow L_2$, $d_1 \leftrightarrow d_2$ and $k_1 \leftrightarrow k_2$ simultaneously does not change the matrix. Hence, we can take the results of case 2 respectively case 3, exchange the variables and get the results for case 4 respectively case 5.

Case 6) $k_1 \neq m_1L_1g, k_2 \neq m_2L_2g, P = 0$

Recall the special structure of a polynomial P in (4.1). It is easy to see, that if $P = 0$ and one of the two summands of P is zero, so does the other. This observation lead us to the first conclusion:

$$P = 0, \text{ if } \frac{L_2}{L_1} = \frac{z_2'}{z_1'} = \frac{d_2'}{d_1'}.$$

Going back to the original variables, it translates into

$$\frac{m_2L_2^2}{m_1L_1^2} = \frac{k_2 - m_2L_2g}{k_1 - m_1L_1g} = \frac{d_2}{d_1}. \tag{4.2}$$

This is especially interesting, since the values, found in [12] for showing the non-generic non-controllability, were $m_1 = m_2 = M_0 = 1$, $L_1 = L_2 = 1$, $d_1 = d_2 = 1$ and $k_1 = k_2 = k$. As we can see, it suffices to set $m_1 = m_2$, $d_1 = d_2$, $k_1 = k_2 \neq m_2L_2g$ and $L_1 = L_2$ for illustrating this phenomenon.

Let us denote by a parameter t the value of the fractions in (4.2). Then,

$$d_2 = t \cdot d_1, \; k_2 = t \cdot k_1 + (m_2L_2 - t \cdot m_1L_1)g, \; m_2L_2^2 = t \cdot m_1L_1^2.$$

We do the substitutions for d_2 and k_2. As a preprocessing before Gröbner bases, we can manipulate the generators. Consider the last generator of a transposed module, that is, the last column of the transposed presentation matrix $(m_1L_1\partial^2, m_2L_2\partial^2)^T$. By multiplying the column with L_2, we can simplify it subject to the substitution to the column $(L_2\partial^2, tL_1\partial^2)^T$. The second generator of the module becomes then $(0, t \cdot (m_1 L_1^2\partial^2 + d_1\partial + z_1))^T$, from which we cancel the parameter t out.

With the following code we perform the controllability and the autonomy analysis for this particular case.

```
ring r6 = (0,g,t,m1,L1,L2,d1,k1),(d),(c,dp);
poly z1 = k1 - m1*L1*g;
module RR =
          [m1*L1^2*d^2+d1*d+z1, 0, L2*d^2],
          [0, m1*L1^2*d^2+d1*d+z1, t*L1*d^2];
module R = transpose(RR);
print(R);
list LC = canonize(control(R));
list LA = canonize(autonom(R));
```

We conclude that this system is neither controllable nor autonomous. In particular, the annihilator of the torsion submodule is the ideal $\langle m_1 L_1^2\partial^2 + d_1\partial + k_1 - m_1 L_1 g\rangle$. Note that in view of the equation (4.2), we obtain the equivalent symmetric annihilator $\langle m_2 L_2^2\partial^2 + d_2\partial + k_2 - m_2 L_2 g\rangle$ by e. g. multiplying the previous annihilator with the constant t.

Now let us assume, that $P = 0$ but neither of its summands vanishes. The polynomial P is quadratic with respect to any of the variables $m_1, m_2, k_1, k_2, d_1, d_2$ and is quartic with respect to L_1 and L_2. Let us fix one of the variables $m_1, m_2, k_1, k_2, d_1, d_2$. Consider the rest of variables as parameters and compute the discriminant of a corresponding quadratic equation. Since the involved variables might take only positive real values, we obtain a condition on the discriminant of a quadratic equation. If we fix m_1, k_1 or d_1, we get

$$\frac{d_2^2}{4m_2 L_2^2} \geq k_2 - m_2 L_2 g. \tag{4.3}$$

For fixed m_2, k_2 or d_2, we obtain, either by a direct computation or via the symmetry, that

$$\frac{d_1^2}{4m_1 L_1^2} \geq k_1 - m_1 L_1 g.$$

Notably, both inequalities cannot become equalities simultaneously.

Provided (4.3), the explicit solution with respect to, say, d_1 gives the following expression (recall, we use the short notation $z_i = k_i - m_i L_i g$):

$$d_1 = \frac{d_2(m_2 L_2^2 z_1 + m_1 L_1^2 z_2) \pm (m_2 L_2^2 z_1 - m_1 L_1^2 z_2)\sqrt{d_2^2 - 4m_2 L_2^2 z_2}}{2m_2 L_2^2 z_2}.$$

Each root corresponds to a separate system. Substituting the roots into our system, we obtain, that as in all previous cases, it is neither controllable nor autonomous. The annihilators of torsion submodules are then $\langle 2m_2 L_2^2\partial + d_2 \pm \sqrt{d_2^2 - 4m_2 L_2^2 z_2}\rangle$. The annihilators with respect to m_1, d_1, L_1, z_1 we obtain by the symmetry.

Finally, we summarize the obtained results.

Proposition 4.1 *The complete stratification of the obstructions to genericity for the generically controllable system with the essential friction is obtained. All the components of the stratification correspond to non-controllable and non-autonomous systems, whose torsion submodules are annihilated by one of the ideals (for $i = 1, 2$)*

$$\langle m_i L_i^2 \partial^2 + d_i \partial \rangle, \quad \langle m_i L_i^2 \partial^2 + d_i \partial + k_i - m_i L_i g \rangle, \quad \langle \partial \rangle,$$

and $\langle 2 m_i L_i^2 \partial + d_i \pm \sqrt{d_i^2 - 4 m_i L_i^2 (k_i - m_i L_i g)} \rangle$, *provided* $k_i \leq \frac{d_i^2}{4 m_i L_i^2} + m_i L_i g$.

5 Conclusion and future work

We have investigated the parameter-dependence of structural properties (such as torsion-freeness) of modules over certain algebras over $\mathbb{K}(p_1, \ldots, p_t)$, where $\mathbb{K}$ is a ground field and p_i are parameters. The central idea is to keep track of all polynomial expressions in the p_i that occur as denominators during Gröbner basis computation. These problems have practical applications in control theory as outlined in the Introduction. We have shown several nontrivial phenomena that arise with these questions in terms of illustrative worked examples. Our goal for the future is to extend this approach to the study of more general parametric module properties, leading to the implementation of systematic procedures for such problems.

In particular, one is interested in working with parameters, on which the involved operators act nontrivially. That is, the parameters may correspond to (q-)differentiable and/or (q-)shiftable functions. Then, the field $\mathbb{K}(p_1, \ldots, p_t)$ must be a differential and/or a difference field. The obstructions to genericity are then presented as systems of differential-difference algebraic equations (DDAE) instead of just algebraic equations treated in this article. Though the main principles remain the same, there is a strong need for specialized techniques and systematic computer-algebraic support for both theoretical and implementational parts of the further research in this area. The case of differentiable parameters was treated in the articles [14, 15], the software package OREMODULES [3] seems to be able to provide computational support for this case.

Yet another important direction of investigation is the analysis of numerical phenomena, namely inexact computations with parameters defined as floating point numbers or as certain inequalities. The generalization of our approach to these domains seems to be possible with the help of e. g. cylindrical algebraic decomposition techniques. Alternatively, one may first obtain an exact symbolic solution to parametric problem, say, in form of the complete stratification, and postprocess it with numerical or symbolical-numerical tools.

Acknowledgements. The authors are grateful to A. Quadrat and J.-F. Pommaret for discussions on theoretical methods as well as particular applications.

Bibliography

[1] M. Becker, V. Levandovskyy, and O. Yena, *A* SINGULAR *2.0 library for computations in control theory* `control.lib`, 2003, Available from `http://www.singular.uni-kl.de`.

[2] F. Chyzak, A. Quadrat, and D. Robertz, *Effective algorithms for parametrizing linear control systems over Ore algebras*, Applicable Algebra in Engineering, Communications and Computing 16 (2005), pp. 319–376.

[3] ———, OREMODULES*: A symbolic package for the study of multidimensional linear systems*, Applications of Time-Delay Systems (J. Chiasson and J.-J. Loiseau, eds.), Springer LNCIS 352, 2007, Available from `http://wwwb.math.rwth-aachen.de/OreModules/`, pp. 233–264.

[4] G.-M. Greuel and G. Pfister with contributions by O. Bachmann, C. Lossen, and H. Schönemann, *A SINGULAR Introduction to Commutative Algebra*, Springer, 2002.

[5] G.-M. Greuel, G. Pfister, and H. Schönemann, SINGULAR *3.0. A Computer Algebra System for Polynomial Computations. Centre for Computer Algebra, University of Kaiserslautern*, 2005, Available from `http://www.singular.uni-kl.de`.

[6] H. Kredel, *Solvable polynomial rings*, Shaker, 1993.

[7] V. Levandovskyy, *Non-commutative Computer Algebra for polynomial algebras: Gröbner bases, applications and implementation*, Doctoral Thesis, Universität Kaiserslautern (2005), Available from `http://kluedo.ub.uni-kl.de/volltexte/2005/1883/`.

[8] V. Levandovskyy and H. Schönemann, *Plural – a computer algebra system for noncommutative polynomial algebras*, Proceedings of the International Symposium on Symbolic and Algebraic Computation (ISSAC'03), ACM Press, 2003.

[9] V. Levandovskyy and E. Zerz, *Computer algebraic methods for the structural analysis of linear control systems*, Proceedings in Applied Mathematics and Mechanics (PAMM) 5 (2005), pp. 717–718, DOI: 10.1002/pamm.200510333.

[10] ———, *Algebraic systems theory and computer algebraic methods for some classes of linear control systems*, Proceedings of the International Symposium on Mathematical Theory of Networks and Systems (MTNS'06), 2006, pp. 536–541.

[11] A. Leykin, *Constructibility of the Set of Polynomials with a Fixed Bernstein-Sato Polynomial: an Algorithmic Approach*, Journal of Symbolic Computation 32 (2001), pp. 663–675, DOI: 10.1006/jsco.2001.0488.

[12] J. Polderman and J. Willems, *Introduction to mathematical systems theory. A behavioral approach*, Texts in Applied Mathematics, vol. 26, Springer New York, 1997.

[13] J.-F. Pommaret, *Partial differential control theory*, Mathematics and its Applications 530, Dordrecht: Kluwer Academic Publishers, 2001.

[14] J.-F. Pommaret and A. Quadrat, *Formal obstructions to the controllability of partial differential control systems*, Proceedings of the 15th IMACS World Congress, Berlin (Germany), vol. 5, 1997, pp. 209–214.

[15] ———, *Formal elimination for multidemsional systems and application to control theory*, Mathematics of Control, Signals and Systems 13 (2000), pp. 193–215.

[16] H. Tsai and A. Leykin, *D-modules package for* MACAULAY *2 – algorithms for D-modules*, 2006, Available from `http://www.ima.umn.edu/~leykin/Dmodules/`.

[17] U. Walther, *Cohomology, stratifications and parametric Gröbner bases in characteristic zero*, Journal of Symbolic Computation 35 (2003), pp. 527–542.

[18] V. Weispfenning, *Canonical comprehensive Gröbner bases*, Journal of Symbolic Computation 36 (2003), pp. 669–683.

Author information

Viktor Levandovskyy, Lehrstuhl D für Mathematik,
RWTH Aachen,
Templergraben 64, D-52062 Aachen, Germany.
Email: `viktor.levandovskyy@math.rwth-aachen.de`

Eva Zerz, Lehrstuhl D für Mathematik,
RWTH Aachen,
Templergraben 64, D-52062 Aachen, Germany.
Email: `eva.zerz@math.rwth-aachen.de`

Radon Series Comp. Appl. Math 3, 151–190

A survey of (BIBO) stability and (proper) stabilization of multidimensional input/output systems

Ulrich Oberst and Martin Scheicher

Key words. Stability, BIBO stability, stabilization, proper stabilization, multidimensional system, multidimensional behavior, input/output system.

AMS classification. 93D15, 93D25, 93C20, 93B52, 35B35

1 Introduction

Stability and stabilization theory is a part of control theory and usually involves the following

Ingredients. [22, p. 60] 1. *Stability*: Select the class of admissible systems and define and characterize the stable systems in this class and, in particular, check BIBO (bounded input/bounded output) stability of stable systems. 2. *Stabilizability*: Which admissible systems can be stabilized by output feedback? 3. *Stabilization*: Construct a stabilizing compensator for a given stabilizable system. 4. *Parametrization*: Classify or construct all stabilizing compensators for a given stabilizable system.

First author: Support by the Austrian FWF through project P18974 is gratefully acknowledged.
Second author: See first author.

In this survey we discuss these problems for continuous and discrete multidimensional *input/output (IO) behaviors* which are described by linear systems of complex partial differential equations on $\mathbb{R}^r$ resp. difference equations on $\mathbb{N}^r$ with constant coefficients, the signals being taken from various function spaces, in particular from those of polynomial-exponential functions. After an introduction into multidimensional behavioral systems theory we give a survey without detailed proofs, but sufficient motivations of [26] and of our talks at the conferences D2 and D3 of the Gröbner semester 2006 in Linz and also present new results on BIBO stability. Complete proofs of the new results on BIBO stability and proper stabilization will appear in [38] and [39].

J. W. Polderman and J. C. Willems in [31, section 10.8] and P. Rocha [36] suggest to renounce the input/output structure and output feedback in favor of more general behavior interconnections. The first stabilization results for multidimensional systems in this generality are due to S. Shankar [41, 42].

In contrast to our approach most multidimensional stability and stabilization papers use a commutative integral domain $\mathcal{S}$ of 'SISO stable plants' and describe the admissible systems, often called plants, by a *transfer operator or input/output map* which is a matrix with coefficients in the quotient field of $\mathcal{S}$. These systems are called structurally [16] or internally [33] stable if their transfer matrix has entries in $\mathcal{S}$. In the one-dimensional case this approach to stabilization theory is due to C. A. Desoer, V. Kucera, M. Vidyasagar, D. C. Youla, and their coworkers. Among the well-known contributors to multidimensional stability and stabilization are M. Bisiacco [2], N. K. Bose [6, 9, 10, 7, 8], D. E. Dudgeon [11], E. Fornasini [13], J. P. Guiver [16], E. I. Jury [21], Z. Lin [22, 23], G. Marchesini [13], R. M. Mersereau [11], A. Quadrat [32, 33, 34], H. C. Reddy [35], S. Shankar [41, 42, 43], V. R. Sule [45], J. Wood [51], L. Xu [52], J. Q. Ying [53], E. Zerz [54] et al. We refer the reader especially to the interesting surveys [6, 7, 8] by N. K. Bose and [21] by E. I. Jury with detailed descriptions of the results, problems, history and further important contributors of the subject and with comprehensive lists of references.

An IO behavior $\mathcal{B}$ gives rise to its autonomous part $\mathcal{B}^0$, its transfer matrix H and to the characteristic varieties $\mathrm{char}(\mathcal{B}) \subseteq \mathrm{char}(\mathcal{B}^0)$ of $\mathcal{B}$ and of $\mathcal{B}^0$. The entries of the transfer matrix are complex rational functions in r indeterminates s_ρ, i. e., are contained in the quotient field $\mathbb{C}(s)$ of the polynomial algebra $A := \mathbb{C}[s] = \mathbb{C}[s_1, \dots, s_r]$. In the one-dimensional theory the elements of $\mathrm{char}(\mathcal{B}^0)$ are called the poles, modes, characteristic values or natural frequencies of the system.

Stability and stabilization of an IO system are defined with respect to a disjoint decomposition $\mathbb{C}^r = \Lambda_1 \uplus \Lambda_2$ of the complex space into a stable region Λ_1 and an unstable region Λ_2, the standard continuous resp. discrete cases being

$$\Lambda_{2,\mathrm{cont}} = \overline{\mathbb{C}_+}^r,\ \overline{\mathbb{C}_+} := \{z \in \mathbb{C};\ \Re(z) \geq 0\} \text{ resp. } \Lambda_{2,\mathrm{dis}} = \{z \in \mathbb{C};\ \mid z \mid \geq 1\}^r. \quad (1.1)$$

A polynomial $t \in A$ is called *stable* if its zeros lie in the stable region, i. e., if it belongs to $T := \{t \in A;\ \forall \lambda \in \Lambda_2 :\ t(\lambda) \neq 0\}$ [45]. A rational function is called stable if its poles lie in the stable region, i. e., if it belongs to the quotient ring $A_T \subseteq \mathbb{C}(s)$ of 'SISO stable plants'. An IO system $\mathcal{B}$ is called stable if the characteristic variety of its autonomous part is contained in the stable region or, equivalently, if all polynomial-exponential trajectories in $\mathcal{B}^0$ involve exponents in the stable region only (Theorem 6.2, 1.(i),(ii)). In [51] this is called a CV-condition and used to define stable autonomous systems. Stability of $\mathcal{B}$ in this sense is equivalent with the stability of the transfer matrix and an additional condition (Theorem 6.2, 1.(iii)). A one-dimensional IO system is stable with respect to (1.1) if and only if its autonomous part is asymptotically stable. A stable IO system is *input/output stable* in the sense that the transfer matrix acts as operator on interesting classes of inputs and generates outputs of the same type as shown in parts 2, 3 and 4 of Theorem 6.2, but BIBO stability does not hold in general.

A rational function in $\mathbb{C}(s)$ is called *proper* if it is a formal and then automatically a convergent power series in $s_1^{-1}, \dots, s_r^{-1}$. Properness of (the entries of) the transfer matrix H is not generally assumed in [26] or in the present survey since standard transfer functions like $(s_1 - s_2)^{-1}$, $(s_1^2 - s_2^2)^{-1}$ (wave equation) or $(s_1 - s_2^2)^{-1}$ (heat equation) are not proper. Properness of the transfer function or matrix is, however, usually assumed in the literature and has often the desirable consequence that the transfer matrix can be considered as an operator or input/output map on large classes of inputs and that a proper stable transfer matrix is also BIBO stable. In part of the engineering literature on stability (compare [21]) stability of a proper rational function is *defined* as BIBO stability. We prove in Theorem 10.2 in the standard discrete case of equation (1.1) that proper stable rational functions in our sense coincide precisely with the structurally stable rational functions of the literature [16, 22] and are BIBO stable in particular. In the continuous case of equation (1.1) stable polynomials in our sense are called strict Hurwitz in the literature [21, Definition 5 on p. 140]. E. I. Jury conjectured BIBO stability of the inverse of a bivariate very strict Hurwitz polynomial [21, Remark on p. 143]. We prove this conjecture in Theorem 11.7. The methods of its proof will also be applicable to higher dimensional cases.

An IO system $\mathcal{B}$ is called *stabilizable* if there is a compensator IO system $\mathcal{B}'$ such that the feedback system of $\mathcal{B}$ and $\mathcal{B}'$ is well-posed [49] and stable, and $\mathcal{B}'$ is then called a *stabilizing compensator*. In Theorem 8.1 we characterize stabilizability of $\mathcal{B}$ and construct all stabilizing compensators (parametrization). In particular, a controllable IO system is stabilizable if and only if its transfer matrix is stabilizable in the usual sense [33]. The famous prototype of the parametrization in the one-dimensional case is that of V. Kucera, D. C. Youla, J. J.Bongiorno and H. A. Jabr and is exposed by M. Vidyasagar in [49, chapter 5]. The proofs employ localization after P. Gabriel as a new mathematical tool in systems theory which is described in B. Stenström's book [44], in [26] and in section 7. At no place the results and proofs need or employ the so-called fractional representation approach, i. e., matrix fraction descriptions of the transfer matrix of various kinds and the, sometimes long [49, 33, 34], ensuing matrix computations, and seem thus simpler and of interest even in the one-dimensional case. The localization technique also avoids the difficulties in [51] with the lack of *ideal-convexity* [43] of the unstable regions Λ_2. Ideal-convexity is characterized by the coincidence of Gabriel localization with the standard localization functor $M \mapsto M_T$ on A-modules M (Theorem 7.2). Algorithm 8.2 discusses tests for stabilizability and the construction of all stabilizing compensators. There remain open problems, however. The difficulties are related to those pointed out in [52].

An IO system is called *proper stabilizable* if it is stabilizable with the additional property that the transfer matrix of the resulting feedback system is proper (and, of course, stable). In Theorem 9.6 we characterize proper stabilizability and in Algorithm 9.7 we describe an algorithmic test for proper stabilizability and construct all proper stabilizing compensators. However, as in Algorithm 8.2, there remain open problems. In the one-dimensional case proper stabilization of transfer matrices is discussed in [49, chapter 5] and of (Rosenbrock) systems in [48, chapter 6].

The table of contents exhibits the plan of the paper in detail.

2 Input/output systems

In engineering mathematics *input/output (IO) systems* are often pictorially represented as

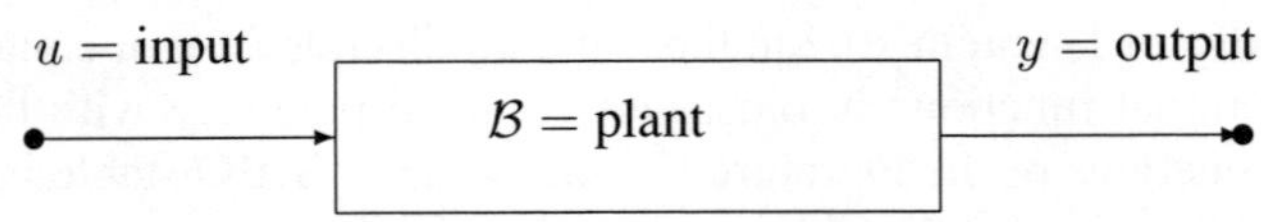

Figure 2.1 IO system

where u resp. y are the *input* resp. the *output signal* of the system and are, in most cases and also in this paper, functions into finite-dimensional complex spaces $\mathbb{C}^m$ resp. $\mathbb{C}^p$. In colloquial terms, the system produces the output y from the input u. In this paper the system is given as a *multidimensional behavior* which will be defined below, hence the letter $\mathcal{B}$. The attribute *plant* refers to a primary system whose behavior is to be steered or controlled.

In general, the assignment $u \mapsto y$ is not a *map (=operator, functional)*, but the system may, for instance, autonomously produce a non-zero output y from a zero input $u = 0$. If this assignment is actually a map $u \mapsto y = H(u)$ then H is called an *input/output map* or *transfer operator*.

In context with stability and stabilization of IO systems by output feedback one has

1. to describe the admissible systems in exact mathematical form, in particular the spaces of admissible signals

2. and to characterize the stable systems among the admissible ones.

As usual we use output feedback systems as in Figure 2.2 with the following interpretation: The plant $\mathcal{B}_1$ is controlled by a *controller or compensator system* $\mathcal{B}_2$ which will often be a computer. For $i = 1, 2$ we assume that the input u_i of $\mathcal{B}_i$ and the output y_{3-i} of $\mathcal{B}_{3-i}$ have the same dimension and can therefore be added. The output y_1 of the plant and a chosen input u_2 of the controller are combined, i. e., summed up, and produce the output y_2 of the controller. This, in turn, is combined with the input u_1 of

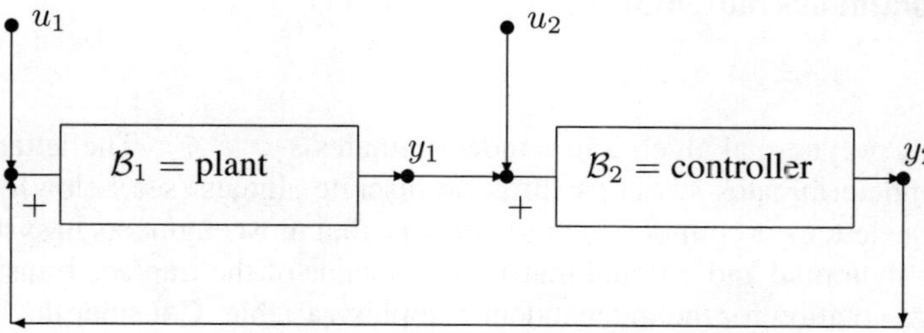

Figure 2.2 Feedback system

the plant and then produces its output y_1. The resulting total system is

$$\mathcal{B} := \text{feedback}(\mathcal{B}_1, \mathcal{B}_2) \quad \text{with input} \begin{pmatrix} u_2 \\ u_1 \end{pmatrix} \quad \text{and output} \begin{pmatrix} y_1 \\ y_2 \end{pmatrix} \tag{2.1}$$

of the same dimension $p + m$. Mathematically the roles of $\mathcal{B}_1$ and $\mathcal{B}_2$ can be interchanged, but in technical realizations there is, of course and in general, a big difference between the primary plant and its controller which is used to steer the plant.

Once the admissible systems and their stability are defined as will be done below we can also define the *stabilization* of a system.

Definition 2.1 An IO systems $\mathcal{B}_1$ is called *stabilizable* if there is another IO system $\mathcal{B}_2$ such that the feedback system $\mathcal{B} := \text{feedback}(\mathcal{B}_1, \mathcal{B}_2)$ is stable. Then $\mathcal{B}_2$ is called a *stabilizing compensator or controller* of $\mathcal{B}_1$. The construction of all stabilizing compensators of one given plant $\mathcal{B}_1$ is usually described by the term *parametrization*. If a stable compensator can be found one talks about *strong stabilization*.

3 Multidimensional signal modules

We are going to discuss systems governed by linear partial differential *resp.* difference *equations with constant coefficients and talk, as usual in engineering mathematics, about the* continuous *resp. the* discrete *case. We start with the description of the continuous signals.*

Consider the simple linear differential system

$$\overset{\bullet}{y} = Ay, \; A \in \mathbb{R}^{n \times n} \; \text{ where } \; y : \mathbb{R} \to \mathbb{R}, \; t \mapsto y(t),$$

with the solution $y(t) = e^{tA}y(0)$. It is a standard result that the stability behavior of this system for $t \to \infty$ is determined by the *complex* eigenvalues of A. In the sequel we therefore consider *complex-valued* functions y.

Let $r \geq 1$. We consider (multidimensional) signals of r independent variables.

Definition 3.1 (Continuous multidimensional signals) Let

$$A := \mathbb{C}[s] := \mathbb{C}[s_1, \ldots, s_r]$$

denote the complex polynomial algebra in r indeterminates $s_1, \ldots, s_r$. The letter s is chosen since the indeterminates s_ρ act as shifts on discrete signals (see below). Another reason why the letter s is common in this setting is, that most engineers in systems theory introduce polynomial and rational matrices by means of the Laplace transform where s is the usual notation for the independent complex variable. Consider the function spaces

$$\begin{aligned} \mathrm{C}^\infty(\mathbb{R}^r) &= \{\text{arbitrarily often differentiable functions}\} \\ &\subset \mathcal{D}' := \mathcal{D}'(\mathbb{R}^r) = \{\text{distributions}\} \end{aligned}$$

of arbitrarily often differentiable, complex-valued functions

$$y = y(t_1, \ldots, t_r),\ t = (t_1, \ldots, t_r) \in \mathbb{R}^r,$$

resp. of distributions. These spaces become $\mathbb{C}[s]$-modules via the action

$$s_\rho \circ y := \partial y / \partial t_\rho$$

by partial differentiation.

In engineering practice C^∞-signals are too restrictive since piecewise continuous signals appear frequently, for instance the one-dimensional Heaviside step function

$$Y(t) := \begin{cases} 1 & \text{if } t \geq 0 \\ 0 & \text{if } t < 0. \end{cases}$$

The space of distributions contains these and is closed under differentiation, i. e., a $\mathbb{C}[s]$−module. These two properties are the reason that distributions, in particular Dirac's $\delta = Y'$ and its derivatives, play such an important part in engineering mathematics.

For the purposes of the present paper the space of *polynomial-exponential* functions is of particular importance.

Lemma and Definition 3.2 (Continuous polynomial-exponential functions) [28, Theorem 6.6] *A distribution y is called* locally finite *if it generates a $\mathbb{C}$-finite dimensional cyclic module $\mathbb{C}[s] \circ y$ consisting of all derivatives of y and their $\mathbb{C}$-linear combinations. The $\mathbb{C}[s]$-submodule of all locally finite distributions is the space of* polynomial exponential *functions*

$$\begin{aligned} \mathcal{F}_{\text{cont}} := \mathcal{D}'_{\text{lf}} := \{y \in \mathcal{D}'(\mathbb{R}^r);\ \dim_{\mathbb{C}}(\mathbb{C}[s] \circ y) < \infty\} = \oplus_{\lambda \in \mathbb{C}^r} \mathbb{C}[t] e^{\lambda \bullet t}, \\ \lambda = (\lambda_1, \ldots, \lambda_r) \in \mathbb{C}^r,\ t = (t_1, \ldots, t_r) \in \mathbb{R}^r, \\ \lambda \bullet t := \lambda_1 t_1 + \ldots + \lambda_r t_r. \end{aligned} \tag{3.1}$$

In particular the locally finite distributions are entire functions.

The preceding $\mathbb{C}[s]$-modules are *injective cogenerators*, see e. g. [37, p. 65ff and p. 79ff] for a broader explanation of these concepts. This fact is of paramount significance for the solution of linear systems of partial differential equations with constant coefficients in the function modules just introduced. To discuss this property in more detail we first introduce some standard objects from polynomial algebra.

Let $\mathrm{Max}(A)$ denote the set of all maximal ideals of the commutative ring A. Hilbert's Nullstellensatz implies the bijection

$$\mathbb{C}^r \cong \mathrm{Max}(\mathbb{C}[s]),\ \lambda \mapsto \mathfrak{m}_\lambda,\ \text{where}$$
$$\mathfrak{m}_\lambda := _{\mathbb{C}[s]}\langle s_1 - \lambda_1, \ldots, s_r - \lambda_r\rangle = \{t \in A;\ t(\lambda) = 0\}.$$

The $\mathbb{C}$-one-dimensional factor modules

$$A/\mathfrak{m}_\lambda \cong \mathbb{C},\ \overline{f} \mapsto f(\lambda),\ \lambda \in \mathbb{C}^r,$$

constitute a representative system of the *irreducible* or *simple* $\mathbb{C}[s]$-modules. Each $\lambda \in \mathbb{C}^r$ gives also rise to the *local ring* of rational functions defined in λ, i. e., to

$$A_{\mathfrak{m}_\lambda} = \left\{\frac{f}{t};\ f, t \in \mathbb{C}[s],\ t(\lambda) \neq 0\right\} \subset \mathbb{C}(s),$$
$$A/\mathfrak{m}_\lambda \cong A_{\mathfrak{m}_\lambda}/(\mathfrak{m}_\lambda)_{\mathfrak{m}_\lambda}.$$

Recall that a $\mathbb{C}[s]$-module $\mathcal{F}$ is *injective* if the contravariant duality functor

$$\mathrm{D} := \mathrm{Hom}_{\mathbb{C}[s]}(-, \mathcal{F}),\ \left(M' \xrightarrow{f} M\right) \mapsto$$
$$\left(\mathrm{D}(M) = \mathrm{Hom}_{\mathbb{C}[s]}(M, \mathcal{F}) \xrightarrow{\mathrm{D}(f)} \mathrm{D}(M'),\ \varphi \mapsto \varphi f\right)$$

is exact on the category of $\mathbb{C}[s]$-modules or transforms monomorphisms into epimorphisms.

Lemma and Definition 3.3 *An injective A-module $\mathcal{F}$ is a cogenerator if it satisfies the following equivalent conditions:*

1. *If M is a non-zero module then so is* $\mathrm{Hom}_{\mathbb{C}[s]}(M, \mathcal{F})$.
2. *The module $\mathcal{F}$ contains all simple modules* $\mathbb{C}[s]/\mathfrak{m}$, $\mathfrak{m} \in \mathrm{Max}(\mathbb{C}[s])$, *up to isomorphism.*
3. *A sequence of A-modules*

$$M' \xrightarrow{f} M \xrightarrow{g} M'' \text{ is exact if and only if the dual sequence}$$
$$\mathrm{D}(M'') \xrightarrow{\mathrm{D}(g)} \mathrm{D}(M) \xrightarrow{\mathrm{D}(f)} \mathrm{D}(M') \quad \text{is exact.}$$

Theorem 3.4 [27, Theorem 2.54] [28, Theorem 6.6] *The three $\mathbb{C}[s]$-modules*

$$\mathcal{F}_{\text{cont}} = \oplus_{\lambda \in \mathbb{C}^r} \mathbb{C}[t] e^{\lambda \bullet t} \subset \mathrm{C}^\infty(\mathbb{R}^r) \subset \mathcal{D}'(\mathbb{R}^r)$$

are injective cogenerators.

The difficult injectivity of the two larger modules was established by L. Ehrenpreis [12], B. Malgrange [25] and V. Palamodov [30] in the beginning 1960s and was called the *fundamental principle* for these (and other) function modules. The result implies the *divisibility* of these modules and therefore the important existence of a *fundamental solution* $y \in \mathcal{D}'$ of $f \circ y = \delta$ for any non-zero polynomial or differential operator f. The injectivity of the module of polynomial-exponential functions can be shown more easily by reduction to the discrete case as explained below.

It is easy to see that $\mathbb{C}e^{\lambda \bullet t}$ and $\mathbb{C}[t]e^{\lambda \bullet t}$ are $\mathbb{C}[s]$-submodules of $\mathcal{D}'$,

$$A/\mathfrak{m}_\lambda \cong \mathbb{C}e^{\lambda \bullet t}, \ \overline{f} \longmapsto f(\lambda)e^{\lambda \bullet t}$$

is a $\mathbb{C}[s]$-isomorphism and the inclusion $\mathbb{C}e^{\lambda \bullet t} \subset \mathbb{C}[t]e^{\lambda \bullet t}$ is essential, i. e.,

$$\forall \text{ non-zero } y \in \mathbb{C}[t]e^{\lambda \bullet t} \ \exists f \in \mathbb{C}[s] \text{ such that } 0 \neq f \circ y \in \mathbb{C}e^{\lambda \bullet t}.$$

Hence for all polynomials $f \in \mathbb{C}[s] \setminus \mathfrak{m}_\lambda$ the maps

$$f\circ : \mathbb{C}e^{\lambda \bullet t} \cong \mathbb{C}e^{\lambda \bullet t} \text{ and } f\circ : \mathbb{C}[t]e^{\lambda \bullet t} \cong \mathbb{C}[t]e^{\lambda \bullet t}$$

are isomorphisms and therefore $\mathbb{C}[t]e^{\lambda \bullet t}$ is an $A_{\mathfrak{m}_\lambda}$-module.

Corollary 3.5 [28] *For each $\lambda \in \mathbb{C}^r$ the module $\mathbb{C}[t]e^{\lambda \bullet t}$, $\lambda \in \mathbb{C}^r$, is injective and indecomposable as A- and as $A_{\mathfrak{m}_\lambda}$-module. More precisely, the modules $\mathcal{F}_{\text{cont}} = \oplus_{\lambda \in \mathbb{C}^r} \mathbb{C}[t]e^{\lambda \bullet t}$ of polynomial-exponential functions resp. $\mathbb{C}[t]e^{\lambda \bullet t}$ are the* minimal injective cogenerators *over A resp. over $A_{\mathfrak{m}_\lambda}$, up to isomorphism. This property will be used frequently later on.*

The following lines contain the analogous definitions for the *discrete case* of partial difference equations.

Lemma and Definition 3.6 (Discrete multidimensional signals) [27, sections 4.12, 4.13] [28, Theorem 1.14] *As signal space in the discrete case we use the space*

$$\mathbb{C}^{\mathbb{N}^r} = \{y = (y_\mu)_{\mu \in \mathbb{N}^r} : \mathbb{N}^r \to \mathbb{C}, \ \mu = (\mu_1, \ldots, \mu_r) \mapsto y(\mu) := y_\mu\}$$

of multisequences or functions y in r independent discrete variables $\mu_\rho \in \mathbb{N}$ with the $\mathbb{C}[s]$-action

$$(s^\mu \circ y)(\nu) := y(\mu + \nu), \ s^\mu := s_1^{\mu_1} * \ldots * s_r^{\mu_r},$$

by left shifts. The $\mathbb{C}[s]$-isomorphism

$$\mathrm{Hom}_{\mathbb{C}[s]}(\mathbb{C}[s], \mathbb{C}) \cong \mathbb{C}^{\mathbb{N}^r}, \ \varphi \longleftrightarrow y, \ y(\mu) = \varphi(s^\mu),$$

easily implies that also $\mathbb{C}^{\mathbb{N}^r}$ is an injective $\mathbb{C}[s]$-cogenerator, and so is the submodule

$$\mathcal{F}_{\text{dis}} := \mathbb{C}_{\text{lf}}^{\mathbb{N}^r} := \left\{ y \in \mathbb{C}^{\mathbb{N}^r}; \ \dim_{\mathbb{C}}(\mathbb{C}[s] \circ y) < \infty \right\}$$

which is again the space of polynomial-exponential sequences.

This definition can be extended to sequences $y = (y_\mu)_{\mu\in\mathbb{Z}^r} \in \mathbb{C}^{\mathbb{Z}^r}$ [27, 29, 11, 47]. However, the stability theory of the present paper has not yet been developed for this case.

The discrete polynomial-exponential sequences have the following form.

Lemma and Definition 3.7 [28, Theorem 1.25] *For $\alpha \in \mathbb{C}$, $k, i \in \mathbb{N}$, $\lambda \in \mathbb{C}^r$, $\mu, \nu \in \mathbb{N}^r$ define $e_{\alpha,k} \in \mathbb{C}^{\mathbb{N}}$ and $e_{\lambda,\mu} \in \mathbb{C}^{\mathbb{N}^r}$ by*

$$e_{\alpha,k}(i) := \begin{cases} \binom{i}{k}\alpha^{i-k} & \text{if } \alpha \neq 0 \\ \delta_{i,k} & \text{if } \alpha = 0 \end{cases}, \quad e_{\lambda,\mu}(\nu) := \prod_{\rho=1}^{r} e_{\lambda_\rho,\mu_\rho}(\nu_\rho).$$

Then

$$(s-\lambda)^\nu \circ e_{\lambda,\mu} = \begin{cases} e_{\lambda,\mu-\nu} & \text{if } \mu \in \nu + \mathbb{N}^r \\ 0 & \text{otherwise} \end{cases}$$

and

$$\mathcal{F}_{\mathrm{dis}} := \mathbb{C}_{\mathrm{lf}}^{\mathbb{N}^r} := \left\{ y \in \mathbb{C}^{\mathbb{N}^r};\ \dim_{\mathbb{C}}(\mathbb{C}[s] \circ y) < \infty \right\} = \bigoplus_{\lambda\in\mathbb{C}^r, \mu\in\mathbb{N}^r} \mathbb{C}e_{\lambda,\mu}.$$

In the one-dimensional case, i. e. $r = 1$, these are the sequences that are polynomial-exponential after finitely many time steps. For $r \in \mathbb{N}$ arbitrary, there exists $\mu \in \mathbb{N}^r$ such that a sequence in $\mathcal{F}_{\mathrm{dis}}$ is polynomial-exponential on $\mu + \mathbb{N}^r$. With $t^\mu := t_1^{\mu_1} * \ldots * t_r^{\mu_r}$ the $\mathbb{C}[s]$-linear isomorphism

$$\mathbb{C}^{\mathbb{N}^r} \cong \mathbb{C}[[t]],\ y = (y_\mu)_{\mu\in\mathbb{N}^r} \mapsto \sum_{\mu\in\mathbb{N}^r} y_\mu \frac{t^\mu}{\mu!},$$

induces the isomorphisms

$$\mathcal{F}_{\mathrm{dis}} = \oplus_{\lambda,\mu}\mathbb{C}e_{\lambda,\mu} \cong \mathcal{F}_{\mathrm{cont}} = \oplus_\lambda \mathbb{C}[t]e^{\lambda\bullet t} \text{ and}$$

$$\oplus_{\mu\in\mathbb{N}^r}\mathbb{C}e_{\lambda,\mu} \cong \mathbb{C}[t]e^{\lambda\bullet t},\ e_{\lambda,\mu} \mapsto \frac{t^\mu}{\mu!}e^{\lambda\bullet t}.$$

The isomorphism $\mathcal{F}_{\mathrm{dis}} \cong \mathcal{F}_{\mathrm{cont}}$ shows that the minimal injective cogenerator property of $\mathcal{F}_{\mathrm{cont}}$ follows from that of $\mathcal{F}_{\mathrm{dis}}$.

4 Stable one-dimensional systems

In this section we recall the stability theory for the simplest one-dimensional systems as a model for the following multidimensional theory and start with the continuous case.

Let

$$r := 1,\ s := s_1,\ t := t_1 \in \mathbb{R},\ \lambda := \lambda_1 \in \mathbb{C},$$

$$A \in \mathbb{C}^{p\times p},\ \mathrm{Spec}(A) := \{\text{eigenvalues of } A\},$$

$$\text{stable region } \Lambda_1 := \{\lambda \in \mathbb{C};\ \Re(\lambda) < 0\}.$$

We consider the simple *differential system*

$$\overset{\bullet}{y} = Ay + u,\ u, y \in \mathcal{D}'^p, \text{ with the solutions}$$

$$\mathcal{B}^0 := \left\{ y \in \mathcal{D}'(\mathbb{R})^p;\ \overset{\bullet}{y} = Ay \right\} \subset \bigoplus_{\lambda \in \mathrm{Spec}(A)} \mathbb{C}[t]^p e^{\lambda \bullet t},$$

$$y(t) = e^{tA} y(0) + \int_0^t e^{(t-\tau)A} u(\tau) d\tau, \text{ if } u \text{ is continuous.}$$

The following theorem is the standard result on the stability of these systems.

Theorem 4.1 *1. The following assertions are equivalent:*

(i) Analysis: $\mathcal{B}^0 := \{y \in \mathcal{D}'^p;\ \overset{\bullet}{y} = Ay\} \subset \oplus_{\lambda \in \Lambda_1} \mathbb{C}[t]^p e^{\lambda \bullet t}$.
(ii) $\mathcal{B}^0$ *ist* asymptotically stable, *i. e., for all* $y \in \mathcal{B}^0 :\ \lim_{t \to \infty} y(t) = 0$.
(iii) Geometry: $\mathrm{Spec}(A) \subset \Lambda_1 = \{\lambda \in \mathbb{C};\ \Re(\lambda) < 0\}$.

2. *If the conditions of the first item are satisfied the system is* BIBO stable *(bounded input/bounded output stable), i. e., if the differential equation*

$$\overset{\bullet}{y} = Ay + u$$

is satisfied and if u *is continuous and bounded for* $t \geq 0$ *then so is* y.

We reformulate the preceding data in order to get a clue for a reasonable multidimensional generalization:

$$P := s\,\mathrm{id}_p - A,\ Q := \mathrm{id}_p \in \mathbb{C}[s]^{p \times p},\ \mathrm{rank}(s\,\mathrm{id}_p - A) = p,$$

$$\overset{\bullet}{y} = Ay + u \Longleftrightarrow P \circ y = P(\frac{d}{dt})(y) = Q \circ u, \tag{4.1}$$

$$\mathrm{Spec}(A) = \{\lambda \in \mathbb{C};\ \mathrm{rank}(P(\lambda)) < p\}.$$

The rank of a polynomial matrix is that as matrix with coefficients in the *field* $\mathbb{C}(s)$ of rational functions and therefore the maximal number of linearly independent columns. In the *discrete* case we consider the stable region $\Lambda_1 := \{\lambda \in \mathbb{C};\ |\lambda| < 1\}$ and the *difference system*

$$y(i+1) = Ay(i) + u(i),\ u, y \in \mathbb{C}^{\mathbb{N}}, \text{ with the solutions}$$

$$\mathcal{B}^0 := \left\{ y \in (\mathbb{C}^p)^{\mathbb{N}};\ y(i+1) = Ay(i) \right\} \subset \bigoplus_{\lambda \in \mathrm{Spec}(A),\ i \in \mathbb{N}} \mathbb{C}^p e_{\lambda,i} \subset \mathcal{F}^p_{\mathrm{dis}},$$

$$y(i) = A^i y(0) + \sum_{j,k,j+k=i-1} A^j u(k).$$

Theorem 4.1 remains valid with the new data

$$\mathcal{B}^0 \subset \bigoplus_{\lambda \in \Lambda_1,\ i \in \mathbb{N}} \mathbb{C}^p e_{\lambda,i} \quad \text{and} \quad \mathrm{Spec}(A) \subset \{\lambda \in \mathbb{C};\ |\lambda| < 1\}.$$

5 Multidimensional IO behaviors

In this section we specify the multidimensional IO systems which we study in this paper, in particular in context with the feedback stabilization of section 2. We generalize the data of equation (4.1). *The behaviors have an algebraic, an analytic and a geometric aspect, and we discuss them all.*

Algebra: Consider a polynomial matrix

$$R \in \mathbb{C}[s]^{k\times l} \text{ of } \operatorname{rank}(R) =: p \quad \text{and} \quad m := l - p.$$

Various choices of p columns of R are linearly independent and are called an *input/output or IO structure* in systems theory. This terminology will be justified below. We make such a choice. After a possible permutation of the columns the first p columns of R are the IO structure and the matrix is written as

$$R = (P, -Q) \in \mathbb{C}[s]^{k\times(p+m)} \quad \text{with} \quad \operatorname{rank}(P) = \operatorname{rank}(R) = p.$$

Hence there is a unique rational matrix

$$H \in \mathbb{C}(s)^{p\times m} \quad \text{with} \quad PH = Q$$

which is called the *transfer matrix* in systems theory.

The matrices $R = (P, -Q)$ and P give rise to their row modules and corresponding factor modules

$$U := \mathbb{C}[s]^{1\times k}R = \sum_{i=1}^{k} \mathbb{C}[s]R_{i-} \subset \mathbb{C}[s]^{1\times(p+m)} \quad \text{and}$$

$$U^0 := \mathbb{C}[s]^{1\times k}P = \sum_{i=1}^{k} \mathbb{C}[s]P_{i-} \subset \mathbb{C}[s]^{1\times p}, \quad \text{as well as}$$

$$M := \mathbb{C}[s]^{1\times(p+m)}/U, \quad \text{and } M^0 := \mathbb{C}[s]^{1\times p}/U^0.$$

The isomorphism

$$\mathbb{C}(s) \otimes_{\mathbb{C}[s]} M \underset{\mathbb{C}(s)}{\cong} \mathbb{C}(s)^{1\times(p+m)}/\mathbb{C}(s)^{1\times k}R$$

motivates to define the factor module's rank as

$$\operatorname{rank}(M) := \dim_{\mathbb{C}(s)}(\mathbb{C}(s) \otimes_{\mathbb{C}[s]} M) = m.$$

Since M^0 is a torsion module we have

$$\mathbb{C}(s) \otimes_{\mathbb{C}[s]} M^0 = 0 \quad \text{and} \quad \operatorname{rank}(M^0) = 0.$$

The last row shows that M^0 is a torsion module or has a non-zero annihilator ideal

$$\operatorname{ann}_{\mathbb{C}[s]}(M^0) := \{f \in \mathbb{C}[s];\ fM^0 = 0 \text{ or } f\mathbb{C}[s]^{1\times p} \subseteq U^0\}. \tag{5.1}$$

Geometry [27, p. 155ff]: We use the algebraic data, i. e., matrices and modules, from above. For every point $\lambda \in \mathbb{C}^r$ the module M and the maximal ideal $\mathfrak{m}_\lambda$ give rise to the quotient module

$$M_{\mathfrak{m}_\lambda} = \left\{\frac{x}{t};\ x \in M,\ t \in \mathbb{C}[s], t(\lambda) \neq 0\right\} \quad \text{over the local ring } \mathbb{C}[s]_{\mathfrak{m}_\lambda}.$$

For any ideal $\mathfrak{a} \subseteq \mathbb{C}[s]$ we consider its variety

$$V(\mathfrak{a}) := \{\lambda \in \mathbb{C}^r;\ \forall f \in \mathfrak{a}:\ f(\lambda) = 0\} \subseteq \mathbb{C}^r.$$

We obtain two characteristic varieties which are essentially connected with M and M^0, viz.

$$\begin{aligned}(\text{the } \textit{characteristic variety} \text{ of } M) &:= \operatorname{char}(M)\\ &:= \{\lambda \in \mathbb{C}^r; \operatorname{rank}(R(\lambda)) < p = \operatorname{rank}(R)\}\\ &= \{\lambda \in \mathbb{C}^r;\ M_{\mathfrak{m}_\lambda} \text{ is not free}\}\end{aligned}$$

and

$$\begin{aligned}(\text{the } \textit{characteristic variety} \text{ of } M^0) &:= \operatorname{char}(M^0)\\ &:= \{\lambda \in \mathbb{C}^r;\ \operatorname{rank}(P(\lambda)) < p = \operatorname{rank}(P)\}\\ &= \{\lambda \in \mathbb{C}^r;\ M^0_{\mathfrak{m}_\lambda} \neq 0\}\\ &= V\left(\operatorname{ann}_{\mathbb{C}[s]}(M^0)\right)\end{aligned}$$

with $\operatorname{char}(M) \subset \operatorname{char}(M^0)$.

In systems theory these varieties are also called the varieties of *rank singularities* for obvious reasons. In the one-dimensional example $P := s\,\mathrm{id}_p - A,\ A \in \mathbb{C}^{p\times p}$, the characteristic variety $\operatorname{char}(M^0)$ coincides with the spectrum $\operatorname{Spec}(A)$. Therefore one expects that the characteristic varieties determine the stability properties of a multidimensional behavior, and this is indeed the case.

Analysis: The data from above are in force. Let $\mathcal{F}$ denote one of the injective $\mathbb{C}[s]$-cogenerators discussed in section 3 in the continuous or discrete case. The *multidimensional $\mathcal{F}$-behavior* defined by R is the solution module of the linear system $R \circ w = 0$ of partial differential or difference equations with constant coefficients, more precisely

$$\begin{aligned}\mathcal{B} &:= \left\{w = (w_1, \dots, w_l)^\top \in \mathcal{F}^l;\ R \circ w = 0\right\}\\ &= \ker\left(R\circ : \mathcal{F}^l \to \mathcal{F}^k,\ w \mapsto R \circ w\right) \subset \mathcal{F}^l.\end{aligned}$$

For obvious reasons J. C. Willems calls the preceding form a *kernel representation* of $\mathcal{B}$. With the standard basis

$$\delta_i := (0, \dots, 0, \overset{i}{1}, 0, \dots, 0) \in \mathbb{C}[s]^{1\times l}, \quad i = 1, \dots, l,$$

one obtains the $\mathbb{C}[s]$-isomorphism, due to B. Malgrange [25],

$$\operatorname{sol}_{\mathcal{F}}(M) := \operatorname{Hom}_{\mathbb{C}[s]}(M, \mathcal{F}) \cong \mathcal{B},\ \varphi \longleftrightarrow w,\ \varphi(\overline{\delta_i}) = w_i.$$

The significance of this simple isomorphism lies in the fact that it connects algebra with analysis and thus was one origin of *Algebraic Analysis*. In the same fashion one obtains the behavior

$$\mathcal{B}^0 := \{y = (y_1, \ldots, y_p)^\top \in \mathcal{F}^p;\ P \circ y = 0\} \quad \text{and}$$
$$\mathrm{sol}_{\mathcal{F}}(M^0) := \mathrm{Hom}_{\mathbb{C}[s]}(M^0, \mathcal{F}) \cong \mathcal{B}^0.$$

The IO structure and the associated decompositions $R = (P, -Q)$ and $w := \binom{y}{u} \in \mathcal{F}^l = \mathcal{F}^{p+m}$ of the matrix R resp. the function vectors yield the alternative *IO representation* of $\mathcal{B}$:

$$\mathcal{B} := \left\{\begin{pmatrix} y \\ u \end{pmatrix} \in \mathcal{F}^{p+m};\ R \circ \begin{pmatrix} y \\ u \end{pmatrix} = 0 \text{ or } P \circ y = Q \circ u\right\},$$
$$\mathcal{B}^0 \cong \mathcal{B} \cap (\mathcal{F}^p \times \{0\}),\ y \longleftrightarrow \begin{pmatrix} y \\ 0 \end{pmatrix}.$$

We interpret $P \circ y = Q \circ u$ as an *inhomogeneous linear system* with right side $Q \circ u$ and $P \circ y = 0$ as the associated *homogeneous system* with solution module $\mathcal{B}^0$.

Theorem and Definition 5.1 [27, Theorem 2.69 and Corollary 2.27]

1. *For each $u \in \mathcal{F}^m$ there is a $y \in \mathcal{F}^p$ such that*

$$P \circ y = Q \circ u \text{ or } \begin{pmatrix} y \\ u \end{pmatrix} \in \mathcal{B},$$

i. e., one may select the component u of a trajectory in $\mathcal{B}$ freely or, in other terms, the sequence

$$\begin{array}{ccccccc} 0 & \to & \mathcal{B}^0 & \xrightarrow{\text{inj}} & \mathcal{B} & \xrightarrow{\text{proj}} & \mathcal{F}^m & \to & 0 \\ & & y & \mapsto & \binom{y}{0},\ \binom{y}{u} & \mapsto & u & & \end{array}$$

is exact. It is this property which justifies to call u the input *and y a corresponding* output *of $\mathcal{B}$ for the chosen IO structure. The components u_i of the input are also called* free.

2. *The module U and therefore also the module M are uniquely determined by the solution module $\mathcal{B}$, indeed*

$$U = \mathbb{C}[s]^{1\times k} R = \mathcal{B}^\perp := \left\{x \in \mathbb{C}[s]^{1\times(p+m)};\ \forall w \in \mathcal{B}:\ x \circ w = 0\right\}.$$

J. F. Pommaret calls the elements of M the observables *of $\mathcal{B}$. We also define the characteristic varieties*

$$\mathrm{char}(\mathcal{B}) := \mathrm{char}(M) \quad \textit{and} \quad \mathrm{char}(\mathcal{B}^0) := \mathrm{char}(M^0).$$

3. *The equality* $\mathrm{ann}_{\mathbb{C}[s]}(\mathcal{B}^0) = \mathrm{ann}_{\mathbb{C}[s]}(M^0) \neq 0$ *shows that there is a non-zero polynomial f such that $f \circ y = 0$ for all $y \in \mathcal{B}^0$, i. e., all solutions of the homogeneous*

system $P \circ y = 0$ *satisfy the same differential resp. difference equation. The behavior* $\mathcal{B}^0$ *is called the* autonomous *part of* $\mathcal{B}$ *for the chosen IO structure. The behavior* $\mathcal{B}$ *is called autonomous if*

$$\mathcal{B} = \mathcal{B}^0 \Longleftrightarrow M = M^0 \Longleftrightarrow M \text{ is a torsion module.}$$

All statements of the preceding theorem follow from the injective cogenerator property of $\mathcal{F}$.

6 Stable multidimensional systems

In this section we are going to define and characterize stable multidimensional systems in the continuous and discrete cases. The data and notations are those of the preceding section.

We need an analogue of the stable region $\{\lambda \in \mathbb{C};\ \Re(\lambda) < 0\}$ of the one-dimensional situation of section 4. For maximal flexibility we define stability for an *arbitrary* disjoint decomposition

$$\mathbb{C}^r = \Lambda_1 \uplus \Lambda_2,\ \Lambda_1 =: \text{stable region},\ \Lambda_2 =: \text{unstable region.} \tag{6.1}$$

The standard example is

$$\Lambda_2 := \begin{cases} \{\lambda \in \mathbb{C}^r;\ \forall \rho = 1, \ldots, r:\ \Re(\lambda_\rho) \geq 0\} \\ \{\lambda \in \mathbb{C}^r;\ \forall \rho = 1, \ldots, r:\ |\lambda_\rho| \geq 1\} \end{cases} \text{in the } \begin{cases} \text{continuous} \\ \text{discrete} \end{cases} \text{case.} \tag{6.2}$$

We define the saturated multiplicatively closed set

$$\begin{aligned} T &:= \{t \in \mathbb{C}[s];\ V(t) \subset \Lambda_1\} = \{t \in \mathbb{C}[s];\ \forall \lambda \in \Lambda_2:\ t(\lambda) \neq 0\} \\ &= \bigcap_{\lambda \in \Lambda_2} (\mathbb{C}[s] \setminus \mathfrak{m}_\lambda) = \{\text{stable polynomials}\} \end{aligned}$$

of *stable polynomials* with respect to the decomposition (6.1) and the ensuing quotient ring

$$\begin{aligned} \mathbb{C}[s]_T &:= \left\{\frac{a}{t};\ a \in \mathbb{C}[s], t \in T\right\} = \bigcap_{\lambda \in \Lambda_2} \mathbb{C}[s]_{\mathfrak{m}_\lambda} \\ &= \{\text{stable rational functions }\} \subseteq \mathbb{C}(s) \end{aligned}$$

of *stable rational functions.*

The next lemma shows that the stable region $\Lambda_1 := \{z \in \mathbb{C};\ \Re(z) < 0\}^r$ is no suitable generalization of the one-dimensional situation and uninteresting.

Lemma 6.1 *For $r > 1$ consider disjoint decompositions*

$$\mathbb{C} = \Lambda_{1,\rho} \uplus \Lambda_{2,\rho},\ \rho = 1, \ldots, r, \quad \textit{with} \quad \text{infinite } \Lambda_{2,\rho},$$

$$\Lambda_1 := \prod_{\rho=1}^{r} \Lambda_{1,\rho},\ \mathbb{C}^r = \Lambda_1 \uplus \Lambda_2.$$

Then the non-zero constants are the only stable polynomials with respect to this decomposition, i. e., $T := \{t \in \mathbb{C}[s];\ V(t) \subseteq \Lambda_1\} = \mathbb{C} \setminus \{0\}$.

Proof. Define $s' = (s_1, \ldots, s_{r-1})$, hence $s = (s', s_r)$, and let $f = f(s) = f(s', s_r)$ be a stable polynomial, i. e., $V(f) \subseteq \Lambda_1 := \prod_{\rho=1}^{r} \Lambda_{1,\rho}$ and $f \neq 0$. The assumption implies that for every $\lambda_{2,r} \in \Lambda_{2,r}$ the polynomial

$$f(s', \lambda_{2,r}) \text{ has no zero, hence is constant, i. e., } f(s', \lambda_{2,r}) = f(0, \lambda_{2,r}), \text{ and}$$
$$\mathbb{C}^{r-1} \times \Lambda_{2,r} \subseteq V(f - f(0, s_r)), \text{ hence } f - f(0, s_r) = 0$$

since this polynomial is zero on a product of infinite subsets of $\mathbb{C}$. Thus

$$f = g(s_r),\ 0 \neq g \in \mathbb{C}[s_r], \qquad \text{hence } V(f) = \mathbb{C}^{r-1} \times V_{\mathbb{C}}(g).$$

Again by assumption this set is empty and hence $f = g(s_r)$ is a non-zero constant. □

For the next theorem we use the injective cogenerator $\mathcal{F}$ of polynomial-exponential functions in the continuous resp. discrete cases from (3.1) and Lemma 3.7, i. e.,

$$\mathcal{F} := \begin{cases} \mathcal{F}_{\text{cont}} = \oplus_{\lambda \in \mathbb{C}^r} \mathbb{C}[t] e^{\lambda \bullet t} \\ \mathcal{F}_{\text{dis}} = \oplus_{\lambda \in \mathbb{C}^r} \oplus_{\mu \in \mathbb{N}^r} \mathbb{C} e_{\lambda,\mu} \end{cases} \text{in the } \begin{cases} \text{continuous} \\ \text{discrete} \end{cases} \text{case}$$

and the submodules

$$\mathcal{F}_i := \begin{cases} \oplus_{\lambda \in \Lambda_i} \mathbb{C}[t] e^{\lambda \bullet t} \\ \oplus_{\lambda \in \Lambda_i} \oplus_{\mu \in \mathbb{N}^r} \mathbb{C} e_{\lambda,\mu} \end{cases} \text{in the } \begin{cases} \text{continuous} \\ \text{discrete} \end{cases} \text{case, } i = 1, 2, \text{ hence}$$

$$\mathcal{F} = \mathcal{F}_1 \oplus \mathcal{F}_2,\ \mathcal{B} := \mathcal{B}_1 \oplus \mathcal{B}_2 \text{ with } \mathcal{B}_i := \mathcal{B} \cap \mathcal{F}_i^{p+m}.$$

With these data we obtain the analogue of Theorem 4.1

Theorem and Definition 6.2 (Stable multidimensional behaviors) [26, Theorems 5.1, 5.3, 5.4] *The data are those from section 5 and from above.*

1. *The following statements are equivalent:*

   (i) Analysis: $\mathcal{B}^0 \subset \mathcal{F}_1^p$.
   (ii) Geometry: $\text{char}(\mathcal{B}^0) \subset \Lambda_1 =$*stable region.*
   (iii) Algebra:

      (a) The transfer matrix is stable, i. e., $H \in \mathbb{C}[s]_T^{p \times m}$, and
      (b) for each $\lambda \in \Lambda_2$ the $\mathbb{C}[s]_{\mathfrak{m}_\lambda}$-module $M_{\mathfrak{m}_\lambda}$ is torsionfree.

Under these conditions $\mathcal{B}$ is called stable *with respect to the decomposition* (6.1).

2. *The differential resp. difference operators $f\circ : \mathcal{F}_2 \to \mathcal{F}_2$, $f \in T$, are isomorphisms, i. e., $\mathcal{F}_2$ is a $\mathbb{C}[s]_T$-module. For a stable system $\mathcal{B}$ as in 1. this implies the isomorphism*

$$\mathcal{F}_2^m \cong \mathcal{B}_2 = \mathrm{graph}(H \circ |\mathcal{F}_2^m),\ u \mapsto \begin{pmatrix} H\circ u \\ u \end{pmatrix},\ \textit{hence}$$

$$\mathcal{B} = \mathcal{B}_1 \oplus \mathrm{graph}(H \circ |\mathcal{F}_2^m) \ni w = \begin{pmatrix} y \\ u \end{pmatrix} = \begin{pmatrix} y_1 \\ u_1 \end{pmatrix} + \begin{pmatrix} H\circ u_2 \\ u_2 \end{pmatrix}.$$

In the usual suggestive language the first resp. the second summand of the trajectory $w \in \mathcal{B}$ are called its stable part *resp. its* steady state.

3. *If $\mathcal{B}$ is stable, $d \in \mathbb{N}$, $\Gamma \subset \Lambda_2$ is finite and $u \in \mathcal{F}_2^m$ has the form*

$$u = \sum_{\gamma\in\Gamma} u_\gamma(t)e^{\gamma \bullet t} \quad \textit{with} \quad u_\gamma(t) \in \mathbb{C}[t]^m \quad \textit{and} \quad \deg_t(u_\gamma) \le d$$

where $\deg_t$ denotes the total degree in $\mathbb{C}[t]$ then the output $H \circ u$ has the same form. This is a certain equivalent for BIBO stability. An analogous result holds in the discrete case.

4. *If in the continuous case $i\mathbb{R}^r \subseteq \Lambda_2$ and if $\mathcal{G}$ denotes one of the spaces*

$$\mathcal{G} := \begin{cases} \mathcal{S} & \textit{of rapidly decreasing } \mathrm{C}^\infty\textit{-functions} \\ \mathcal{O} & \textit{of slowly increasing } \mathrm{C}^\infty\textit{-functions} \\ \mathcal{S}' & \textit{of temperate distributions} \\ \mathcal{O}' & \textit{of rapidly decreasing distributions} \end{cases}$$

then $\mathcal{G}$ is also a $\mathbb{C}[s]_T$-module. The stability of $\mathcal{B} := \mathcal{B}_{\mathcal{F}_{\mathrm{cont}}}$ then implies the $\mathbb{C}[s]_T$-monomorphism

$$\mathcal{G}^m \to \mathcal{B}_{\mathcal{G}} := \left\{ \begin{pmatrix} y \\ u \end{pmatrix} \in \mathcal{G}^{p+m};\ P \circ y = Q \circ u \right\},\ u \mapsto \begin{pmatrix} H\circ u \\ u \end{pmatrix}, \tag{6.3}$$

and this is another variant of BIBO stability in the multidimensional situation. The map (6.3) *is even an isomorphism if the set Λ_2 is* ideal-convex *(see section 7).*

It is not clear (to us) how the condition 1.(i) of the preceding theorem can be generalized to other function spaces $\mathcal{G}$ like those of C^∞-functions or distributions instead of polynomial-exponential functions. In this context the so-called *integral representation theorem* of L. Ehrenpreis and V. Palamodov [3, 12, 30] could be useful. This theorem expresses solutions in $\mathcal{B}_{\mathcal{G}}^0$ as integrals over polynomial-exponential solutions with respect to measures which are supported in the characteristic variety. The difficulties in this approach are of an analytic nature. Due to these difficulties we have applied the space of polynomial-exponential functions for the formulation of Theorem 6.2.

7 Gabriel localization

P. Gabriel developed his categorical localization theory for the construction of *non-commutative* quotient rings 45 years ago. The theory was exposed by B. Stenström in [44] and is also explained in [26, §3]. There we observed that the theory is also non-trivial for the Noetherian polynomial algebra $A = \mathbb{C}[s]$ and that it is significant for the stabilization of IO systems. *Here we give a short introduction into this special application.*

The assumptions and notations of sections 5 and 6 are in force. It is a standard and simple observation that the quotient module M_T of an A-module M can be constructed as a colimit or direct limit over the ordered set $\{At;\ t \in T\}$ of ideals which is directed downwards via

$$At_1t_2 \subseteq At_1 \cap At_2.$$

The construction is

$$M_T \cong \operatorname{colim}_{At;\, t\in T} \operatorname{Hom}_A(At, M) \quad \text{with} \quad \operatorname{Hom}_A(At, M) \cong M,\ \alpha \mapsto \alpha(t).$$

Recall that the functor $M \mapsto M_T$ is exact. In the situation of the present paper we replace the set $\{At;\ t \in T\}$ by the downward directed set

$$\begin{aligned}\mathfrak{T} &:= \{\mathfrak{a} \subseteq A;\ \forall\lambda \in \Lambda_2 :\ (A/\mathfrak{a})_{\mathfrak{m}_\lambda} = 0\}\\ &= \{\mathfrak{a} \subseteq A;\ \forall\lambda \in \Lambda_2 :\ \mathfrak{a} \cap (A \setminus \mathfrak{m}_\lambda) \neq \emptyset\} = \{\mathfrak{a} \subseteq A;\ V(\mathfrak{a}) \cap \Lambda_2 = \emptyset\},\end{aligned}$$

$$\text{hence } At \in \mathfrak{T} \text{ for all } t \in T = \bigcap_{\lambda\in\Lambda_2} (A \setminus \mathfrak{m}_\lambda).$$

The set $\mathfrak{T}$ is the basis of neighborhoods of 0 of a unique *linear* topology on A which is called a *Gabriel topology*. It induces the new left-exact *Gabriel localization functor* on the category of A-modules

$$M \mapsto M_{\mathfrak{T}} = \operatorname{colim}_{\mathfrak{a}\in\mathfrak{T}} \operatorname{Hom}_A(\mathfrak{a}, M).$$

The standard colimit construction of $M_{\mathfrak{T}}$ is

$$M_{\mathfrak{T}} = \biguplus_{\mathfrak{a}\in\mathfrak{T}} \operatorname{Hom}_A(\mathfrak{a}, M)/\sim\ \ni [\alpha : \mathfrak{a} \to M] \quad \text{where}$$

$$\begin{gathered}(\alpha_1 : \mathfrak{a}_1 \to M) \sim (\alpha_2 : \mathfrak{a}_2 \to M) :\Longleftrightarrow\\ \exists\mathfrak{b} \in \mathfrak{T} \text{ such that } \mathfrak{b} \subseteq \mathfrak{a}_1 \cap \mathfrak{a}_2 \text{ and } \alpha_1|\mathfrak{b} = \alpha_2|\mathfrak{b}.\end{gathered}$$

The module $M_{\mathfrak{T}}$ is an A_T-module, i. e., for all $t \in T$ the map

$$t : M_{\mathfrak{T}} \cong M_{\mathfrak{T}},\ x \mapsto tx,$$

is an isomorphism. There are canonical linear maps

$$\begin{array}{ccccccc} M & \longrightarrow & M_T & \longrightarrow & M_{\mathfrak{T}} & \longrightarrow & M_{\mathfrak{m}_\lambda}\\ x & \mapsto & \frac{x}{1},\ \frac{x}{t} & \mapsto & [\alpha_{\frac{x}{t}}],\ [\alpha : \mathfrak{a} \to M] & \mapsto & \frac{\alpha(t_\lambda)}{t_\lambda}\end{array}$$

where $\lambda \in \Lambda_2,\ \alpha_{\frac{x}{t}} : At \to M,\ t \mapsto x$, and $t_\lambda \in \mathfrak{a},\ t_\lambda(\lambda) \neq 0$, in particular $A_T \cong A_{\mathfrak{T}}$ or $A_T = A_{\mathfrak{T}}$ by identification, which induce canonical isomorphisms

$$(M_T)_{\mathfrak{T}} \cong M_{\mathfrak{T}},\ M_{\mathfrak{m}_\lambda} \cong (M_T)_{\mathfrak{m}_\lambda} \cong (M_{\mathfrak{T}})_{\mathfrak{m}_\lambda},$$
$$\mathbb{C}(s) \otimes_A M \cong \mathbb{C}(s) \otimes_A M_T \cong \mathbb{C}(s) \otimes_A M_{\mathfrak{T}} \cong \mathbb{C}(s) \otimes_A M_{\mathfrak{m}_\lambda}.$$

For any finitely generated module ${}_AM = A^{1\times l}/U$ and the injective modules

$$\mathcal{F}_\lambda := \mathbb{C}[t]e^{\lambda\bullet t} \cong \bigoplus_{\mu\in\mathbb{N}^r} \mathbb{C}e_{\lambda,\mu},\quad \mathcal{F}_2 = \bigoplus_{\lambda\in\Lambda_2} \mathcal{F}_\lambda$$

there are canonical isomorphisms

$$\mathcal{B}_2 := \mathcal{B} \cap \mathcal{F}_2^l \cong \mathrm{Hom}_A(M, \mathcal{F}_2) \cong \mathrm{Hom}_{A_T}(M_T, \mathcal{F}_2) \cong$$
$$\mathrm{Hom}_{A_T}(M_{\mathfrak{T}}, \mathcal{F}_2) \cong \bigoplus_{\lambda\in\Lambda_2} \mathrm{Hom}_{A_{\mathfrak{m}_\lambda}}(M_{\mathfrak{m}_\lambda}, \mathcal{F}_\lambda).$$

The decisive advantage of Gabriel's localization $M_{\mathfrak{T}}$ over the usual quotient module M_T is that the canonical map

$$M_{\mathfrak{T}} \xrightarrow{\mathrm{can}} \prod_{\lambda\in\Lambda_2} M_{\mathfrak{m}_\lambda} \text{ is injective, and hence}$$
$$M_{\mathfrak{T}} = 0 \Longleftrightarrow \forall \lambda \in \Lambda_2 : M_{\mathfrak{m}_\lambda} = 0 \Longleftrightarrow \mathcal{B}_2 = \mathcal{B} \cap \mathcal{F}_2^l = 0. \tag{7.1}$$

The left exactness of $(-)_{\mathfrak{T}}$ on A-modules, the monomorphism (7.1) and $A_T = A_{\mathfrak{T}}$ also imply the representation

$$U_{\mathfrak{T}} = A_T^{1\times l} \bigcap \bigcap_{\lambda\in\Lambda_2} U_{\mathfrak{m}_\lambda} \subseteq A_T^{1\times l} \subset \mathbb{C}(s)^{1\times l}$$

for any submodule $U \subseteq A^{1\times l}$ and then $U_{\mathfrak{T}} = U_{\mathrm{st},T}$ where

$$U_{\mathrm{st}} := A^{1\times l} \bigcap \bigcap_{\lambda\in\Lambda_2} U_{\mathfrak{m}_\lambda} = A^{1\times k_{\mathrm{st}}} R_{\mathrm{st}},\ R_{\mathrm{st}} \in A^{k_{\mathrm{st}}\times(p+m)},$$

or, in other terms, the rows of R_{st} are A-generators of U_{st} and A_T-generators of $U_{\mathfrak{T}}$. In general, it is difficult to compute these, but in important special cases they can be obtained in the following fashion. The data are again those of sections 5 and 6, i. e.,

$$U = A^{1\times k}(P, -Q) \subseteq A^{1\times(p+m)},\quad M = A^{1\times(p+m)}/U,\quad PH = Q.$$

Define

$$U_{\mathrm{cont}} := \{(\xi, \eta) \in A^{1\times(p+m)}; \xi H + \eta = 0\} = A^{1\times k_{\mathrm{cont}}} R_{\mathrm{cont}}.$$

Then, $t(M) := U_{\mathrm{cont}}/U \subseteq M$ is the *torsion submodule* of M. Let

$$\mathfrak{a} := \mathrm{ann}_A(t(M))$$

be its annihilator ideal. Both U_{cont} and $\mathfrak{a}$ can be computed via Gröbner bases. The IO behavior

$$\mathcal{B}_{\mathrm{cont}} := \left\{ \begin{pmatrix} y \\ u \end{pmatrix} \in \mathcal{F}^{p+m};\ P_{\mathrm{cont}} \circ y = Q_{\mathrm{cont}} \circ u \right\} \text{ with}$$

$$(P_{\mathrm{cont}}, -Q_{\mathrm{cont}}) := R_{\mathrm{cont}} \in A^{k_{\mathrm{cont}} \times (p+m)}$$

is the least or the unique controllable IO behavior with transfer matrix H.

Theorem 7.1 [26, Theorem, 5.14] *Data as just introduced.*

1. *The following properties are equivalent:*

   (i) All localized modules $M_{\mathfrak{m}_\lambda}$, $\lambda \in \Lambda_2$, are torsionfree.
   (ii) $V(\mathfrak{a}) \cap \Lambda_2 = \emptyset$ or, equivalently, $\mathfrak{a} \in \mathfrak{T}$.

   If these equivalent properties hold then

   $$U_{\mathfrak{T}} = U_{\mathrm{cont},T}, \quad \textit{hence} \quad U_{\mathrm{st}} = U_{\mathrm{cont}} \quad \textit{and} \quad R_{\mathrm{st}} := R_{\mathrm{cont}}.$$

2. *The module M_T is torsionfree if and only if $\mathfrak{a} \cap T \neq \emptyset$. If this is the case then*

   $$U_{\mathfrak{T}} = U_{\mathrm{cont},T} = U_T \textit{ and } M_T \cong \mathcal{M} := A_T^{1\times p} H + A_T^{1\times m} \subset \mathbb{C}(s)^{1\times m}. \tag{7.2}$$

   The module $\mathcal{M}$ is an A_T-lattice in $\mathbb{C}(s)^{1+m}$.

For the following notion of *ideal convexity* we make two simple observations.

1. For any ideal $\mathfrak{a} \subseteq A = \mathbb{C}[s]$ we get

   $$\mathfrak{a} \cap T \neq \emptyset \Rightarrow V(\mathfrak{a}) \cap \Lambda_2 = \emptyset. \tag{7.3}$$

2. If $\mathfrak{m}$ is a maximal ideal of A with $\mathfrak{m} \cap T = \emptyset$ then $\mathfrak{m}_T$ is a maximal ideal of A_T and therefore

   $$\{(\mathfrak{m}_\lambda)_T;\ \lambda \in \Lambda_2\} \subseteq \operatorname{Max}\left(\mathbb{C}[s]_T\right).$$

Theorem and Definition 7.2 (Perfect localization and ideal-convexity) [44, p. 231] [43, p. 25] [26, Theorem 5.6] [51]

1. *Under the preceding assumptions the following properties are equivalent for the polynomial ring $A = \mathbb{C}[s]$ and an unstable region $\Lambda_2 \subseteq \mathbb{C}^r$:*

   (i) For every A-module M the canonical map $M_T \to M_{\mathfrak{T}}$ is an isomorphism and hence $M_T = M_{\mathfrak{T}}$ by identification.
   (ii) For every ideal $\mathfrak{a}$ of A the reverse implication of (7.3) holds, i. e., for any ideal $\mathfrak{a} \subseteq A = \mathbb{C}[s]$ we get

   $$V(\mathfrak{a}) \cap \Lambda_2 = \emptyset \Rightarrow \mathfrak{a} \cap T \neq \emptyset.$$

   (iii) The preceding condition holds for all prime *ideals of $\mathbb{C}[s]$.*

(iv) $\mathrm{Max}\,(\mathbb{C}[s]_T) = \{(\mathfrak{m}_\lambda)_T;\ \lambda \in \Lambda_2\}$.

If these conditions are satisfied the region Λ_2 *is called* ideal-convex.

2. *If* Λ_2 *is ideal-convex and* $\mathcal{B} = \left\{\binom{y}{u} \in \mathcal{F}^{p+m};\ P \circ y = Q \circ u\right\}$ *is an IO behavior then* $\mathcal{B}$ *is stable if and only if the matrix* $P \in A^{k\times p}$ *has a left inverse in* $A_T^{p\times k}$.

Remark that the condition (ii) of the preceding theorem holds for all cyclic ideals Af and especially for all prime ideals Af of height one with irreducible f since

$$V(f) \cap \Lambda_2 = \emptyset \ \Leftrightarrow\ f \in T.$$

Lemma 7.3 *The condition 1.(iii) of the preceding theorem holds for all maximal ideals* $\mathfrak{m}_\mu$ *of* $\mathbb{C}[s]$ *or, equivalently, for all Krull-zero-dimensional ideals* $\mathfrak{a}$ *(with finite* $V(\mathfrak{a})$*) if and only if* $\Lambda_1 = \cup_{t\in T} V(t)$, *i. e., if* Λ_1 *is a (possibly infinite) union of hyper-surfaces. This is the case if* Λ_2 *is a Cartesian product* $\Lambda_2 = \prod_{\rho=1}^r \Lambda_{2,\rho}$. *Indeed, for*

$$\mu \in \Lambda_1 \quad \text{and} \quad \mu_\rho \notin \Lambda_{2,\rho}: \quad t := s_\rho - \mu_\rho \in T \quad \text{and} \quad \mu \in V(t) \subset \Lambda_1.$$

In dimension 2 *this condition is equivalent with ideal-convexity since then all prime ideals are either principal or maximal.*

Theorem and Definition 7.4 [43, Proposition 3.1.20 on p. 28] *A* compact *subset* $\Lambda_2 \subset \mathbb{C}^r$ *is ideal-convex if it is* polynomially convex [17, p. 53], *i. e., if*

$$\Lambda_2 = \left\{\lambda \in \mathbb{C}^r;\ \forall f \in \mathbb{C}[\lambda]:\ |f(\lambda)| \le \max_{z\in\Lambda_2} |f(z)|\right\}.$$

A compact *set*

$$\Lambda_2 = \left\{\lambda \in \mathbb{C}^r;\ \ \sup_{i=1,\dots,l}\ |\, f_i(\lambda)| \le 1\right\} \qquad \text{with } f_i \in \mathbb{C}[\lambda],\ i = 1,\dots,l,$$

is polynomially convex and thus ideal-convex.

The proof of this theorem in [43] uses the Theorems A and B by Oka, Cartan, Serre on coherent analytic sheaves.

Examples 7.5 1. According to Theorem 7.4 the closed unit polydisc

$$\{z \in \mathbb{C};\ |z| \le 1\}^r$$

is ideal-convex. This is the standard example in the stabilization theory of discrete multidimensional input/output maps, compare [16] and [22].

2. Every unstable region in $\mathbb{C}$ is ideal-convex.

3. Lemma 7.3 implies that the two-dimensional unstable regions

$$\{z \in \mathbb{C};\ \Re(z) \ge 0\}^2 \quad \text{resp.} \quad \{z \in \mathbb{C};\ |z| \ge 1\}^2$$

are ideal-convex [51]. For higher dimensions this is an open problem.

4. Every compact polyrectangle

$$\prod_{\rho=1}^{r}\{\lambda_\rho \in \mathbb{C}; a_{\rho,1} \leq \Re(\lambda_\rho) \leq b_{\rho,1},\ a_{\rho,2} \leq \Im(\lambda_\rho) \leq b_{\rho,2}\}$$

with real numbers $a_{\rho,i} \leq b_{\rho,i},\ i = 1, 2$, is ideal-convex. The proof is analogous to that of Theorem 7.4.

5. For any stable region $\Lambda_1 = \prod_{\rho=1}^{r} \Lambda_{1,\rho}$ as in Lemma 6.1 we have $T = \mathbb{C} \setminus \{0\}$ and therefore $\cup_{t\in T} V(t) = \emptyset$. Hence Λ_2 is ideal-convex only if $\Lambda_2 = \mathbb{C}^r$.

Open Problem 7.6 *Decide* ideal convexity for a given unstable region Λ_2, for instance for the standard unstable regions

$$\Lambda_2 := \{z \in \mathbb{C};\ \Re(z) \geq 0\}^r \text{ resp. } \Lambda_2 := \{z \in \mathbb{C};\ |z| \geq 1\}^r,\ r \geq 3$$

and *construct*

$$t \in \mathfrak{a} \cap T \quad \text{if} \quad V(\mathfrak{a}) \cap \Lambda_2 = \emptyset.$$

8 Stabilization

Stabilization is defined according to section 2 with respect to the class of multidimensional IO systems from section 5 with the stable ones from section 6. In particular we require that the feedback system from equation (2.1) is *well-posed*. This signifies that the combined input $\binom{u_2}{u_1} \in \mathcal{F}^{p+m}$ is indeed an input according to section 5.

The assumptions and notations are those from sections 5, 6 and 7, in particular

$$U_{\mathfrak{T}} = U_{\mathrm{st},T} = A^{1\times k_{\mathrm{st}}} R_{\mathrm{st}}.$$

Theorem 8.1 (Stabilizability and stabilization) [26, Theorems 4.4, 4.6, 5.8 and 5.14]

1. The following properties of an IO system $\mathcal{B} := \left\{\binom{y}{u} \in \mathcal{F}^{p+m};\ P \circ y = Q \circ u\right\}$ are equivalent:

(i) The system is stabilizable.

(ii) The submodule $U_{\mathfrak{T}} \subseteq A_T^{1\times(p+m)}$ is a direct summand.

(iii) There is a matrix $G_{\mathrm{st}} \in A_T^{(p+m)\times k_{\mathrm{st}}}$ such that $R_{\mathrm{st}} G_{\mathrm{st}} R_{\mathrm{st}} = R_{\mathrm{st}}$.

Then $E_1 := G_{\mathrm{st}} R_{\mathrm{st}} = E_1^2$ is an idempotent matrix or projection in $A_T^{(p+m)\times(p+m)}$ with $U_{\mathfrak{T}} = A_T^{1\times(p+m)} E_1$, and $\widetilde{U} := A_T^{1\times(p+m)}(\mathrm{id}_{p+m} - E_1)$ is a direct complement of $U_{\mathfrak{T}}$.

2. *If the equivalent conditions of 1. are satisfied there are bijections*

$$\left\{X \in A_T^{(p+m)\times k_{\mathrm{st}}};\ R_{\mathrm{st}} X R_{\mathrm{st}} = 0\right\} / \left\{X \in A_T^{(p+m)\times k_{\mathrm{st}}};\ X R_{\mathrm{st}} = 0\right\} \cong$$
$$\left\{E \in A^{(p+m)\times(p+m)};\ E = E^2,\ U_{\mathfrak{T}} = A_T^{1\times(p+m)} E\right\} \cong$$
$$\left\{\widetilde{U} \leq_{A_T} A_T^{1\times(p+m)};\ U_{\mathfrak{T}} \oplus \widetilde{U} = A_T^{1\times(p+m)}\right\},$$
$$\overline{X} \leftrightarrow E \leftrightarrow \widetilde{U},\ E := E_1 + X R_{\mathrm{st}},\ \widetilde{U} = A_T^{1\times(p+m)}(\mathrm{id}_{p+m} - E).$$

Generically the matrix $(\mathrm{id}_{p+m} - E)\binom{0}{\mathrm{id}_m}$ *has rank* m. *Generators of the first* A_T*-module in 2. can be computed via Gröbner bases.*

3. Parametrization of stabilizing compensators*: Let* $E = E^2$ *be a matrix as in 2. such that* $(\mathrm{id}_{p+m} - E)\binom{0}{\mathrm{id}_m}$ *has rank* m *and let* $t \in T$ *be a common denominator of the entries of* E. *Define*

$$(-\widetilde{Q}, \widetilde{P}) := t(\mathrm{id}_{p+m} - E) \in A^{(p+m)\times(p+m)}. \quad \textit{Then}$$
$$\widetilde{\mathcal{B}} := \left\{\begin{pmatrix}\widetilde{u}\\ \widetilde{y}\end{pmatrix} \in \mathcal{F}^{p+m};\ \widetilde{P} \circ \widetilde{y} = \widetilde{Q} \circ \widetilde{u}\right\}$$

is an IO behavior with input $\widetilde{u}$ *and a stabilizing compensator of the given behavior* $\mathcal{B}$. *It satisfies*

$$(\mathcal{B} \cap \mathcal{F}_2^{p+m}) \bigoplus (\widetilde{\mathcal{B}} \cap \mathcal{F}_2^{p+m}) = \mathcal{F}_2^{p+m} = \bigoplus_{\lambda\in\Lambda_2} \mathcal{F}_\lambda^{p+m} \quad \textit{where}$$
$$\mathcal{F}_\lambda := \mathbb{C}[t]e^{\lambda\bullet t} \cong \bigoplus_{\mu\in\mathbb{N}^r} \mathbb{C}e_{\lambda,\mu}.$$

The steady state part $\widetilde{\mathcal{B}} \cap \mathcal{F}_2^{p+m}$ *of each stabilizing compensator* $\widetilde{\mathcal{B}}$ *of* $\mathcal{B}$ *is obtained in this fashion.*

4. *If the unstable region* Λ_2 *is ideal-convex the behavior* $\mathcal{B}$ *is stabilizable if and only if the characteristic variety* $\mathrm{char}(\mathcal{B})$ *is contained in the stable region* Λ_1. *Recall in this context that* $\mathcal{B}$ *is stable if and only if* $\mathrm{char}(\mathcal{B}^0) \subseteq \Lambda_1$*, and that* $\mathrm{char}(\mathcal{B}) \subseteq \mathrm{char}(\mathcal{B}^0)$.

5. *If* M_T *is torsionfree the behavior* $\mathcal{B}$ *is stabilizable if and only if* $M_T \cong \mathcal{M}$ *(compare (7.2)) is a projective* A_T*-module. This is exactly the condition for the stabilizability of the transfer matrix* H *according to* [33, Corollary 3].

Algorithm 8.2 1. At various instances of the stabilization theory one has to check whether a variety meets Λ_2. All relevant unstable regions $\Lambda_2 \subset \mathbb{C}^r = \mathbb{R}^{2r}$ are real semi-algebraic [5, p. 21]. For these the equation $V(\mathfrak{a}) \cap \Lambda_2 = \emptyset$ can be constructively decided by means of the *Tarski-Seidenberg-principle* and *cylindrical algebraic decomposition.* This fact is of highest significance for stabilization theory. *We* learned it from S. Tsarev at the D2-conference of the Gröbner semester in Linz and have yet to study it in detail. In the mean-time we realized that the application of such methods in stabilization theory has already a longer history, for instance in several papers by N. K. Bose et al. [1, 6, 8, 24].

Whether under the necessary condition $V(\mathfrak{a}) \cap \Lambda_2 = \emptyset$ the inequality $\mathfrak{a} \cap T \neq \emptyset$ can likewise be decided and whether even a particular element $t \in \mathfrak{a} \cap T$ can be constructed is open in the moment. This problem is obviously related to the Open Problem 7.6 concerning ideal-convexity and was also discussed in [52] for the closed unit polydisc.

2. The annihilator ideal $\mathfrak{a} := \operatorname{ann}_A(M^0)$ (compare (5.1)) can be computed via Gröbner bases. Its variety $V(\mathfrak{a}) = \operatorname{char}(M^0) = \operatorname{char}(\mathcal{B}^0)$ is the characteristic variety of the autonomous part $\mathcal{B}^0$ of $\mathcal{B}$. According to Theorem 6.2 the behavior $\mathcal{B}$ is stable if and only if $V(\mathfrak{a}) \cap \Lambda_2 = \emptyset$, and this can be decided due to item 1.

3. Stabilizability is checked via the condition 1.(iii) of Theorem 8.1, i. e., by solving the inhomogeneous linear equation

$$R_{\mathrm{st}} G_{\mathrm{st}} R_{\mathrm{st}} = R_{\mathrm{st}} \quad \text{in} \quad A_T^{(p+m)\times k_{\mathrm{st}}}. \tag{8.1}$$

In particular, one needs the matrix R_{st} which is known in the following case. The annihilator $\mathfrak{a} = \operatorname{ann}_A(t(M))$ of the torsion module $t(M)$ of M can be computed via Gröbner bases and the equation $V(\mathfrak{a}) \cap \Lambda_2 = \emptyset$ can be checked constructively according to item 1. If this is the case Theorem 7.1 implies $R_{\mathrm{st}} = R_{\mathrm{cont}}$ and therefore R_{st} is known in this important case.

4. Assume, more generally, that a matrix $R_{\mathrm{st}} \in A^{k_{\mathrm{st}}\times(p+m)}$ is known such that $U_{\mathfrak{T}} = A_T^{1\times k_{\mathrm{st}}} R_{\mathrm{st}}$. Recall that the matrix G_{st} in equation (8.1) has denominators in T. Instead of solving this inhomogeneous equation we compute generators of the solution module L of the homogeneous linear system $R_{\mathrm{st}} G R_{\mathrm{st}} = x R_{\mathrm{st}}$, i. e.,

$$L := \{(G, x) \in A^{k_{\mathrm{st}}\times(p+m)} \times A;\ R_{\mathrm{st}} G R_{\mathrm{st}} - x R_{\mathrm{st}} = 0\},$$

via Gröbner bases for the polynomial algebra $A = \mathbb{C}[s]$ and then obtain generators of the ideal

$$\mathfrak{a} := \operatorname{proj}(L) = \{x \in A;\ \exists (G, x) \in L\}.$$

The equation (8.1) has a solution

$$G_{\mathrm{st}} \in A_T^{(p+m)\times k_{\mathrm{st}}} \text{ if and only if there is } t \in \mathfrak{a} \cap T, \text{ and then}$$

$$G_{\mathrm{st}} := \frac{G}{t} \text{ with } (G, t) \in L \text{ is one solution.}$$

Recall that for the existence of $t \in \mathfrak{a} \cap T$ the equation $V(\mathfrak{a}) \cap \Lambda_2 = \emptyset$ is necessary and can be checked constructively according to item 1. The decision of $\mathfrak{a} \cap T \neq \emptyset$ and the eventual construction of $t \in \mathfrak{a} \cap T$ are, however, still open as explained in item 1.

5. If *one* solution G_{st} and its ensuing idempotent matrix

$$E_1 := G_{\mathrm{st}} R_{\mathrm{st}} \in A_T^{(p+m)\times(p+m)}$$

have been computed the construction of all other matrices E according to Theorem 8.1, (2), is simple. Via Gröbner bases one computes A-generators of the solution module

$$L_0 := \{X \in A^{(p+m)\times k_{\rm st}};\ R_{\rm st} X R_{\rm st} = 0\}$$

and these are also A_T-generators of the solution module

$$L_{\rm st} := \{X \in A_T^{(p+m)\times k_{\rm st}};\ R_{\rm st} X R_{\rm st} = 0\} = L_{0,T}$$

which appears in the theorem.

6. The preceding considerations show that important parts of the algorithms are constructive, but that difficult open problems remain for the constructive checking of stabilizability of a given system. The construction of *all* stabilizing compensators is then simple.

9 Proper stabilization

As explained in the introduction with many references most papers on multidimensional stabilization treat the stabilization of *discrete proper* transfer matrices. The main advantage of the latter is that they can often be naturally interpreted as input/output *maps* or *operators*, see, for instance, [10, chapter 1 and 3] and [27, §6]. *In this section we characterize those IO systems which can be stabilized according to the preceding section with the additional property that the resulting feedback system is proper (and, of course, stable).*

The assumptions and notations of the previous sections remain in force. First we will define the ring of proper and stable rational functions which plays the part of the ring A_T of stable rational functions in the previous sections.

Definition and Corollary 9.1 [27, Theorem 6.50] *Let* $\deg_{s_\rho}(f)$ *denote the univariate degree of* $f \in \mathbb{C}[s] = \mathbb{C}[s_1, \ldots, \widehat{s_\rho}, \ldots, s_r][s_\rho]$. *The* multivariate *or* componentwise degree *of a polynomial is the function*

$$\deg : \mathbb{C}[s] \setminus \{0\} \to \mathbb{N}^r,\ f \mapsto \left(\deg_{s_1}(f), \ldots, \deg_{s_r}(f)\right).$$

It satisfies $\deg(fg) = \deg(f) + \deg(g)$ *and can therefore be extended to the field of rational functions* $\mathbb{C}(s) := \left\{\frac{a}{b},\ a, b \in \mathbb{C}[s],\ b \neq 0\right\}$ *via*

$$\deg : \mathbb{C}(s) \setminus \{0\} \to \mathbb{Z}^r,\ \frac{a}{b} \mapsto \deg\left(\frac{a}{b}\right) := \deg(a) - \deg(b).$$

In contrast to the degree-function induced by a term well-order on $\mathbb{N}^r$ *the componentwise degree* $\deg(f)$ *of a polynomial* $f = \sum_{\mu \in \mathbb{N}^r} a_\mu s^\mu \in \mathbb{C}[s]$ *does not, in general, belong to its support* $\operatorname{supp}(f) := \{\mu \in \mathbb{N}^r;\ a_\mu \neq 0\}$, *for instance*

$$\deg(s_1 + s_2) = (1, 1) \notin \operatorname{supp}(s_1 + s_2) = \{(1, 0), (0, 1)\}.$$

The polynomials satisfying $\deg(f) \in \operatorname{supp}(f)$ *are called* componentwise unital *(cw-unital).*

Theorem and Definition 9.2 [27, Theorem 6.60] *A rational function in* $\mathbb{C}(s)$ *is called* proper *if it is a power series in* $s^{-1} := (s_1^{-1}, \dots, s_r^{-1})$. *Thus we define the algebra of* proper rational functions

$$\mathcal{P} := \mathbb{C}(s) \cap \mathbb{C}[[s^{-1}]]. \tag{9.1}$$

It can also be described as

$$\mathcal{P} = \left\{ \frac{a}{t} \in \mathbb{C}(s);\ a, t \in \mathbb{C}[s],\ t \text{ is cw-unital},\ \deg(a) \leqslant_{\mathrm{cw}} \deg(t) \right\} \tag{9.2}$$

where $\leqslant_{\mathrm{cw}}$ *is the partial ordering of* $\mathbb{Z}^r$ *defined by*

$$\mu \leqslant_{\mathrm{cw}} \nu : \Longleftrightarrow \forall \rho = 1, \dots, r :\ \mu_\rho \leq \nu_\rho : \Longleftrightarrow \exists \gamma \in \mathbb{N}^r :\ \nu = \mu + \gamma.$$

Then

$$\mathcal{S} := \mathbb{C}[s]_T \cap \mathcal{P} = \mathbb{C}[s]_T \cap \mathbb{C}[[s^{-1}]] \tag{9.3}$$

is the algebra of proper stable *rational functions.*

For the interpretation of a proper transfer matrix as discrete IO operator we identify

$$\mathbb{C}[[s^{-1}]] = \mathbb{C}^{\mathbb{N}^r},\ u = (u_\mu)_{\mu \in \mathbb{N}^r} = \sum_{\mu \in \mathbb{N}^r} u_\mu s^{-\mu}.$$

The action of $\mathbb{C}[s]$ on $\mathbb{C}^{\mathbb{N}^r}$ is that by left shifts, i. e., $(s^\lambda \circ u)_u = u_{\lambda+\mu}$. The multiplication or convolution $*$ of $\mathbb{C}[[s^{-1}]]$ induces the action on itself by right shifts. Then [27, (6.37)]

$$P \circ (H * u) = Q \circ u \quad \text{for} \quad P, Q \in \mathbb{C}[s], \quad u \in \mathbb{C}^{\mathbb{N}^r} \quad \text{if} \quad H := P^{-1}Q \in \mathcal{P}.$$

In particular, the unit impulse $\delta_0 = 1 = s^{-0}$ gives rise to the *impulse response* $H = H * \delta_0$. More generally, a proper transfer matrix H of a discrete IO system $\mathcal{B}$ can be interpreted as the IO map $u \mapsto H * u$ and as its impulse response, in detail

$$\mathcal{B} = \left\{ \begin{pmatrix} y \\ u \end{pmatrix} \in (\mathbb{C}^{\mathbb{N}^r})^{p+m};\ P \circ y = Q \circ u \right\},\ \operatorname{rank}(P) = p,\ PH = Q,\ H \in \mathcal{P}^{p \times m}$$

$$\begin{pmatrix} H * u \\ u \end{pmatrix} \in \mathcal{B} \text{ for all } u \in (\mathbb{C}^{\mathbb{N}^r})^m, \text{ especially } H_{-j} = H * (1, \dots, 0, 1, 0, \dots, 0)^\top.$$

The following weak, but important condition on $\mathbb{C}^r = \Lambda_1 \uplus \Lambda_2$ is satisfied by all standard stability decompositions, for example those in (6.2). Let

$$\operatorname{proj}_\rho : \mathbb{C}^r \to \mathbb{C},\ \lambda \mapsto \lambda_\rho,$$

denote the projection on the ρth component for $\rho = 1, \dots, r$. In the sequel we assume

$$\operatorname{proj}_\rho(\Lambda_2) \subsetneq \mathbb{C}\ \text{, i.e., } \mathbb{C} \setminus \operatorname{proj}_\rho(\Lambda_2) \neq \emptyset, \quad \text{for all} \quad \rho = 1, \dots, r,$$

$$\text{choose} \quad \alpha \in \prod_{\rho=1}^{r} (\mathbb{C} \setminus \operatorname{proj}_\rho(\Lambda_2)) \quad \text{and define}$$

$$p_\rho := s_\rho - \alpha_\rho, \quad p := (p_1, \dots, p_r) = s - \alpha \in \mathbb{C}[s]^r, \quad \text{hence} \quad p_\rho^{-1} \in \mathcal{S}.$$

Theorem 9.3 *We give two descriptions of the ring of proper and stable rational functions:*

1. $\mathcal{S} = \left\{\frac{a}{t} \in \mathbb{C}(s);\ a \in \mathbb{C}[s],\ t \in T \text{ and cw-unital},\ \deg\left(\frac{a}{t}\right) \leqslant_{\mathrm{cw}} 0\right\}$. *This description is derived from that of* $\mathcal{P}$ *in equation* (9.2).
2. $\mathcal{S}_1 := \mathbb{C}[p^{-1}] := \mathbb{C}\left[\frac{1}{s_1-\alpha_1}, \ldots, \frac{1}{s_r-\alpha_r}\right] = \left\{\frac{a}{p^\mu},\ a \in \mathbb{C}[p],\ \deg(a) \leqslant_{\mathrm{cw}} \mu\right\}$ *is the polynomial algebra in* $p_1^{-1}, \ldots, p_r^{-1}$ *and a subring of* $\mathcal{S}$ *where*
$$\mathbb{C}[p] := \mathbb{C}[p_1, \ldots, p_r] = \mathbb{C}[s-\alpha] = \mathbb{C}[s], \quad \text{but}$$
$$\mathbb{C}[p^{-1}] \neq \mathbb{C}[s^{-1}] \quad \text{since, for instance,} \quad p_1^{-1} = \sum_{k=0}^{\infty} \alpha_1^k (s_1^{-1})^{k+1} \notin \mathbb{C}[s^{-1}].$$
3. *The set* $T_1 := \left\{\frac{t}{p^{\deg(t)}};\ t \in T \text{ and cw-unital}\right\}$ *is a multiplicatively closed saturated subset of* $\mathcal{S}_1$ *and* $\mathcal{S} = \mathcal{S}_{1,T_1}$ *is the associated quotient ring.*

Theorem 9.4 *As a quotient ring of the polynomial algebra* $\mathcal{S}_1 = \mathbb{C}[p^{-1}]$ *the ring* $\mathcal{S}$ *of proper stable rational functions is a Noetherian factorial integral domain. Its prime elements are the irreducible polynomials in* $\mathcal{S}_1 \setminus T_1$.

With these preparations we can now treat the *proper stabilizability* of IO systems.

Definition and Corollary 9.5 *An IO system* $\mathcal{B}$ *as in section 5 is called* proper *if its transfer matrix* H *belongs to* $\mathcal{P}^{p\times m}$. *If it is proper and stable according to section 6 then* $H \in \mathcal{S}^{p\times m}$.

The following theorem is the analogue of Theorem 8.1 for proper stabilization.

Theorem and Definition 9.6 (Proper stabilizability) *With the notations from Theorem 8.1 the following properties are equivalent for an IO system*
$$\mathcal{B} = \left\{\left(\begin{smallmatrix} y \\ u \end{smallmatrix}\right) \in \mathcal{F}^{p+m},\ P \circ y = Q \circ u\right\}:$$

1. *The system* $\mathcal{B}$ *is* proper stabilizable, *i. e., it is stabilizable in the sense of Definition 2.1 and section 8 with the additional property that the resulting feedback system is proper.*
2. *The* $\mathcal{S}$*-module* $U_{\mathfrak{T}} \cap \mathcal{S}^{1\times l}$ *is a direct summand of* $\mathcal{S}^{1\times l}$.

Algorithm 9.7 We use the preceding theorem to check algorithmically whether an IO behaviour $\mathcal{B}$ is proper stabilizable and, if so, to construct all compensators. This algorithm is subdivided into four main steps:

1. *Find* $\mathbb{C}[s]$*-generators of* U_{st}: In many cases this can be solved according to Algorithm 8.2, item 3.

2. *Find an $\mathcal{S}$-generating system of* $U_{\mathfrak{T}} \bigcap \mathcal{S}^{1\times l}$, $U_{\mathfrak{T}} = U_{\mathrm{st},T}$: Recall that

$$T = \bigcap_{\lambda\in\Lambda_2} (\mathbb{C}[s] \setminus \mathfrak{m}_\lambda)$$

and that therefore

$$U_{\mathrm{st}} = \mathbb{C}[s]^{1\times l} \bigcap \bigcap_{\lambda\in\Lambda_2} U_{\mathfrak{m}_\lambda}$$

is T-closed, i. e.,

$$u \in \mathbb{C}[s]^{1\times l}, \quad t \in T, \quad tu \in U_{\mathrm{st}} \quad \text{imply} \quad u \in U_{\mathrm{st}}.$$

This property implies the quotient representation

$$U_{\mathrm{st},T} \bigcap \mathcal{S}^{1\times l} = \widehat{U}_{T_1} \quad \text{where}$$

$$\widehat{U} := \mathbb{C}[p^{-1}]^{1\times l} \cap \mathbb{C}[p, p^{-1}]U_{\mathrm{st}} \subseteq \mathbb{C}[p^{-1}]^{1\times l}.$$

Hence $\mathbb{C}[p^{-1}]$-generators of $\widehat{U}$ are also $\mathcal{S}$-generators of $U_{\mathrm{st},T} \bigcap \mathcal{S}^{1\times l}$. These can be computed by means of the following argument from [55]. Let

$$\mathbb{C}[p, q] := \mathbb{C}[p_1, \dots, p_r, q_1, \dots, q_r]$$

be the polynomial ring in $2r$ indeterminates with the ideal

$$I := \sum_{\rho=1}^{r} \mathbb{C}[p, q](p_\rho q_\rho - 1).$$

The algebra isomorphism

$$\mathbb{C}[p, p^{-1}] \cong \mathbb{C}[p, q]/I, \; p_\rho \leftrightarrow \overline{p_\rho}, \; p_\rho^{-1} \leftrightarrow \overline{q_\rho},$$

makes the algebra $\mathbb{C}[p, p^{-1}]$ of Laurent polynomials accessible to the Gröbner calculus. We compute a Gröbner basis of $\mathbb{C}[p, q]U_{\mathrm{st}} + I\mathbb{C}[p, q]^{1\times l}$ with respect to an elimination term order for $p_1, \dots, p_r$. With the identification $\mathbb{C}[q] = \mathbb{C}[\overline{q}] = \mathbb{C}[p^{-1}]$ its elements in $\mathbb{C}[q]^{1\times l}$ are a Gröbner basis and hence a generating system of $\widehat{U}$ and of $U_{\mathfrak{T}} \bigcap \mathcal{S}^{1\times l}$.

3. *Check whether* $U_{\mathfrak{T}} \bigcap \mathcal{S}^{1\times l}$, $U_{\mathfrak{T}} = U_{\mathrm{st},T}$, *is a direct summand of* $\mathcal{S}^{1\times l}$: Part 4 of Algorithm 8.2 can be applied in this situation with the polynomial algebra $\mathbb{C}[p^{-1}]$ and $U_{\mathfrak{T}} \bigcap \mathcal{S}^{1\times l}$ instead of $\mathbb{C}[s]$ and $U_{\mathfrak{T}}$. Notice, however, that part 4 of Algorithm 8.2 requires the algorithmic construction of $t_1 \in T_1 \bigcap \mathfrak{a}_1$ for an ideal $\mathfrak{a}_1 \subset \mathbb{C}[p^{-1}]$ which is an open problem in general.

4. *Construct the stabilizing compensators*: Assume that the IO behavior $\mathcal{B}$ is proper stabilizable. According to parts 4 and 5 of Algorithm 8.2 we construct one and then all direct complements

$$V_2 = \mathcal{S}^{1\times k_2} R_2^s, \; R_2^s = (-Q_2^s, P_2^s) \in \mathcal{S}^{k_2\times(p+m)}, \; \text{of } U_{\mathfrak{T}} \bigcap \mathcal{S}^{1\times l} \text{ in } \mathcal{S}^{1\times l}.$$

Let $t \in T$ be a common denominator of the entries of R_2^s, i. e.,

$$(-Q_2, P_2) := t(-Q_2^s, P_2^s) \in \mathbb{C}[s]^{k_2 \times (p+m)}.$$

Then generically

$$\mathrm{rank}(P_2) = m \text{ and } \mathcal{B}_2 := \left\{ \begin{pmatrix} u_2 \\ y_2 \end{pmatrix} \in \mathcal{F}^{p+m};\, P_2 \circ y_2 = Q_2 \circ u_2 \right\}$$

is a proper stabilizing compensator of $\mathcal{B}$. All such compensators are obtained in the same fashion.

10 Discrete BIBO stability

In this section we discuss stability in the discrete case for the signal space $\mathbb{C}^{\mathbb{N}^r}$ *and the stability decomposition*

$$\mathbb{C}^r = \Lambda_1 \uplus \Lambda_2,\ \Lambda_2 := \{w \in \mathbb{C};\ |w| \geq 1\}^r. \tag{10.1}$$

Then the ring $\mathcal{S}$ *of proper stable rational functions in the sense of* (9.3) *coincides precisely with ring of causal structurally stable rational functions as discussed, for instance, in* [16, Definition 3.47] *and* [22, p. 60]. *This implies in particular that the transfer matrix of a proper stable system is BIBO stable.*

To see this in detail we need some preparations. The notations are the same as in section 9. We use various rings of power series and the abbreviation

$$z_\rho := s_\rho^{-1}, \text{ hence } \mathbb{C}[[s^{-1}]] = \mathbb{C}[[z]] = \mathbb{C}[[z_1, \ldots, z_r]] = \mathbb{C}^{\mathbb{N}^r}.$$

For a vector $R = (R_1, \ldots, R_r) \in \mathbb{R}^r_{>0}$, i. e., $R_\rho > 0$ for all ρ, we consider the open resp. closed polydisc

$$U(R) := \prod_{\rho=1}^{r} \{z_\rho \in \mathbb{C};\ |z_\rho| < R_\rho\} \subset \overline{U}(R) := \prod_{\rho=1}^{r} \{z_\rho \in \mathbb{C};\ |z_\rho| \leq R_\rho\}$$

in particular the unit polydiscs $U(\mathbf{1}) \subset \overline{U}(\mathbf{1})$ with $\mathbf{1} := (1, \ldots, 1)$.

If X is an arbitrary subset of $\mathbb{C}^r$ a function is called holomorphic in X if it is *defined in an open neighborhood* of X and is there holomorphic. If X is open f need, of course, only be defined on X. The algebra $\mathcal{O}(X)$ of all holomorphic functions on X is a subalgebra of the algebra $\mathrm{C}^0(X)$ of all continuous functions. If f is holomorphic in $U(R)$, $R \in \mathbb{R}^r_{>0}$, it has a Taylor series expansion $f(z) = \sum_{\mu \in \mathbb{N}^r} a_\mu z^\mu$ which is compactly convergent in $U(R)$, i. e., uniformly convergent on compact sets of $U(R)$. The spaces of power series

$$B_R := \left\{ f = \sum_{\mu \in \mathbb{N}^r} a_\mu z^\mu \in \mathbb{C}[[z]];\ \|f\|_R := \sum_{\mu \in \mathbb{N}^r} |a_\mu| R^\mu < \infty \right\} \tag{10.2}$$

and in particular

$$\ell^1(\mathbb{N}^r) := B_{\mathbf{1}} = \left\{ f = \sum_{\mu\in\mathbb{N}^r} a_\mu z^\mu;\ \|f\|_{\mathbf{1}} = \sum_{\mu\in\mathbb{N}^r} |a_\mu| < \infty \right\}$$

are subalgebras of $\mathbb{C}[[z]]$ and indeed Banach algebras with respect to $\|f\|_R$ [15, Satz 1 on p. 16], and moreover

$$B_R \subset \mathcal{O}(U(R)) \cap \mathrm{C}^0(\overline{U}(R)).$$

Let $\ell^\infty(\mathbb{N}^r)$ denote the space of all bounded multisequences, i. e.,

$$\ell^\infty(\mathbb{N}^r) := \left\{ u = (u_\mu)_{\mu\in\mathbb{N}^r} = \sum_{\mu\in\mathbb{N}^r} u_\mu z^\mu \in \mathbb{C}^{\mathbb{N}^r};\ \exists M > 0 \forall \mu \in \mathbb{N}^r : |u_\mu| \leq M \right\}.$$

Corollary 10.1 *The algebra $B_{\mathbf{1}} = \ell^1(\mathbb{N}^r)$ and the space $\ell^\infty(\mathbb{N}^r)$ are $B_{\mathbf{1}}$-submodules of $\mathbb{C}[[z]]$ with respect to the convolution multiplication of $\mathbb{C}[[z]]$. In other terms,*

$$\text{if} \quad H \in \ell^1(\mathbb{N}^r),\ u = (u_\mu)_{\mu\in\mathbb{N}^r} \in \ell^i(\mathbb{N}^r) \quad \text{then} \quad H * u \in \ell^i(\mathbb{N}^r),\ i = 1, \infty.$$

The first resp. the second of these properties are called the ℓ^1, ℓ^1- resp. BIBO stability *of the transfer function H.*

Let now $\mathcal{S}_{ss}$ denote the ring of causal structurally stable rational functions according to [16, Definition 3.47], i. e.,

$$\mathcal{S}_{ss} := \left\{ \frac{a(z)}{b(z)} \in \mathbb{C}(z) = \mathbb{C}(s);\ \forall z \in \overline{U}(\mathbf{1}) : b(z) \neq 0 \right\} \subset \mathbb{C}[z]_{\mathfrak{m}} = \mathbb{C}(z) \cap \mathbb{C}[[z]]$$

where $\mathfrak{m}$ is the maximal ideal $\mathfrak{m} := \sum_{\rho=1}^r \mathbb{C}[z]z_\rho \subset \mathbb{C}[z]$, and let

$$\mathcal{S} := \mathcal{P} \cap \mathbb{C}[s]_T = \mathbb{C}[z]_{\mathfrak{m}} \cap \mathbb{C}[s]_T \subset \mathbb{C}(s) = \mathbb{C}(z)$$

be the ring of proper stable rational functions with respect to the stability decomposition (10.1) from (9.3).

Theorem 10.2 *These two rings of stable rational functions coincide, i. e.,*

$$\mathcal{S} = \mathcal{S}_{ss}, \quad \text{in particular} \quad \mathcal{S} \subset \mathcal{O}(\overline{U}(\mathbf{1})) \subset B_{\mathbf{1}} = \ell^1(\mathbb{N}^r).$$

Therefore proper stable rational functions are ℓ^1, ℓ^1-stable and BIBO stable.

The proof of the preceding theorem uses multivariate *Laurent series* [50, p. 38ff]. In [26, Remark 5.12, (1)] we called it an open problem.

Remark 10.3 A causal structurally stable rational function

$$H = \frac{a(z)}{b(z)} \quad \text{with} \quad b(z) \neq 0 \quad \text{for all} \quad z \in \overline{U}(\mathbf{1}) \quad \text{and relatively prime} \quad a, b$$

belongs to $B_{\mathbf{1}} = \ell^1(\mathbb{N}^r) \subset \mathcal{O}(U(\mathbf{1})) \cap \mathrm{C}^0(\overline{U}(\mathbf{1}))$, i. e., H is BIBO stable. If $a = 1$ the converse is true as is easily seen. In general, BIBO stability of H implies that $b(z)$ has no zeros in $U(\mathbf{1})$ and that any zero of b in $\overline{U}(\mathbf{1})$ is also one of a. It may happen that there are infinitely many such zeros in $\overline{U}(\mathbf{1}) \setminus U(\mathbf{1})$.

Consult [10, section 1.2, Problem 1 on p. 245], [8, section 1.4.1], [11, pp. 190–198], [21], [47, pp. 492–496] and the original papers quoted there for further discussions of multidimensional discrete BIBO stable transfer functions and various stability tests.

11 Continuous BIBO stability

In this section we discuss BIBO stability in the continuous case of partial differential equations for various signal spaces and the stability decomposition

$$\mathbb{C}^r = \Lambda_1 \uplus \Lambda_2, \; \Lambda_2 := \{w \in \mathbb{C};\; \Re(w) \geq 0\}^r.$$

Stability of polynomials, rational functions, the ring $\mathcal{S}$ etc. always refer to this decomposition. In [7, p. 12], [21, Definition 5 on p. 140] and [35, p. 79] the stable polynomials in our sense are called *strict Hurwitz polynomials.*

In the last chapter 8 on 'Stability of Multidimensional Continuous Systems' of the survey paper [21] E. I. Jury conjectures with an algebraic, non-analytic reduction to the discrete case that the inverse of a bivariate very strict Hurwitz polynomial (VSHP) is BIBO stable in the sense that its impulse response is an absolutely integrable function [21, Definition 6, Theorem 23 and Remark]. However, the exact nature of the continuous systems is not specified and the (usually difficult) existence of the impulse response is not addressed in that survey. In [35, Theorem 6.1] Jury's conjecture is quoted as a *theorem.* Notice that there are other types of continuous systems than those of the present paper, for instance delay-differential systems.

In this section we show that the impulse response of a proper rational function exists, but is a measure *and not a function. We derive necessary and sufficient conditions for BIBO stability and in particular prove Jury's conjecture.*

If X is a set,

$$x = (x_1, \ldots, x_r) \in X^r \quad \text{and} \quad S \subseteq \{1, \ldots, r\}, \quad S' := \{1, \ldots, r\} \setminus S$$

we define the subvector $x_S := (x_\rho)_{\rho \in S} \in X^S$, for instance

$$\mu = (\mu_1, \ldots, \mu_r), \quad \mathbf{1} := (1, \ldots, 1) \in \mathbb{N}^r, \quad \mu_S = (\mu_\rho)_{\rho \in S}, \quad \mathbf{1}_S \in \mathbb{N}^S,$$

and identify $x = (x_S, x_{S'}) \in X^r = X^S \times X^{S'}$. As before we use the variables

$$s = (s_1, \ldots, s_r), \quad z = (z_1, \ldots, z_r) := s^{-1}, \quad z_\rho := s_\rho^{-1}, \quad t = (t_1, \ldots, t_r)$$

which we consider both as indeterminates and as vectors in $\mathbb{C}^r$.

We need certain preparations concerning distributions from [40, pp. 170–180] and consider the subspace $\mathcal{D}'_+ := \mathcal{D}'_+(\mathbb{R}^r) \subset \mathcal{D}' := \mathcal{D}'(\mathbb{R}^r)$ of all distributions with left bounded support, i. e., whose support is contained in a subset $t + \mathbb{R}^r_{\geq 0}$, $t \in \mathbb{R}^r$. The space $\mathcal{D}'_+$ is a commutative integral domain with respect to the convolution product $*$ [40, Theorem XIV on p. 173]. If $\varphi_1, \dots, \varphi_r$ are univariate functions in $\mathcal{D}(\mathbb{R}) = \mathrm{C}_0^\infty(\mathbb{R})$ and $T_1, \dots, T_r$ are univariate distributions in $\mathcal{D}'(\mathbb{R})$ then their tensor products are defined [40, chapter IV]:

$$T := \otimes_{\rho=1}^r T_\rho \in \mathcal{D}'(\mathbb{R}^r), \quad \varphi := \otimes_{\rho=1}^r \varphi_\rho \in \mathcal{D}(\mathbb{R}^r),$$

$$\varphi(t_1, \dots, t_r) := \prod_{\rho=1}^r \varphi_\rho(t_\rho), \quad T(\varphi) := \prod_{\rho=1}^r T_\rho(\varphi_\rho).$$

If the T_ρ have even left bounded support, i. e., are contained in $\mathcal{D}'_+(\mathbb{R})$, then so has their tensor product. With the Heaviside step function in one variable $t_1 \in \mathbb{R}$

$$Y(t_1) := \begin{cases} 1 & \text{if } t_1 > 0 \\ 0 & \text{otherwise} \end{cases} \quad \text{and its derivative} \quad \delta := Y', \ \delta(\varphi) = \varphi(0),$$

we define [40, (II,2;31)]

$$Y_k(t_1) := \begin{cases} \frac{t_1^{k-1}}{(k-1)!} Y(t_1) & \text{if } k \in \mathbb{Z}, \ k > 0 \\ \delta^{(-k)} = s_1^{-k} \circ \delta & \text{if } k \leq 0 \end{cases} \qquad Y_k \in \mathcal{D}'_+(\mathbb{R}),$$

where $f(s)\circ$ denotes the action of a polynomial as differential operator. For $k > 0$ the distribution Y_k is a function with

$$Y_k(\varphi) = \int_0^\infty \frac{t_1^{k-1}}{(k-1)!} \varphi(t_1) dt_1, \quad \varphi \in \mathcal{D}(\mathbb{R}).$$

The distributions Y_k are defined for all complex numbers k, but we do not need this here. The Y_k satisfy [40, p. 43, (VI,5;6)]

$$Y_k' = s_1 \circ Y_k = Y_{k-1} \quad \text{and} \quad Y_k * Y_l = Y_{k+l}, \quad k, l \in \mathbb{Z}.$$

These univariate distributions give rise to their r-dimensional counter-parts

$$Y_\mu := Y_{\mu_1} \otimes \dots \otimes Y_{\mu_r} \in \mathcal{D}'_+(\mathbb{R}^r), \ \mu \in \mathbb{Z}^r,$$

in particular for $\mu \in \mathbb{N}^r$, $\varphi \in \mathcal{D}(\mathbb{R}^r)$ and $S := S(\mu) := \mathrm{supp}(\mu) = \{\rho; \ \mu_\rho \neq 0\}$:

$$Y_\mu = \frac{t_{S(\mu)}^{\mu_{S(\mu)} - \mathbf{1}_{S(\mu)}}}{(\mu_{S(\mu)} - \mathbf{1}_{S(\mu)})!} \otimes_{\rho \in S(\mu)} Y(t_\rho) \otimes \otimes_{\rho \in S(\mu)'} \delta(t_\rho)$$

$$Y_\mu(\varphi) = \int_{0_S}^{\infty_S} \frac{t_S^{\mu_S - \mathbf{1}_S}}{(\mu_S - \mathbf{1}_S)!} \varphi(t_S, 0_{S'}) dt_S,$$

where

$$t_S^{\mu_S-\mathbf{1}_S}=\prod_{\rho\in S}t_\rho^{\mu_\rho-1},\quad (\mu_S-\mathbf{1}_S)!=\prod_{\rho\in S}(\mu_\rho-1)!,$$

$$\int_{0_S}^{\infty_S}(-)\,dt_S:=\int_0^\infty\cdots\int_0^\infty(-)\prod_{\rho\in S}dt_\rho.$$

It is easy to see that the identities

$$Y_\mu * Y_\nu = Y_{\mu+\nu}\quad\text{for}\quad \mu,\nu\in\mathbb{Z}^r\quad\text{and}$$
$$s^\nu\circ Y_\mu = Y_{-\nu}*Y_\mu = Y_{\mu-\nu}\quad\text{for}\quad \mu\in\mathbb{Z}^r,\ \nu\in\mathbb{N}^r$$

hold. With B_R from (10.2) let

$$\mathbb{C}\langle z\rangle := \bigcup_{R\in\mathbb{R}^r_{>0}} B_R\subset\mathbb{C}[[z]]$$

denote the algebra of *(locally) convergent* power series. We also consider the algebras of *Laurent* polynomials resp. power series

$$\mathbb{C}[z,s]:=\mathbb{C}[z]_z=\oplus_{\mu\in\mathbb{Z}^r}\mathbb{C}z^\mu,\ \mathbb{C}[[z]]_z=\{s^\mu H(z);\ H\in\mathbb{C}[[z]],\ \mu\in\mathbb{N}^r\},$$
$$\mathbb{C}\langle z\rangle_z=\{s^\mu H(z);\ H\in\mathbb{C}\langle z\rangle,\ \mu\in\mathbb{N}^r\}.$$

The equations $Y_\mu * Y_\nu = Y_{\mu+\nu}$ imply that the map

$$\mathbb{C}[z]_z\to\mathcal{D}'_+(\mathbb{R}^r),\ f=\sum_{\mu\in\mathbb{Z}^r}f_\mu z^\mu\mapsto f(Y):=\sum_{\mu\in\mathbb{Z}^r}f_\mu Y_\mu,$$

is an algebra homomorphism, and it is indeed a monomorphism. This monomorphism can be extended to $\mathbb{C}\langle z\rangle_z$ as follows. A convergent power series $H=\sum_{\mu\in\mathbb{N}^r}H_\mu z^\mu$ can be considered as an analytic functional on $\mathbb{C}^r$ [17, chapter 4, §5], [27, pp. 64–70], i. e., as the continuous linear map

$$H:\mathcal{O}(\mathbb{C}^r)\to\mathbb{C},\ u=\sum_{\mu\in\mathbb{N}^r}u_\mu z^\mu\mapsto H(u):=\sum_{\mu\in\mathbb{N}^r}H_\mu u_\mu,$$
$$\text{where}\quad \mathcal{O}(\mathbb{C}^r):=\bigcap_{R\in\mathbb{R}^r_{>0}}B_R\subset\mathbb{C}\langle z\rangle=\bigcup_{R\in\mathbb{R}^r_{>0}}B_R$$

is the algebra of entire functions or everywhere convergent power series in z. Its *Laplace transform* [17, Definition 4.5.2]

$$\widehat{H}(t):=H_z(\exp(z\bullet t))=\sum_{\mu\in\mathbb{N}^r}H_\mu\frac{t^\mu}{\mu!}$$

is contained in $\mathcal{O}(\mathbb{C}^r,\exp)$, i. e., is an entire holomorphic function of *exponential type*, and the map

$$\mathbb{C}\langle z\rangle\cong\mathcal{O}(\mathbb{C}^r,\exp),\quad H=\sum_{\mu\in\mathbb{N}^r}H_\mu z^\mu\mapsto\widehat{H}(t)=H_z(\exp(z\bullet t))=\sum_{\mu\in\mathbb{N}^r}H_\mu\frac{t^\mu}{\mu!},$$

is an isomorphism [27, (4.28)].

Theorem 11.1 *The algebra monomorphism* $\mathbb{C}[z] \to \mathcal{D}'_+$, $z^\mu \mapsto Y_\mu$, *can be uniquely extended to algebra monomorphisms*

$$\mathbb{C}\langle z\rangle \to \mathcal{D}'_+, \quad H = \sum_{\mu\in\mathbb{N}^r} H_\mu z^\mu \mapsto H(Y) := \sum_{\mu\in\mathbb{N}^r} H_\mu Y_\mu, \quad \text{and}$$

$$\mathbb{C}\langle z\rangle_z \to \mathcal{D}'_+, \quad a = s^\mu H \mapsto a(Y) := Y_{-\mu} * H(Y) = s^\mu \circ H(Y), \quad \mu \in \mathbb{N}^r,$$

the first one being continuous in the sense that it transforms analytically convergent sequences in $\mathbb{C}\langle z\rangle$ *into convergent sequences of distributions.*

The algebra homomorphism $\mathbb{C}\langle z\rangle_z \to \mathcal{D}'_+(\mathbb{R}^r)$ *makes* $\mathcal{D}'_+(\mathbb{R}^r)$ *a* $\mathbb{C}\langle z\rangle_z$*-module via* $a \circ T := a(Y) * T$.

Each $H(Y)$, $H \in \mathbb{C}\langle z\rangle$, *is indeed a measure, i.e., a continuous linear function on* $\mathrm{C}_0^0(\mathbb{R}^r)$. *If* $\varphi(t)$ *is a continuous function with compact support in*

$$K := \{t \in \mathbb{R}^r;\ \forall\rho:\ -R_\rho \le t_\rho \le R_\rho\}, \quad R \in \mathbb{R}^r_{>0},$$

and $\|\varphi\|_K := \sup_{t\in K} |\varphi(t)|$ *is its maximum norm then*

$$\left|\sum_{\mu\in\mathbb{N}^r} H_\mu Y_\mu(\varphi)\right| \le \sum_{\mu\in\mathbb{N}^r} |H_\mu| \frac{R^\mu}{\mu!} \|\varphi\|_K,$$

where $\sum_{\mu\in\mathbb{N}^r} |H_\mu| \frac{R^\mu}{\mu!} < \infty$ *since* $\widehat{H}(t) = \sum_{\mu\in\mathbb{N}^r} H_\mu \frac{t^\mu}{\mu!}$ *is entire. Here a sequence of (locally) convergent power series is* analytically convergent [15, p. 31] *if it is convergent in some Banach algebra* B_R, $R \in \mathbb{R}^r_{>0}$.

Corollary and Definition 11.2 (Impulse response and BIBO stability) *For a proper rational function*

$$H = \frac{f(s)}{q(s)} = \sum_{\mu\in\mathbb{N}^r} H_\mu z^\mu \in \mathcal{P} \subset \mathbb{C}\langle z\rangle$$

and an input $u \in \mathcal{D}'_+(\mathbb{R}^r)$ *the output*

$$y := H \circ u = H(Y) * u = (H \circ \delta) * u$$

is the unique solution of the partial differential equation $q \circ y = f \circ u$ *in* $\mathcal{D}'_+(\mathbb{R}^r)$. *For obvious reasons the solution*

$$H \circ \delta = H(Y) = \sum_{\mu\in\mathbb{N}^r} H_\mu \frac{t_{S(\mu)}^{\mu_{S(\mu)} - \mathbf{1}_{S(\mu)}}}{(\mu_{S(\mu)} - \mathbf{1}_{S(\mu)})!} \otimes_{\rho\in S(\mu)} Y(t_\rho) \otimes \otimes_{\rho\in S(\mu)'} \delta(t_\rho)$$

is called the fundamental solution *or the* impulse response *of the differential equation. The transfer function* H *is called* BIBO stable *if for a bounded continuous function* u *the output* $y := H \circ (uY_{\mathbf{1}})$ *with support in* $\mathbb{R}^r_{\ge 0}$ *is also bounded.*

Remark 11.3 The injectivity and therefore divisibility of the $\mathbb{C}[s]$-module $\mathcal{D}'(\mathbb{R}^r)$ of distributions according to Theorem 3.4 implies that for every rational function $H =$

$\frac{f}{q} \in \mathbb{C}(s)$ and every distribution u the differential equation $q \circ y = f \circ u$ has a distributional solution y, in particular $q \circ y_0 = f \circ \delta$. In general and in contrast to the preceding corollary, however, the convolution $y_0 * u$ and the equation $q \circ (y_0 * u) = f \circ (\delta * u) = f \circ u$ make no sense for u with non-compact support and cannot be used to solve $q \circ y = f \circ u$. The search for fundamental solutions *with good properties* is one of the basic tasks of the theory of partial differential equations [18, 19].

Definition and Corollary 11.4 *1. For $\mu \in \mathbb{Z}^r$ and $\alpha \in \mathbb{C}^r$ define* [40, VI,5;13]

$$Y_\mu^\alpha := e^{\alpha \bullet t} Y_\mu \in \mathcal{D}'_+(\mathbb{R}^r).$$

Then,

$$Y_\mu^\alpha * Y_\nu^\alpha = Y_{\mu+\nu}^\alpha \quad \text{for } \mu, \nu \in \mathbb{Z}^r$$

and

$$(s-\alpha)^\nu \circ Y_\mu^\alpha = Y_{\mu-\nu}^\alpha \quad \text{for } \mu \in \mathbb{Z}^r,\ \nu \in \mathbb{N}^r.$$

In particular, for $\mu \in \mathbb{N}^r$ the transfer function

$$H := (s-\alpha)^{-\mu} = z^\mu (1-\alpha z)^{-\mu},\ (1-\alpha z)^{-\mu} := \prod_{\rho=1}^{r} (1-\alpha_\rho z_\rho)^{-\mu_\rho},$$

has the impulse response Y_μ^α. If the components of α have negative real parts the rational function $(s-\alpha)^{-\mu}$ is proper stable and BIBO stable.

2. *The equation*

$$H \circ Y_{\mathbf{1}} = H(Y) * Y_{\mathbf{1}} = \sum_{\mu \in \mathbb{N}^r} H_\mu Y_\mu * Y_{\mathbf{1}} = \sum_{\mu \in \mathbb{N}^r} H_\mu Y_{\mu+\mathbf{1}} = \sum_{\mu \in \mathbb{N}^r} H_\mu \frac{t^\mu}{\mu!} Y_{\mathbf{1}}$$
$$= \widehat{H}(t) Y_{\mathbf{1}}$$

shows that BIBO stability of H implies the boundedness of $\widehat{H}(t)$ on $\mathbb{R}^r_{\geq 0}$.

The preceding corollary and partial fraction decomposition imply the well-known and important fact that each *univariate* proper stable rational function is BIBO stable.

Sufficient and necessary conditions for BIBO stability can be derived from the disjoint decomposition

$$\mathbb{N}^r = \biguplus_{S \subseteq \{1,\ldots,r\}} \{\mu \in \mathbb{N}^r;\ \mathrm{supp}(\mu) = S\} = \biguplus_S (\mathbf{1}_S + \mathbb{N}^S)$$

which induces the direct decomposition $\mathbb{C}[[z]] = \oplus_S \mathbb{C}[[z_S]] z_S^{\mathbf{1}_S}$. Hence, any $H \in \mathbb{C}[[z]]$ has a unique finite sum representation

$$H(z) = \sum_{\mu \in \mathbb{N}^r} H_\mu z^\mu = \sum_{S \subseteq \{1,\ldots,r\}} H_S(z_S)$$

with

$$H_S(z_S) := \sum_{\mu \in \mathbb{N}^r, \mathrm{supp}(\mu) = S} H_\mu z_S^{\mu_S} \in \mathbb{C}[[z_S]].$$

If H is convergent and u is continuous on $\mathbb{R}^r$ this implies

$$H(Y) = \sum_S \widetilde{H_S}(t_S) \otimes_{\rho\in S} Y(t_\rho) \otimes \otimes_{\rho\in S'} \delta(t_\rho)$$

with

$$\widetilde{H_S}(t_S) := (s_S^{\mathbf{1}_S} \circ \widehat{H_S})(t_S) = \sum_{\mu\in\mathbb{N}^r, \mathrm{supp}(\mu)=S} H_\mu \frac{t_S^{\mu_S-\mathbf{1}_S}}{(\mu_S-\mathbf{1}_S)!} \in \mathcal{O}(\mathbb{C}^S, \exp),$$

and

$$(H \circ (uY_{\mathbf{1}}))(t) = (H(Y) * (uY_{\mathbf{1}}))(t) = \sum_S \int_{0_S}^{t_S} \widetilde{H_S}(\tau_S) u(t_S - \tau_S, t_{S'})\, d\tau_S.$$

Corollary 11.5 *A proper rational function $H \in \mathcal{P} \subset \mathbb{C}\langle z\rangle$ is BIBO stable if the 2^r entire functions $\widetilde{H_S}(t_S)$, $S \subseteq \{1,\dots,r\}$, are absolutely integrable over $\mathbb{R}^S_{\geq 0}$. If, in particular,*

$$q(s) = s^d + \sum_{\mu<_{\mathrm{cw}} d} q_\mu s^\mu, \quad d_\rho > 0 \quad \textit{for } \rho = 1,\dots,r, \quad \textit{and } f(s) = \sum_{\mu\leq_{\mathrm{cw}} d-\mathbf{1}} f_\mu s^\mu,$$

the strictly *proper rational function*

$$H := \frac{f(s)}{q(s)} = \left(\sum_{\mu\leq_{\mathrm{cw}} d-\mathbf{1}} f_\mu z^{d-\mu}\right)\left(\sum_{k=0}^{\infty}(-1)^k \left(\sum_{\mu<_{\mathrm{cw}} d} q_\mu z^{d-\mu}\right)^k\right) = \sum_{\mu\geq_{\mathrm{cw}}\mathbf{1}} H_\mu z^\mu = H_{\{1,\dots,r\}}$$

is BIBO stable if and only if the function $\widetilde{H}(t)Y_{\mathbf{1}}$ with

$$\widetilde{H}(t) := \left((s_1\dots s_r)\circ \widehat{H}\right)(t) := \sum_{\mu\geq_{\mathrm{cw}}\mathbf{1}} H_\mu \frac{t^{\mu-\mathbf{1}}}{(\mu-\mathbf{1})!}$$

is absolutely integrable, i. e., if

$$\int_0^\infty \cdots \int_0^\infty |\widetilde{H}(\tau)|\, d\tau_1 \dots d\tau_r < \infty.$$

With the notations of the preceding corollary we derive integral representations of the functions $\widetilde{H}(t)$ and $\widehat{H}(t)$ as in [17, (4.5.7)]. Let Γ_ρ with $\rho = 1,\dots,r$ be simple piecewise smooth closed curves around the origin in $\mathbb{C}$ and $\Gamma := \Gamma_1 \times \dots \times \Gamma_r$. Then the standard equations

$$\frac{1}{(2\pi i)^r}\int_\Gamma s^\mu ds = \delta_{\mu,-\mathbf{1}}, \quad ds := ds_1 \dots ds_r, \quad \text{for } \mu \in \mathbb{Z}^r, \quad \text{hence}$$

$$\frac{t^\mu}{\mu!} = \frac{1}{(2\pi i)^r}\int_\Gamma e^{t\bullet s} s^{-\mu-\mathbf{1}} ds, \quad t \in \mathbb{C}^r, \quad \mu \in \mathbb{N}^r,$$

hold. Now choose $R \in \mathbb{R}^r_{>0}$ and the curves Γ_ρ such that the proper rational function H satisfies

$$\frac{f(s)}{q(s)} = H(z) = \sum_{\mu\in\mathbb{N}^r} H_\mu s^{-\mu} \in B_{R^{-1}}, \quad z := s^{-1}, \quad \text{i.e.,}$$
$$\sum_{\mu\in\mathbb{N}^r} |H_\mu| R^{-\mu} < \infty, \quad \text{and} \quad \Gamma_\rho \subset \{s_\rho \in \mathbb{C};\ |s_\rho| \geq R_\rho\} \quad \text{for all } \rho. \tag{11.1}$$

Corollary 11.6 *1. If* $H(s) = \frac{f(s)}{q(s)} = \sum_{\mu\in\mathbb{N}^r} H_\mu s^{-\mu}$ *is* strictly *proper as in Corollary 11.5 and if the conditions of* (11.1) *are satisfied then the integral representations*

$$\widetilde{H}(t) = \sum_{\mu\geq_{\mathrm{cw}}\mathbf{1}} H_\mu \frac{t^{\mu-\mathbf{1}}}{(\mu-\mathbf{1})!} = \frac{1}{(2\pi i)^r}\int_{\Gamma_1}\cdots\int_{\Gamma_r} e^{t\bullet s} H(s)ds_1\ldots ds_r$$

and

$$\widehat{H}(t) = \sum_{\mu\geq_{\mathrm{cw}}\mathbf{1}} H_\mu \frac{t^\mu}{\mu!} = \frac{1}{(2\pi i)^r}\int_{\Gamma_1}\cdots\int_{\Gamma_r} e^{t\bullet s} \frac{H(s)}{s_1\cdots s_r} ds_1\ldots ds_r$$

are valid.

2. *If*

$$H(s) = \frac{f(s)}{q(s)} = \sum_{S\subseteq\{1,\ldots,r\}} \sum_{\mu\in\mathbb{N}^r,\mathrm{supp}(\mu)=S} H_\mu s^{-\mu}$$

is an arbitrary proper rational function and if the conditions of (11.1) *are satisfied then for each subset* $S \subseteq \{1,\ldots,r\}$ *with* $|S|$ *elements*

$$\widetilde{H_S}(t_S) = \sum_{\mu\in\mathbb{N}^r,\mathrm{supp}(\mu)=S} H_\mu \frac{t_S^{\mu_S-\mathbf{1}_S}}{(\mu_S-\mathbf{1}_S)!}$$
$$= \frac{1}{(2\pi i)^{|S|}}\int_{\prod_{\rho\in S}\Gamma_\rho} e^{t_S\bullet s_S}\left(\sum_{\mu\in\mathbb{N}^r,\mathrm{supp}(\mu)=S} H_\mu s^{-\mu}\right) ds_S.$$

The preceding integral representation of $\widetilde{H}(t)$ permits to prove Jury's conjecture. For this purpose let $H(s) = \frac{f(s)}{q(s)} \in \mathbb{C}(s) = \mathbb{C}(s_1, s_2)$ be a strictly proper bivariate rational function with $\deg_{s_\rho}(f) < \deg_{s_\rho}(q)$ for $\rho = 1, 2$ where $q(s_1, s_2)$ is a *very strict Hurwitz polynomial* (VSHP) in the sense of [21, Definition 6]. This signifies that q is proper and stable with the following additional property. Let

$$q = s^d + \sum_{\mu<_{\mathrm{cw}}d} q_\mu s^\mu = \sum_{\mu_1=0}^{d_1} b^1_{\mu_1}(s_2)s_1^{\mu_1} = \sum_{\mu_2=0}^{d_2} b^2_{\mu_2}(s_1)s_2^{\mu_2}$$

with polynomials $b^1_{\mu_1}(s_2) \in \mathbb{C}[s_2]$ and $b^2_{\mu_2}(s_1) \in \mathbb{C}[s_1]$ where $b^1_{d_1}$ and $b^2_{d_2}$ are, of course, not zero. The additional condition on q is that the univariate polynomials $b^1_{d_1}$, $b^2_{d_2}$ are strict Hurwitz polynomials, i. e., have no zeros in the closed right half-plane of $\mathbb{C}$.

Theorem 11.7 (Jury's conjecture) [21, Theorem 23 and Remark] *If* $H(s) = \frac{f(s)}{q(s)}$ *is a bivariate and strictly proper rational function with a very strict Hurwitz polynomial* q *then there is a vector* $(a_1, a_2) \in \mathbb{R}^2_{>0}$ *and a constant* $C > 0$ *such that*

$$|\widetilde{H}(t)| \leq Ce^{-a \bullet t} = Ce^{-a_1 t_1 - a_2 t_2} \quad \textit{for } t \in \mathbb{R}^2_{\geq 0}.$$

In particular, $\widetilde{H}(t)$ *is absolutely integrable over* $\mathbb{R}^2_{\geq 0}$ *and* H *is BIBO stable.*

Example 11.8 The rational functions

$$H_1(s) := (s_1 s_2 + s_1 + 2s_2 + 1)^{-1}$$

$$\text{and} \quad H_2(s) := (s_1 s_2 + s_1 + s_2 + 1)^{-1} = (s_1 + 1)^{-1}(s_2 + 1)^{-1}$$

satisfy the assumptions of the preceding theorem. BIBO stability of H_2 also follows from Corollary 11.4.

In the future we will study further applications to BIBO stability of Corollaries 11.5 and 11.6, in particular to higher dimensions.

Acknowledgement: We thank the referee for his/her critical reading of the article and the suggestions, in particular for additional research. Discussions at the conferences D2 and D3 of the Gröbner semester in Linz in May 2006, arranged by Professors B. Buchberger and H. Engl and their coworkers at RISC and RICAM, led to improvements of our work.

Bibliography

[1] B. D. O. Anderson, N. K. Bose, and E. I. Jury, *Output feedback stabilization and related problems–solution via decision methods*, IEEE Trans. Automatic Control AC-20 (1975), pp. 53–66.

[2] M. Bisiacco, E. Fornasini, and G. Marchesini, *2D systems feedback compensation: an approach based on commutative linear transformations*, Linear Algebra Appl. 121 (1989), pp. 135–150.

[3] J.-E. Björk, *Rings of Differential Operators*, North Holland, Amsterdam, 1979.

[4] V. D. Blondel and A. Megretski (eds.), *Unsolved Problems in Mathematical Systems and Control Theory*, Princeton University Press, Princeton, NJ, 2004.

[5] J. Bochnak, M. Coste, and M.-F. Roy, *Géométrie algébrique réelle*, Ergebnisse der Mathematik und ihrer Grenzgebiete (3) [Results in Mathematics and Related Areas (3)], vol. 12, Springer-Verlag, Berlin, 1987.

[6] N. K. Bose, *New techniques and results in multidimensional problems*, in [9], pp. 136–154.

[7] ———, *Trends in multidimensional systems theory*, in [10], pp. 1–40.

[8] ———, *Two decades of multidimensional systems research and future trends*, in [14], pp. 5–27.

[9] N. K. Bose (ed.), *Multidimensional systems: theory and applications*, IEEE Press, New York, 1979.

[10] N. K. Bose, J. P. Guiver, E. W. Kamen, H. M. Valenzuela, and B. Buchberger, *Multidimensional systems theory*, Mathematics and its Applications, vol. 16, D. Reidel Publishing Co., Dordrecht, 1985.

[11] D. E. Dudgeon and R. M. Mersereau, *Multidimensional digital signal processing*, Prentice-Hall, Englewood Cliffs, 1984.

[12] L. Ehrenpreis, *Fourier analysis in several complex variables*, Pure and Applied Mathematics, Vol. XVII, Wiley-Interscience Publishers, New York-London-Sydney, 1970.

[13] E. Fornasini and G. Marchesini, *Structure and properties of two-dimensional systems*, in [46], pp. 37–88.

[14] K. Gałkowski and J. Wood (eds.), *Multidimensional Signals, Circuits and Systems*, Taylor and Francis, London, 2001.

[15] H. Grauert and R. Remmert, *Analytische Stellenalgebren*, Die Grundlehren der mathematischen Wissenschaften, vol. 176, Springer-Verlag, Berlin, 1971.

[16] J. P. Guiver and N. K. Bose, *Causal and weakly causal 2-D filters with application in stabilization*, in [10], pp. 52–100.

[17] L. Hörmander, *An introduction to complex analysis in several variables*, revised. ed., North-Holland Publishing Co., Amsterdam, 1973.

[18] ______, *The analysis of linear partial differential operators. I*, Grundlehren der Mathematischen Wissenschaften [Fundamental Principles of Mathematical Sciences], vol. 256, Springer-Verlag, Berlin, 1983, Distribution theory and Fourier analysis.

[19] ______, *The analysis of linear partial differential operators. II*, Grundlehren der Mathematischen Wissenschaften [Fundamental Principles of Mathematical Sciences], vol. 257, Springer-Verlag, Berlin, 1983, Differential operators with constant coefficients.

[20] R. Jeltsch and M. Mansour (eds.), *Stability theory*, International Series of Numerical Mathematics, vol. 121, Birkhäuser Verlag, Basel, 1996, Papers from the International Conference "Centennial Hurwitz on Stability Theory" held in Ascona, May 21–26, 1995.

[21] E. I. Jury, *Stability of multidimensional systems and related problems*, in [46], pp. 89–159.

[22] Z. Lin, *Output feedback stabilizability and stabilization of linear n-D systems*, in [14], pp. 59–76.

[23] Z. Lin, J. Lam, K. Gałkowski, and S. Xu, *A constructive approach to stabilizability and stabilization of a class of nD systems*, Multidimensional Syst. Signal Process. 12 (2001), pp. 329–343, Special issue: Applications of Gröbner bases to multidimensional systems and signal processing.

[24] R. Liska and S. Steinberg, *Solving stability problems using quantifier elimination*, in [20], pp. 205–210.

[25] B. Malgrange, *Systèmes différentiels à coefficients constants*, Séminaire Bourbaki, Soc. Math. France, 1962–1963, pp. 246.01–246.11.

[26] U. Oberst, *Stability and stabilization of multidimensional input/output systems*, SIAM J. Control and Optim. 45 (2006), pp. 1467–1507.

[27] ______, *Multidimensional constant linear systems*, Acta Appl. Math. 20 (1990), pp. 1–175.

[28] ______, *Variations on the fundamental principle for linear systems of partial differential and difference equations with constant coefficients*, Appl. Algebra Engrg. Comm. Comput. 6 (1995), pp. 211–243.

[29] U. Oberst and F. Pauer, *The constructive solution of linear systems of partial difference and differential equations with constant coefficients*, Multidimensional Syst. Signal Process. 12 (2001), pp. 253–308, Special issue: Applications of Gröbner bases to multidimensional systems and signal processing.

[30] V. P. Palamodov, *Linear differential operators with constant coefficients*, Die Grundlehren der mathematischen Wissenschaften, Band 168, Springer-Verlag, New York, 1970.

[31] J. W. Polderman and J. C. Willems, *Introduction to mathematical systems theory*, Texts in Applied Mathematics, vol. 26, Springer-Verlag, New York, 1998, A behavioral approach.

[32] A. Quadrat, *An algebraic interpretation to the operator-theoretic approach to stabilizability. I. SISO systems*, Acta Appl. Math. 88 (2005), pp. 1–45.

[33] ———, *A lattice approach to analysis and synthesis problems*, Math. Control Signals Systems 18 (2006), pp. 147–186.

[34] ———, *On a generalization of the Youla-Kučera parametrization. II. The lattice approach to MIMO systems*, Math. Control Signals Systems 18 (2006), pp. 199–235.

[35] H. C. Reddy, G. Rojan, and G. S. Moschytz, *Two-dimensional Hurwitz polynomials*, in [20], pp. 75–85.

[36] P. Rocha, *Feedback control of multidimensional behaviors*, Systems Control Lett. 45 (2002), pp. 207–215.

[37] Joseph J. Rotman, *An introduction to homological algebra*, Academic Press Inc., New York, 1979.

[38] M. Scheicher and U. Oberst, *Multidimensional BIBO stability and Jury's conjecture*, to appear in Math. Control, Signals and Systems.

[39] ———, *Proper stabilization of multidimensional input/output behaviours*, submitted to Multidimensional Syst. Signal Process.

[40] L. Schwartz, *Théorie des distributions*, Publications de l'Institut de Mathématique de l'Université de Strasbourg, No. IX-X. Nouvelle édition, entiérement corrigée, refondue et augmentée, Hermann, Paris, 1966.

[41] S. Shankar, *Can one control the vibrations of a drum?*, Multidimensional Syst. Signal Process. 11 (2000), pp. 67–81, Recent progress in multidimensional control theory and applications.

[42] ———, *The lattice structure of behaviors*, SIAM J. Control Optim. 39 (2001), pp. 1817–1832.

[43] S. Shankar and V. R. Sule, *Algebraic geometric aspects of feedback stabilization*, SIAM J. Control Optim. 30 (1992), pp. 11–30.

[44] B. Stenström, *Rings of quotients*, Springer-Verlag, New York, 1975, Die Grundlehren der Mathematischen Wissenschaften, Band 217, An introduction to methods of ring theory.

[45] V. R. Sule, *Feedback stabilization over commutative rings: the matrix case*, SIAM J. Control Optim. 32 (1994), pp. 1675–1695.

[46] S. G. Tzafestas (ed.), *Multidimensional systems. Techniques and applications*, Marcel Dekker, New York, 1986.

[47] R. Unbehauen, *Systemtheorie*, Oldenbourg, München, 1990.

[48] A. I. G. Vardulakis, *Linear multivariable control*, John Wiley & Sons Ltd., Chichester, 1991, Algebraic analysis and synthesis methods.

[49] M. Vidyasagar, *Control system synthesis*, MIT Press Series in Signal Processing, Optimization, and Control, 7, MIT Press, Cambridge, MA, 1985, A factorization approach.

[50] V. S. Vladimirov, *Methods of the theory of functions of many complex variables*, The M.I.T. Press, Cambridge, Mass.-London, 1966.

[51] J. Wood, V. R. Sule, and E. Rogers, *Causal and stable input/output structures on multidimensional behaviors*, SIAM J. Control Optim. 43 (2004/05), pp. 1493–1520.

[52] L. Xu, Z. Lin, J.-Q. Ying, O. Saito, and Y. Anazawa, *Open problems in control of linear discrete multidimensional systems*, in [4], pp. 221–228.

[53] J.-Q. Ying, *On the strong stabilizability of MIMO n-dimensional linear systems*, SIAM J. Control Optim. 38 (1999), pp. 313–335.

[54] E. Zerz, *Topics in multidimensional linear systems theory*, Lecture Notes in Control and Information Sciences, vol. 256, Springer-Verlag London Ltd., London, 2000.

[55] E. Zerz and U. Oberst, *The canonical Cauchy problem for linear systems of partial difference equations with constant coefficients over the complete r-dimensional integral lattice $\mathbb{Z}^r$*, Acta Appl. Math. 31 (1993), pp. 249–273.

Author information

Ulrich Oberst, Institut für Mathematik der Universität Innsbruck,
Technikerstraße 25, A-6020 Innsbruck, Austria.
Email: `ulrich.oberst@uibk.ac.at`

Martin Scheicher, Institut für Mathematik der Universität Innsbruck,
Technikerstraße 25, A-6020 Innsbruck, Austria.
Email: `martin.scheicher@uibk.ac.at`

Radon Series Comp. Appl. Math 3, 191–214

Applications of filter coefficients and wavelets parametrized by moments

Georg Regensburger

Key words. Orthonormal wavelets, parametrized filter coefficients, moments, regularity, Hölder and Sobolev exponent, least asymmetric filters, rational filter coefficients.

AMS classification. 42C40, 65T60, 94A12, 68W30

1 Introduction

Wavelets and their generalizations are used in many areas of mathematics ranging from harmonic analysis over numerical analysis to signal and image processing, see for example Daubechies [11], Mallat [29], and Strang and Nguyen [42]. A function $\psi \in L^2(\mathbb{R})$ is an *orthonormal wavelet* if the family

$$\psi_{jk}(x) = 2^{j/2}\psi(2^j x - k) \quad \text{for } j, k \in \mathbb{Z}$$

of translated an dilated versions of ψ is an orthonormal basis of the Hilbert space $L^2(\mathbb{R})$. Alfred Haar gave in his dissertation from 1909, published in [17], the first example of an orthonormal wavelet

$$\psi(x) = \begin{cases} 1, & \text{for } 0 \le x < \frac{1}{2}, \\ -1, & \text{for } \frac{1}{2} \le x < 1, \\ 0, & \text{otherwise,} \end{cases}$$

which is now known as the *Haar wavelet*. Daubechies introduced in her seminal paper [10] a general method to construct compactly supported wavelets. Her construction is

This work was supported by the Austrian Science Fund (FWF) under the SFB grant F1322.

based on *scaling functions*, satisfying a *dilation equation*

$$\phi(x) = \sum_{k=0}^{N} h_k \phi(2x - k) \tag{1.1}$$

given by a linear combination of real *filter coefficients* h_k and dilated and translated versions of the scaling function; see the next section for an outline.

Imposing conditions on the scaling function gives, via the dilation equation (1.1), constraints on the filter coefficients. Orthonormality implies quadratic equations and vanishing moments of the associated wavelet and normalization linear constraints. Daubechies wavelets [10] have the maximal number of vanishing moments for a fixed number of filter coefficients, and so there are only finitely many solutions. Parametrizing all possible filter coefficients that correspond to compactly supported orthonormal wavelets has been studied by several authors [21, 28, 33, 38, 41, 46, 47, 49]. All parametrizations express the filter coefficients in terms of trigonometric functions, and there is no natural interpretation of the angular parameters for the resulting scaling function. Furthermore, one has to solve transcendental constraints for the parameters to find wavelets with more than one vanishing moment.

We gave parametrizations of filter coefficients such that the corresponding wavelets have several vanishing moments and that use the first discrete moments as parameters first in [36] and then simplified in [35]. See section 3 for the parametrizations of four to eight filter coefficients with one parameter and at least one, two, and three vanishing moments, respectively. To compute these parametrizations we used symbolic computation and for the more involved equations in particular Gröbner bases, which were introduced by Buchberger in [3], see also [4]. Other applications of Gröbner bases to the design of wavelets and filter coefficients are for example discussed in [6, 7, 16, 25, 26, 31, 32, 39].

As a first application of parametrized wavelets, we discussed in [36] how they can by used for compression by computing an optimal parameter for a given signal, see also [18]. In this paper, we describe several other applications. In section 4, we discuss the regularity of the scaling functions and wavelets corresponding to our parametrizations. We construct wavelets that have a higher Hölder exponent than the Daubechies wavelets. Filter design is another possible application of our parametrizations. We deal with the construction of least asymmetric orthonormal wavelets in section 5. Finally, we address the existence of rational filter orthogonal filter coefficients in section 6. For example, we show that there are no orthogonal filters with six nonzero filter coefficients and at least two sum rules. A Maple worksheet with all computations, several MATLAB functions to produce the figures and a GUI to compute with and illustrate parametrized wavelets are available on request from the author.

2 Wavelets and moments

We outline the construction of orthonormal wavelets based on scaling functions and recall the polynomial equations for the filter coefficients, see for example Daubechies [11] or Strang and Nguyen [42].

Orthonormality of the integer translates $\{\phi(x-l)\}_{l\in\mathbb{Z}}$ in $L^2(\mathbb{R})$, that is,

$$\int \phi(x)\phi(x-l)dx = \delta_{0,l}$$

implies, using the dilation equation (1.1), the quadratic equations

$$\sum_{k\in\mathbb{Z}} h_k h_{k-2l} = 2\delta_{0,l} \quad \text{for } l \in \mathbb{Z} \tag{2.1}$$

where we set $h_k = 0$ for $k < 0$ and $k > N$. We can assume that $h_0 h_N \neq 0$. Then with equation (2.1) we see that N must be odd and the number of filter coefficients even.

If the filter coefficients satisfy the necessary conditions for orthogonality (2.1) and the normalization

$$\sum_{k=0}^{N} h_k = 2, \tag{2.2}$$

there exists a unique solution of the dilation equation (1.1) in $L^2(\mathbb{R})$ with support $[0, N]$ and for which $\int \phi = 1$, see Lawton [23]. For almost all such scaling functions the integer translates $\{\phi(x-l)\}_{l\in\mathbb{Z}}$ are orthogonal, and then

$$\psi(x) = \sum_{k=0}^{N} (-1)^k h_{N-k}\phi(2x-k) \tag{2.3}$$

is an orthonormal wavelet.

Necessary and sufficient conditions for orthonormality were given by Cohen [8] and Lawton [24], see also Daubechies [11, section 6.3]. The only example with four filter coefficients that satisfies the equations (2.1) and (2.2) and where the integer translates of the corresponding scaling are not orthogonal is $h_0 = h_3 = 1$ and $h_1 = h_2 = 0$ with the scaling function

$$\phi(x) = \begin{cases} 1/3, & \text{for } 0 \leq x < 3, \\ 0, & \text{otherwise.} \end{cases} \tag{2.4}$$

The corresponding scaling function for the Haar wavelet is the box function

$$\phi(x) = \begin{cases} 1, & \text{for } 0 \leq x < 1, \\ 0, & \text{otherwise,} \end{cases}$$

with the filter coefficients $h_0 = h_1 = 1$. In general, there is no closed analytic form for the scaling function, and for computations with scaling functions and wavelets only the filter coefficients are used.

Vanishing moments of the associated wavelet are related to several properties of the scaling function and wavelet. For example, to regularity, the polynomial reproduction and the approximation order of the scaling function, and the decay of the wavelet coefficients for smooth functions, see Strang and Nguyen [42] and the survey [43] by Unser and Blu for details. The condition that the first p moments of the wavelet ψ vanish

$$\int x^l \psi(x)\, dx = 0 \quad \text{for } l = 0, \ldots, p-1$$

is using equation (2.3) equivalent to the *sum rules*

$$\sum_{k=0}^{N}(-1)^k k^l h_k = 0 \quad \text{for } l = 0, \dots, p-1. \tag{2.5}$$

One then says that ψ has p *vanishing moments* or the filter coefficients satisfy p sum rules.

Since we use *discrete moments*

$$m_n = \sum_{k=0}^{N} h_k k^n$$

of the filter coefficients as a parameters, we recall a well-known recursive relation between discrete and *continuous moments*

$$M_n = \int x^n \phi(x)\, dx$$

of the scaling function. Let ϕ be a scaling function satisfying $M_0 = \int \phi = 1$. Then $m_0 = 2$ and

$$M_n = \frac{1}{2^{n+1}-2}\sum_{i=1}^{n}\binom{n}{i} m_i M_{n-i},$$

$$m_n = \left(2^{n+1}-2\right) M_n - \sum_{i=1}^{n-1}\binom{n}{i} m_i M_{n-i} \quad \text{for } n > 0,$$

see for example Strang and Nguyen [42, p. 396]. Using the recursion we obtain for the first moments

$$\begin{aligned} M_1 &= 1/2\, m_1, \\ M_2 &= 1/6\, m_1^2 + 1/6\, m_2, \\ M_3 &= 1/28\, m_1^3 + 1/7\, m_1 m_2 + 1/14\, m_3 \end{aligned}$$

and

$$\begin{aligned} m_1 &= 2\, M_1, \\ m_2 &= -4\, M_1^2 + 6\, M_2, \\ m_3 &= 12\, M_1^3 - 24\, M_1 M_2 + 14\, M_3. \end{aligned}$$

Explicit formulas expressing the discrete moments in terms of the continuous and vice versa are given in [36].

3 Parametrizations

We discuss the parametrizations from [35] of four, six, and eight filter coefficients corresponding respectively to orthonormal wavelets with at least one, two, and three vanishing moments. All families depend on the first discrete moment

$$m = m_1 = \sum_{k=0}^{N} h_k k$$

of the filter coefficients.

3.1 Four filter coefficients

We have the following parametrization of filter coefficients with at least one vanishing moment:

$$\begin{aligned} h_0 &= 1/2 - 1/4\,a - 1/4\,w, \\ h_1 &= 1/2 - 1/4\,a + 1/4\,w, \\ h_2 &= 1/2 + 1/4\,a + 1/4\,w, \\ h_3 &= 1/2 + 1/4\,a - 1/4\,w \end{aligned} \tag{3.1}$$

with $w = \sqrt{4 - a^2}$ and $a = m - 3 \in [-2, 2]$.

Note that for $a = -a$ we obtain the flipped filter coefficients. For $a = 0$ we get the filter coefficients $(0, 1, 1, 0)$, which correspond to a translated Haar scaling function and wavelet. The parameter values $a = -2, 2$ give also Haar scaling functions with the filter coefficients $(1, 1, 0, 0)$ and $(0, 0, 1, 1)$. The *Daubechies wavelet* has two vanishing moments, so we have one more sum rule

$$2\,h_0 - h_1 + h_3 = 0.$$

Substituting the parametrized filter coefficients into this equations and solving for a, we get the two solutions $a = -\sqrt{3}, \sqrt{3}$ with the first discrete moments $m = 3-\sqrt{3}, 3+\sqrt{3}$. The first solution gives the famous Daubechies filters [10]

$$1/4\,(1 + \sqrt{3}, 3 + \sqrt{3}, 3 - \sqrt{3}, 1 - \sqrt{3}) \tag{3.2}$$

and the second the flipped version. See Figure 3.1 for plots of scaling functions for various parameter values.

We have a second parametrization of filter coefficients with at least one vanishing moment:

$$\begin{aligned} h_0 &= 1/2 - 1/4\,a + 1/4\,w, \\ h_1 &= 1/2 - 1/4\,a - 1/4\,w, \\ h_2 &= 1/2 + 1/4\,a - 1/4\,w, \\ h_3 &= 1/2 + 1/4\,a + 1/4\,w \end{aligned} \tag{3.3}$$

with $w = \sqrt{4 - a^2}$ and $a = m - 3 \in [-2, 2]$.

Comparing this solution with the parametrized filter coefficients (3.1), we see that w is replaced by $-w$ and so the two first and the two last filter coefficients are swapped. Note that again for $a = -a$ we obtain the flipped filters.

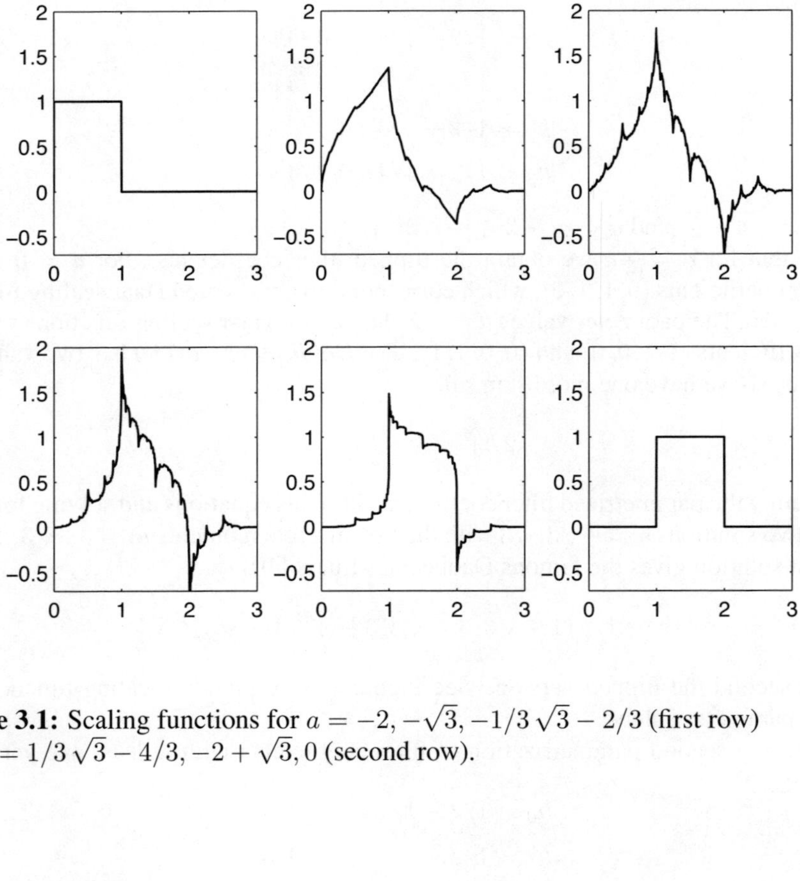

Figure 3.1: Scaling functions for $a = -2, -\sqrt{3}, -1/3\,\sqrt{3} - 2/3$ (first row) and $a = 1/3\,\sqrt{3} - 4/3, -2 + \sqrt{3}, 0$ (second row).

For $a = 0$ we now get the filter coefficients $(1, 0, 0, 1)$, which give the scaling function (2.4) where the integer translates of the scaling function are not orthogonal. The parameter values $a = -2, 2$ also give Haar scaling functions with the filter coefficients $(1, 1, 0, 0)$ and $(0, 0, 1, 1)$. This parametrization does not contain filter coefficients with a second vanishing moment. The corresponding scaling functions are, compared to the parametrization (3.1), irregular, see section 4 for details.

3.2 Six filter coefficients

We have the following parametrization of filter coefficients with with at least two vanishing moments:

$$\begin{aligned} h_0 &= -3/32 - 1/8\,a + 1/32\,a^2 - 1/32\,w, \\ h_1 &= 5/32 - 1/8\,a + 1/32\,a^2 + 1/32\,w, \\ h_2 &= 15/16 - 1/16\,a^2 + 1/16\,w \\ h_3 &= 15/16 - 1/16\,a^2 - 1/16\,w, \\ h_4 &= 5/32 + 1/8\,a + 1/32\,a^2 - 1/32\,w, \\ h_5 &= -3/32 + 1/8\,a + 1/32\,a^2 + 1/32\,w \end{aligned} \tag{3.4}$$

with $w = \sqrt{-a^4 + 14\,a^2 + 15}$ and $a = m - 5 \in [-\sqrt{15}, \sqrt{15}]$.

The Daubechies wavelet has one more vanishing moment, that is, it satisfies the sum rule

$$-9\,h_0 + 4\,h_1 - h_2 - h_4 + 4\,h_5 = 0.$$

Substituting the parametrized filter coefficients into this equations and solving for a, we get one real solution $a = -\sqrt{5 + 2\sqrt{10}}$, which gives the filter coefficients

$$\begin{aligned} 1/16\,(1 + \sqrt{10} + w, 5 + \sqrt{10} + 3\,w, 10 - 2\sqrt{10} + 2\,w, \\ 10 - 2\sqrt{10} - 2\,w, 5 + \sqrt{10} - 3\,w, 1 + \sqrt{10} - w) \end{aligned} \tag{3.5}$$

with $w = \sqrt{5 + 2\sqrt{10}}$. The Daubechies filters with four nonzero filter coefficients (3.2) satisfy two sum rules and are therefore contained in this parametrization. Their first discrete moment is $m = 3 - \sqrt{3}$. So here the corresponding parameter is $a = -2 - \sqrt{3}$. We get a translated version for $a = -\sqrt{3}$. For $a = -\sqrt{15}$ we obtain

$$1/8\,(3 + \sqrt{15}, 5 + \sqrt{15}, 0, 0, 5 - \sqrt{15}, 3 - \sqrt{15}).$$

The parameter $a = -1$ gives the first coiflet

$$1/16\,(1 - \sqrt{7}, 5 + \sqrt{7}, 14 + 2\sqrt{7}, 14 - 2\sqrt{7}, 1 - \sqrt{7}, -3 + \sqrt{7}),$$

see Daubechies [12] and [11, section 8.2]. For $a = 0$ we get

$$1/32\,(-3 - \sqrt{15}, 5 + \sqrt{15}, 30 + 2\sqrt{15}, 30 - 2\sqrt{15}, 5 - \sqrt{15}, -3 + \sqrt{15}).$$

See Figure 3.2 for plots of scaling functions for various parameter values. The corresponding scaling functions and wavelets for $a > 0$ become increasingly irregular, see section 4 for details.

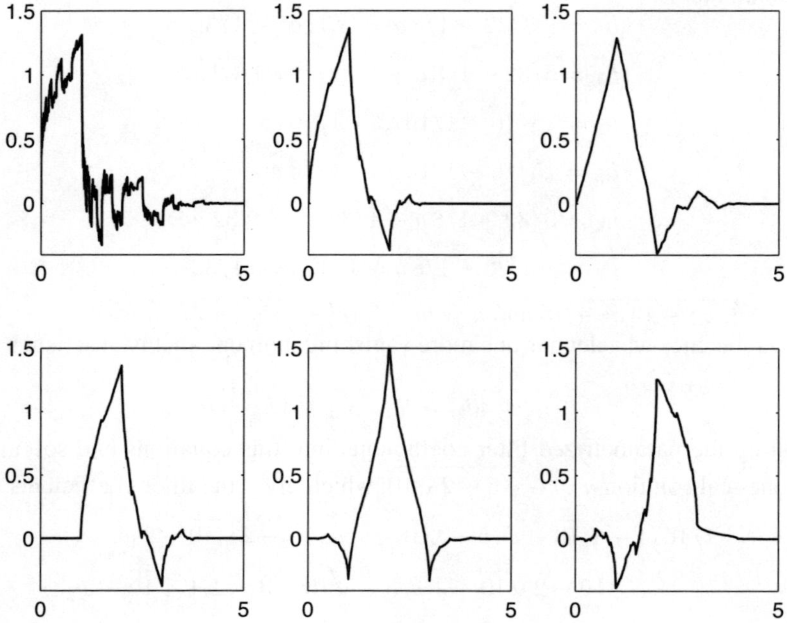

Figure 3.2: Scaling functions for $a = -\sqrt{15}, -2 - \sqrt{3}, -\sqrt{5 + 2\sqrt{10}}$ (first row) and $a = -\sqrt{3}, -1, 0$ (second row).

3.3 Eight filter coefficients

We have the following parametrization of filter coefficients with at least three vanishing moments:

$$
\begin{aligned}
h_0 &= -\frac{1}{512}\,\frac{a^5 - 7\,a^4 - 2\,a^3 + 30\,a^2 - 55\,a - 15 + (1-a)w}{a^2+1},\\
h_1 &= -\frac{1}{512}\,\frac{a^5 - 9\,a^4 + 30\,a^3 + 2\,a^2 - 23\,a + 63 + (1+a)w}{a^2+1},\\
h_2 &= \frac{1}{512}\,\frac{3\,a^5 - 5\,a^4 - 102\,a^3 + 186\,a^2 - 261\,a + 35 + 3(1-a)w}{a^2+1},\\
h_3 &= \frac{1}{512}\,\frac{3\,a^5 - 11\,a^4 - 70\,a^3 + 358\,a^2 - 229\,a + 525 + 3(1+a)w}{a^2+1},\\
h_4 &= -\frac{1}{512}\,\frac{3\,a^5 + 11\,a^4 - 70\,a^3 - 358\,a^2 - 229\,a - 525 + 3(1-a)w}{a^2+1},\\
h_5 &= -\frac{1}{512}\,\frac{3\,a^5 + 5\,a^4 - 102\,a^3 - 186\,a^2 - 261\,a - 35 + 3(1+a)w}{a^2+1},\\
h_6 &= \frac{1}{512}\,\frac{a^5 + 9\,a^4 + 30\,a^3 - 2\,a^2 - 23\,a - 63 + (1-a)w}{a^2+1},\\
h_7 &= \frac{1}{512}\,\frac{a^5 + 7\,a^4 - 2\,a^3 - 30\,a^2 - 55\,a + 15 + (1+a)w}{a^2+1}
\end{aligned}
\tag{3.6}
$$

with

$$w = \sqrt{-a^8 + 36\,a^6 - 182\,a^4 + 1540\,a^2 - 945},$$

$a = m - 7$ and a in the intervals

$$[-\sqrt{\beta}, -\sqrt{\alpha}] \quad \text{or} \quad [\sqrt{\alpha}, \sqrt{\beta}],$$

where α denotes the smaller and β the larger real root of

$$x^4 - 36\,x^3 + 182\,x^2 - 1540\,x + 945,$$

with numerical approximations

$$\sqrt{\alpha} = 0.8113601077\ldots \quad \text{and} \quad \sqrt{\beta} = 5.636256558\ldots .$$

The Daubechies wavelet satisfies one more sum rule

$$64\,h_0 - 27\,h_1 + 8\,h_2 - h_3 + h_5 - 8\,h_6 + 27\,h_7 = 0.$$

Substituting the parametrized filter coefficients (3.6) into this equation and solving for a, we get two real solution $a = -\sqrt{\beta}, -\sqrt{\alpha}$, where α denotes the smaller and β the larger real root of

$$x^4 - 28\,x^3 + 126\,x^2 - 1260\,x + 1225$$

or numerically

$$a = -4.989213573\ldots, -1.029063869\ldots .$$

The first parameter gives the Daubechies wavelet with extremal phase [11, p. 195] and the second the "least asymmetric" [11, p. 198]. The Daubechies wavelet with six nonzero filter coefficients (3.5) has the first discrete moment

$$m = 5 - \sqrt{5 + 2\sqrt{10}},$$

so the corresponding parameter value for the parametrization (3.6) is

$$a = -2 - \sqrt{5 + 2\sqrt{10}} = -5.365197664\ldots .$$

See Figure 3.3 for plots of scaling functions for various parameter values.

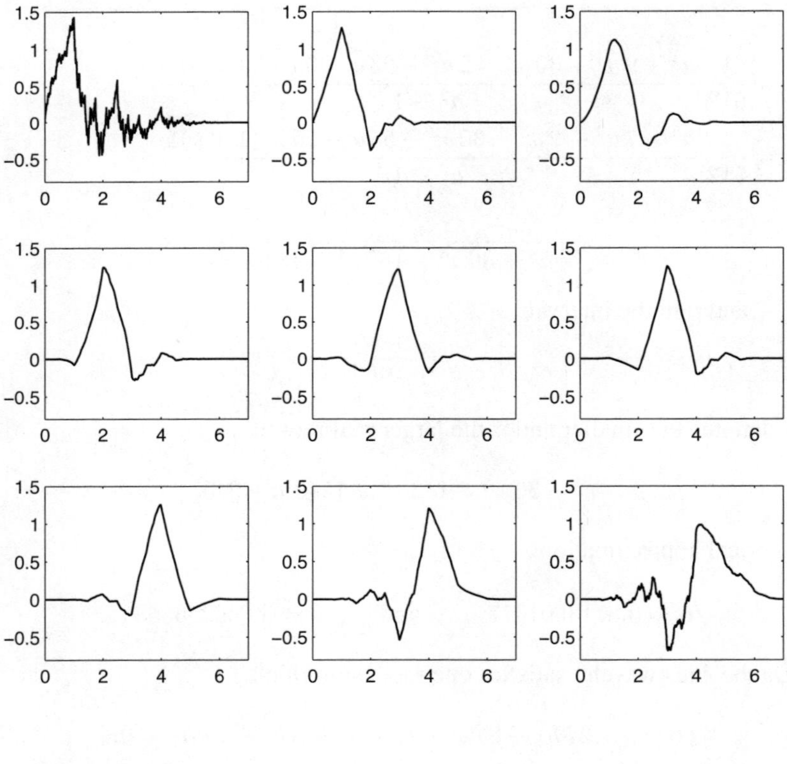

Figure 3.3: Scaling functions for $a = -5.636256559, -5.365197664, -4.989213573$ (first row), $a = -3.009138721, -1.029063869, -0.8113601077$ (second row), and $a = 0.8113601077, 2, 3$ (third row).

4 Regularity of scaling functions and wavelets

In this section, we discuss the regularity or smoothness of the scaling functions and wavelets corresponding to the parametrized filter coefficients from the previous section. The regularity of a function can be measured in different ways, we consider here the Hölder and Sobolev exponent.

We first recall the definitions. For $\alpha = n + \beta$, where $n \in \mathbb{N}$ and $0 \leq \beta < 1$, the set $C^\alpha = C^\alpha(\mathbb{R})$ is defined as the set of all functions f that are n times continuously differentiable and such that the nth derivative $f^{(n)}$ is uniformly *Hölder continuous* with exponent β, that is,

$$|f^{(n)}(x+h) - f^{(n)}(x)| \leq C|h|^\beta \quad \text{for all } x, h \in \mathbb{R}$$

where C is a constant. For $s \geq 0$ the *Sobolev space* $H^s = H^s(\mathbb{R})$ consists of all functions $f \in L^2(\mathbb{R})$ such that $(1 + |\xi|^2)^{s/2}\hat{f}(\xi) \in L^2(\mathbb{R})$, where $\hat{f}$ denotes the Fourier transform of f.

To measure the regularity or smoothness of a scaling function ϕ, one is interested respectively in the (optimal) *Sobolev*

$$s_{\max} = \sup\{s : \phi \in H^s\}$$

and *Hölder exponent*

$$\alpha_{\max} = \sup\{\alpha : \phi \in C^\alpha\}.$$

For a scaling function the Hölder exponent satisfies [44]

$$\alpha_{\max} \in [s_{\max} - 1/2, s_{\max}]. \tag{4.1}$$

The regularity of scaling functions is also related to vanishing moments of the corresponding wavelet. Villemoes [44] proved that if $\phi \in H^n$ with $n \in \mathbb{N}$, the filter coefficients satisfy $n+1$ sum rules or equivalently the corresponding wavelet has $n+1$ vanishing moments. So in particular if $\phi \in C^n$, then the filter coefficients satisfy $n+1$ sum rules, see also [11, pp. 153–156].

Eirola [14] and Villemoes [44] independently showed how the optimal Sobolev exponent can be computed from the spectral radius of a matrix depending on the filter coefficients, see also Strang and Nguyen [42] for further details. To find the optimal Hölder exponent is much more involved, see for example [9, 11, 13, 37], but Rioul [37] gave an algorithm to compute good lower bounds for the Hölder exponent. The algorithm produces monotonically increasing lower bounds with an increasing number of iterations, but the storage and computational costs approximately double for each additional iteration.

In Figures 4.1, 4.2 and 4.3 you can see plots of the Sobolev exponent of the corresponding scaling functions and wavelets depending on one parameter. For four filter coefficients the Sobolev exponents range from 0.5 to 1 (parametrization (3.1)) and from 0 to 0.5 (parametrization (3.3)). The maximum 1 is attained for the Daubechies wavelet since all other filter coefficients satisfy only one sum rule and hence their Sobolev exponent is necessarily less than one. We obtain numerically the maximal Sobolev exponent

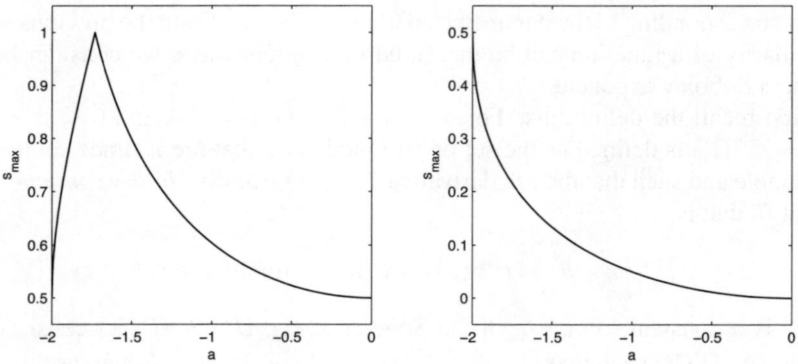

Figure 4.1: Sobolev exponent for scaling functions with four filter coefficients from equation (3.1) (left) and (3.3) (right).

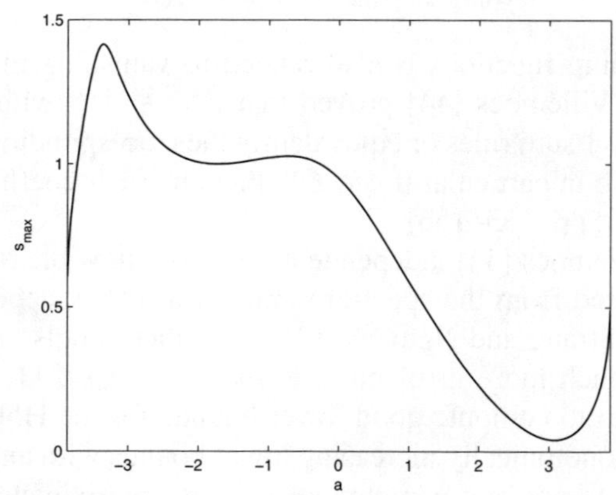

Figure 4.2: Sobolev exponent for scaling functions with six filter coefficients from equation (3.4).

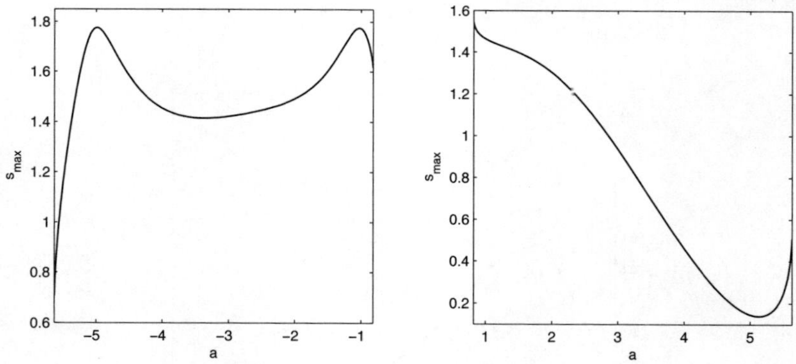

Figure 4.3: Sobolev exponent for scaling functions with eight filter coefficients from equation (3.6).

for respectively six and eight filter coefficients

$$s_{\max} = 1.4150, 1.7756,$$

at the parameter values for the Daubechies wavelets and the minimum is

$$s_{\max} = 0.0399, 0.1393$$

with parameter values

$$a = 3.077681946, 5.131603420.$$

For more than six filter coefficients it is possible to construct wavelets with a higher Sobolev exponents than the Daubechies wavelets by omitting more than one sum rule, see [27, 30, 45].

In Figures 4.4, 4.5 and 4.6 you can see plots of lower bounds for the Hölder exponent of the corresponding scaling functions and wavelets depending on one parameter, with the bounds from equation (4.1). We used 24 iteration in the algorithm from [37].

Note that for most, and for eight filter coefficients for all, parameters the computed lower bound is higher than the lower bound $s_{\max} - 1/2$. The negative lower bound in Figure 4.5 indicates that the corresponding scaling function is not continuous. We obtain numerically the maximal lower bound for the Hölder exponent for respectively four, six and eight filter coefficients

$$\alpha_{24} = 0.5776, 1.1386, 1.6344$$

with parameters

$$a = -1.66260325442517, -3.28211108661493, -4.93905744197576$$

and filter coefficients

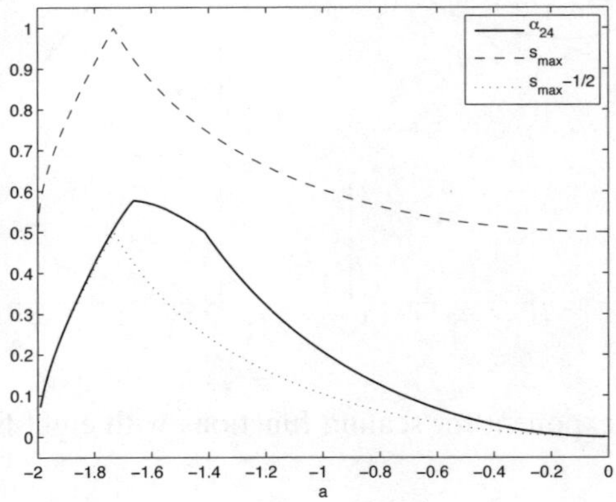

Figure 4.4: Lower bound for Hölder exponent for scaling functions with four filter coefficients from equation (3.1).

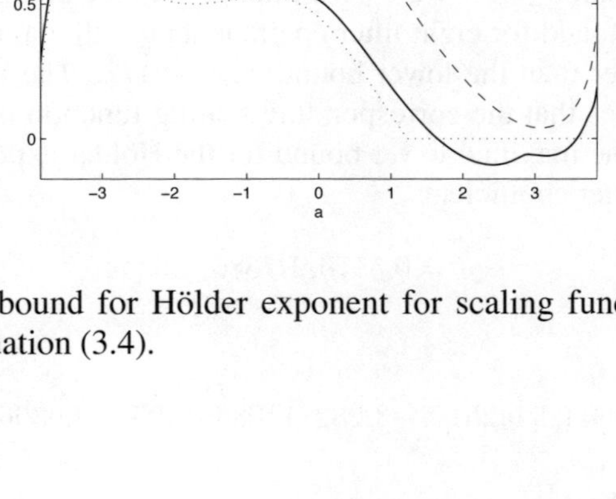

Figure 4.5: Lower bound for Hölder exponent for scaling functions with six filter coefficients from equation (3.4).

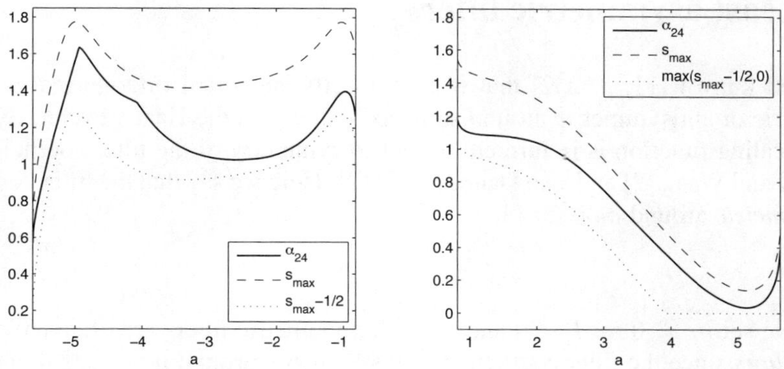

Figure 4.6: Lower bound for Hölder exponent for scaling functions with eight filter coefficients from equation (3.6).

```
 0.31887001724554, 0.59678079636075,
 0.18112998275446,-0.09678079636075

 0.21634225649014, 0.56180454136425, 0.35257937284659,
-0.08834519690163,-0.06892162933673, 0.02654065553738

 0.15488273436983, 0.49644876596501, 0.45767418856225,
-0.00833281609981,-0.13761439998701, 0.01970151455156,
 0.02505747705493,-0.00781746441676.
```

Daubechies and Lagarias [13] obtained the optimal Hölder exponents for the Daubechies wavelets with a different method (four, six, and eight filter coefficients)

$$\alpha_{\max} = 0.5500, 1.0878, 1.6179,$$

where the last one is for the Daubechies wavelet with extremal phase. So we obtained in all cases wavelets that have a higher Hölder exponent than the Daubechies wavelets.

Daubechies addressed in [12] and [11, p. 242] the question of finding wavelets with more regularity. For four filter coefficients she obtained the rational filter coefficients $(3/5, 6/5, 2/5, -1/5)$, which corresponds to $a = -8/5$ in (3.1), see also section 6. With the methods from [13] she found that the Hölder exponent of the corresponding scaling function is at least 0.5864.

Lang and Heller [22] discussed the general optimization problem of maximizing the Hölder exponent for a fixed number of filter coefficients. They found smoother wavelets than the Daubechies wavelets for more than eight filter coefficients, but the numerical method failed to find the more regular wavelets that we obtained using the explicit parametrizations of the filter coefficients. This might be due to the fact that Lang and Heller used a general purpose optimization routine while we could directly apply the golden section search for finding the maximum of a univariate function.

5 Least asymmetric filters

It is well known [11, p. 252] that if a compactly supported orthonormal wavelet is symmetric or antisymmetric around some axis, then it is the Haar wavelet. Symmetry of the scaling function is in turn equivalent to symmetry of the filter coefficients, see Belogay and Wang [2] and also Daubechies [12]. Here we say that the filter coefficients are *symmetric* around $n_0 \in \mathbb{Z}/2$ if

$$h_n = h_{2n_0-n},$$

where we set $h_k = 0$ for $k < 0$ and $k > N$. Symmetric filters are often called *linear phase filters* since the filter coefficients are symmetric around $n_0 \in \mathbb{Z}/2$ if and only if the phase of the *frequency response*

$$h(\xi) = \sum_n h_n e^{in\xi}$$

is a linear function of ξ, that is, if

$$h(\xi) = e^{in_0\xi}|h(\xi)|.$$

So we know that complete symmetry and orthogonality is not possible, and one can only try to find the least asymmetric filter coefficients out of a fixed family. For example, Daubechies discussed in [11] and [12] how to choose the least asymmetric out of the finitely many wavelets with a maximal number of vanishing moments. Another possibility is to omit some vanishing moments and use the additional degrees of freedom to find filters with partial symmetry. Several authors [1, 25, 40] discussed the use of Gröbner bases to find orthogonal filter coefficients with partial symmetry where several pairs of filters are equal. Zhao and Swamy [48] designed least asymmetric orthogonal wavelets with several vanishing moments via numerical optimization.

An immediate application of our parametrized filter coefficients is to find symbolically the least asymmetric filter coefficients using some criteria to measure symmetry. In the following, we discuss some examples, where we minimize the sum of squares error as in [48].

We want to find six filter coefficients satisfying two sum rules such that they are almost symmetric around 2, so that

$$h_0 \approx h_4, \quad h_1 \approx h_3, \quad h_6 \approx 0.$$

Using Maple, we find the minimum of the sum of squares error is attained at $a = \alpha$, where α denotes the largest negative real root of

$$25\,x^{10}-30\,x^9-702\,x^8+652\,x^7+5866\,x^6-3256\,x^5-13140\,x^4-1036\,x^3+5797\,x^2-2730\,x-5190$$

or numerically

$$a = -1.102986298\dots\,.$$

The filter coefficients are:

```
-0.090589559870111, 0.504872307867382, 1.206925694336121,
 0.516001958861136,-0.116336134466010,-0.020874266728517.
```

See Figure 5.1 for the corresponding scaling function, which has a Sobolev exponent $s_{\max} = 1.0180$ and a lower bound for the Hölder exponent $\alpha_{24} = 0.5370$.

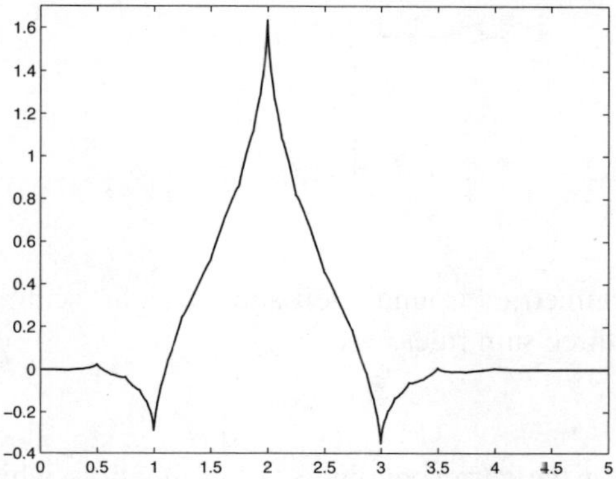

Figure 5.1: Least asymmetric (around 2) scaling function with six filter coefficients and two sum rules.

Now we consider eight filter coefficients. First we want to find filter coefficients that are almost symmetric around 3, so that

$$h_0 \approx h_6, \quad h_1 \approx h_5, \quad h_2 \approx h_4, \quad h_7 \approx 0.$$

The minimum of the sum of squares error is attained at $a = \alpha$, where α denotes the largest negative real root of

$$11025\,x^{24} - 21000\,x^{23} - 901900\,x^{22} + 1407480\,x^{21} + 25484946\,x^{20} - 23935800\,x^{19} - 280989500\,x^{18}$$
$$-149785464\,x^{17} + 837190927\,x^{16} + 6460372400\,x^{15} + 4612440168\,x^{14} - 53422512976\,x^{13}$$
$$-69302308420\,x^{12} + 344858640016\,x^{11} - 84085760856\,x^{10} - 294800719088\,x^{9} + 2435452393919\,x^{8}$$
$$-1913025285928\,x^{7} - 18887356576348\,x^{6} + 10024351195096\,x^{5} + 51733811048402\,x^{4}$$
$$-17259269191640\,x^{3} - 57876449779820\,x^{2} + 8466676099560\,x + 21625605062145$$

or numerically

$$a = -0.8395579286\ldots.$$

The filter coefficients are:

```
-0.073484394510424,-0.071424517120364, 0.556147092523951,
 1.154912201440016, 0.568048480655853,-0.135661369346454,
-0.050711178669381, 0.052173685026802.
```

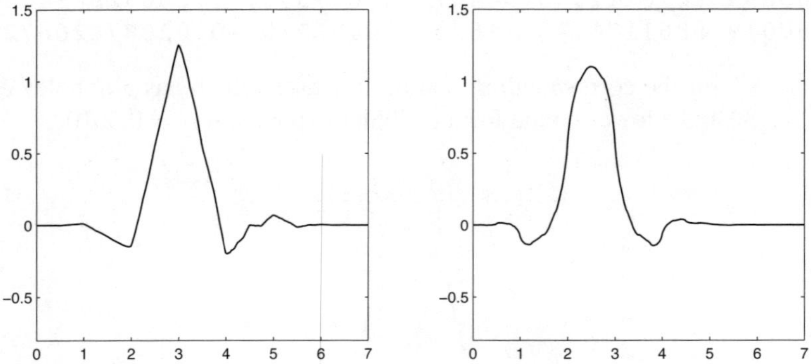

Figure 5.2: Least asymmetric (around 3 left and 2.5 right) scaling function with eight filter coefficients and three sum rules.

See Figure 5.2 (left) for the corresponding scaling function, which has a Sobolev exponent $s_{\max} = 1.6569$ and a lower bound for the Hölder exponent $\alpha_{24} = 1.3080$.

Finally, we want to design filters that are almost symmetric around 2.5, so that

$$h_0 \approx h_5, \quad h_1 \approx h_4, \quad h_2 \approx h_3, \quad h_6 \approx 0, \quad h_7 \approx 0.$$

This is related to the example considered in [1, 25], where the authors constructed using Gröbner bases eight orthogonal filters with two sum rules such that $h_0 = h_5$, $h_1 = h_4$ and $h_2 = h_3$. The minimum of the sum of squares error is attained at $a = \alpha$, where α denotes the second largest negative real root of

$$2025\,x^{24} - 9000\,x^{23} - 168020\,x^{22} + 823000\,x^{21} + 4733434\,x^{20} - 27869720\,x^{19} - 46538164\,x^{18}$$
$$+437384872\,x^{17} - 40684609\,x^{16} - 3591330192\,x^{15} + 3105046936\,x^{14} + 20835868016\,x^{13}$$
$$-35438686580\,x^{12} - 64147246896\,x^{11} + 233849168056\,x^{10} - 48135550128\,x^{9} - 894126414729\,x^{8}$$
$$+1033511750456\,x^{7} + 2682874758716\,x^{6} - 4634966862792\,x^{5} - 4762513155302\,x^{4}$$
$$+10857513198280\,x^{3} + 182957235580\,x^{2} - 6268723929720\,x + 2258107786305$$

or numerically

$$a = -1.927469761\ldots .$$

The filter coefficients are:

```
-0.114678365799638, 0.127976021526492, 0.977783792709255,
 0.990754350911186, 0.120334952341046,-0.133569326041206,
 0.016559620749336, 0.014838953603528.
```

See Figure 5.2 (right) for the corresponding scaling function, which has a Sobolev exponent $s_{\max} = 1.5026$ and a lower bound for the Hölder exponent $\alpha_{24} = 1.0633$.

6 Rational filter coefficients

In this section, we address the existence of rational orthogonal filter coefficients. We know from section 2 that filter coefficients are determined by quadratic equations for orthonormality (2.1) and linear equations for normalization (2.2) and vanishing moments (2.5). Note that all these equations have integer coefficients, and we want to find a rational solution. This leads to “Hilbert’s 10th Problem over $\mathbb{Q}$”, which asks if there exists an algorithm for deciding the existence of rational points for a system of polynomial equations with integer coefficients. The answer is not known, and despite centuries of effort, even for curves it is an open problem although many results and computational methods are known, see for example Poonen [34] for an introduction and further references. Using our parametrizations, we can reduce the question of rational filter coefficients to finding rational points on curves and give some answers.

The case of four filter coefficients is not difficult. Daubechies [10] already gave a rational parametrization of all orthogonal filter coefficients

$$h_0 = \frac{t\,(t-1)}{t^2+1}, \quad h_1 = \frac{1-t}{t^2+1}, \quad h_2 = \frac{t+1}{t^2+1}, \quad h_3 = \frac{t\,(t+1)}{t^2+1}$$

with $t \in \mathbb{R}$. Note that for $t = -t$ we obtain the flipped filter coefficients. The interval $-1 \le t \le 1$ corresponds to the filter coefficients from (3.1) and $t \le -1, 1 \le t$ to (3.3), except for $(1, 0, 0, 1)$, which are approached for $t \to \infty$ and $t \to -\infty$.

The Daubechies wavelet corresponds to $t = -1/\sqrt{3}$. Computing the continued fraction expansion of $-1/\sqrt{3}$, we obtain the periodic expansion

$$-\frac{1}{\sqrt{3}} = [-1;\, 2,\, \overline{2,\, 1}] = -1 + \cfrac{1}{2 + \cfrac{1}{2 + \cfrac{1}{1 + \cfrac{1}{2 + \cfrac{1}{1 + \cdots}}}}}$$

with the first convergents

$$-1,\ -1/2,\ -3/5,\ -4/7,\ -\frac{11}{19},\ -\frac{15}{26},\ -\frac{41}{71},\ -\frac{56}{97},\ -\frac{153}{265},\ -\frac{209}{362}.$$

For further details on continued fractions see for example Khinchin [19] or Knuth [20].

Taking $t = -209/362$, we get a good rational approximation

$$1/174725\,(119339, 206702, 55386, -31977)$$

for the Daubechies filters. Surprisingly, we obtain the filter coefficients corresponding to the most regular scaling function found by Daubechies for the second convergent $t = -1/2$, see section 4.

In parametrization (3.4) for six filter coefficients there appears only the square root

$$w = \sqrt{-a^4 + 14\,a^2 + 15}.$$

So the question of the existence of rational filters reduces to finding a rational point $(a, b) \in \mathbb{Q}^2$ on the (hyperelliptic) algebraic curve defined by the equation

$$y^2 = -x^4 + 14\,x^2 + 15 = -(x^2 + 1)(x^2 - 15). \tag{6.1}$$

Proposition 6.1 *There are no rational points on the curve defined by equation* (6.1).

Proof. Substituting $x = X/Z$ and $y = Y/Z^2$ in (6.1) and multiplying by Z^4, we obtain

$$Y^2 = -(X^2 + Z^2)(X^2 - 15\,Z^2),$$

and we equivalently would have to find integers a, b, c with a and c coprime satisfying this equation. Suppose that we had integers a, b, c satisfying

$$b^2 = -(a^2 + c^2)(a^2 - 15\,c^2). \tag{6.2}$$

Then

$$b^2 \equiv (a^2 + c^2)^2 \pmod 2$$

and hence

$$b \equiv (a + c) \pmod 2.$$

This implies that either

$$a \equiv 1,\ c \equiv 0 \pmod 2 \quad \text{or} \quad a \equiv 0,\ c \equiv 1 \pmod 2$$

or, since a and c are coprime,

$$a \equiv c \equiv 1 \pmod 2.$$

In the first case, we get

$$(a^2 + c^2)^2 \equiv 1 \pmod 4.$$

But then by equation (6.2)

$$b^2 \equiv -1 \equiv 3 \pmod 4,$$

which is not possible since the only quadratic residues modulo 4, that is, the integers d for which

$$x^2 \equiv d \pmod 4$$

has a solution, are

$$d \equiv 0, 1 \pmod 4.$$

In the second case, we get

$$(a^2 + c^2)^2 \equiv 4 \pmod{16}.$$

But then by equation (6.2)

$$b^2 \equiv -4 \equiv 12 \pmod{16},$$

which is not possible since the only quadratic residues modulo 16 are

$$d \equiv 0, 1, 4, 9 \pmod{16},$$

and the proposition is proved. □

Corollary 6.2 *There are no rational orthogonal filters with six nonzero filter coefficients and at least two sum rules.*

In parametrization (3.6) for eight filter coefficients, we have the square root

$$w = \sqrt{-a^8 + 36\,a^6 - 182\,a^4 + 1540\,a^2 - 945}.$$

So we would have to find a rational point on the algebraic curve defined by the equation

$$y^2 = -x^8 + 36\,x^6 - 182\,x^4 + 1540\,x^2 - 945.$$

This is a nonsingular curve with genus 3. Hence by Falting's theorem [15] it has only finitely many rational points, and so there are at most finitely many rational orthogonal filters with eight nonzero filter coefficients and at least three sum rules. So far we could neither find rational points on this curve nor prove that there do not exist any.

Acknowledgements. I would like to thank Josef Schicho for his comments and help with the proof of Proposition 6.1 and the reviewers for useful remarks.

Bibliography

[1] A. F. Abdelnour and I. W. Selesnick, *Symmetric nearly orthogonal and orthogonal nearly symmetric wavelets*, Arab. J. Sci. Eng. Sect. C Theme Issues 29 (2004), pp. 3–16.

[2] E. Belogay and Y. Wang, *Compactly supported orthogonal symmetric scaling functions*, Appl. Comput. Harmon. Anal. 7 (1999), pp. 137–150.

[3] B. Buchberger, *An Algorithm for Finding the Bases Elements of the Residue Class Ring Modulo a Zero Dimensional Polynomial Ideal (German)*, Ph.D. thesis, Univ. of Innsbruck, 1965, English translation published in [5].

[4] ———, *Ein algorithmisches Kriterium für die Lösbarkeit eines algebraischen Gleichungssystems*, Aequationes Math. 4 (1970), pp. 374–383.

[5] ———, *Bruno Buchberger's PhD thesis 1965: An algorithm for finding the basis elements of the residue class ring of a zero dimensional polynomial ideal*, J. Symbolic Comput. 41 (2006), pp. 475–511, Translated from the 1965 German original by Michael P. Abramson.

[6] C. Charoenlarpnopparut and N. K. Bose, *Multidimensional FIR filter bank design using Gröbner bases.*, IEEE Trans. Circuits Syst., II, Analog Digit. Signal Process. 46 (1999), pp. 1475–1486.

[7] F. Chyzak, P. Paule, O. Scherzer, A. Schoisswohl, and B. Zimmermann, *The construction of orthonormal wavelets using symbolic methods and a matrix analytical approach for wavelets on the interval*, Experiment. Math. 10 (2001), pp. 67–86.

[8] A. Cohen, *Ondelettes, analyses multirésolutions et filtres miroirs en quadrature*, Ann. Inst. H. Poincaré Anal. Non Linéaire 7 (1990), pp. 439–459.

[9] D. Colella and C. Heil, *Characterizations of scaling functions: continuous solutions*, SIAM J. Matrix Anal. Appl. 15 (1994), pp. 496–518.

[10] I. Daubechies, *Orthonormal bases of compactly supported wavelets*, Comm. Pure Appl. Math. 41 (1988), pp. 909–996.

[11] ———, *Ten lectures on wavelets*, Society for Industrial and Applied Mathematics (SIAM), Philadelphia, PA, 1992.

[12] ———, *Orthonormal bases of compactly supported wavelets. II. Variations on a theme*, SIAM J. Math. Anal. 24 (1993), pp. 499–519.

[13] I. Daubechies and J. C. Lagarias, *Two-scale difference equations. II. Local regularity, infinite products of matrices and fractals*, SIAM J. Math. Anal. 23 (1992), pp. 1031–1079.

[14] T. Eirola, *Sobolev characterization of solutions of dilation equations*, SIAM J. Math. Anal. 23 (1992), pp. 1015–1030.

[15] G. Faltings, *Endlichkeitssätze für abelsche Varietäten über Zahlkörpern*, Invent. Math. 73 (1983), pp. 349–366.

[16] J.-C. Faugère, F. Moreau de Saint-Martin, and F. Rouillier, *Design of regular nonseparable bidimensional wavelets using Gröbner basis techniques*, IEEE Trans. Signal Process. 46 (1998), pp. 845–856.

[17] A. Haar, *Zur Theorie der orthogonalen Funktionensysteme. (Erste Mitteilung.)*, Math. Ann. 69 (1910), pp. 331–371.

[18] J. M. Hereford, D. W. Roach, and R. Pigford, *Image compression using parameterized wavelets with feedback*, vol. 5102, SPIE, 2003, pp. 267–277.

[19] A. Ya. Khinchin, *Continued fractions*, russian. ed., Dover Publications Inc., Mineola, NY, 1997, With a preface by B. V. Gnedenko, Reprint of the 1964 translation.

[20] D. E. Knuth, *The art of computer programming. Vol. 2: Seminumerical algorithms. 3rd ed.*, Bonn: Addison-Wesley. xiii, 1998.

[21] M.-J. Lai and D. W. Roach, *Parameterizations of univariate orthogonal wavelets with short support*, Approximation theory, X (St. Louis, MO, 2001), Innov. Appl. Math., Vanderbilt Univ. Press, Nashville, TN, 2002, pp. 369–384.

[22] M. Lang and P. N. Heller, *The design of maximally smooth wavelets*, IEEE Proc. Int. Conf. Acoust., Speech, Signal Processing (Atlanta, GA), vol. 3, 1996, pp. 1463–1466.

[23] W. M. Lawton, *Tight frames of compactly supported affine wavelets*, J. Math. Phys. 31 (1990), pp. 1898–1901.

[24] ———, *Necessary and sufficient conditions for constructing orthonormal wavelet bases*, J. Math. Phys. 32 (1991), pp. 57–61.

[25] J. Lebrun and I. W. Selesnick, *Gröbner bases and wavelet design*, J. Symbolic Comput. 37 (2004), pp. 227–259.

[26] J. Lebrun and M. Vetterli, *High-order balanced multiwavelets: theory, factorization, and design*, IEEE Trans. Signal Process. 49 (2001), pp. 1918–1930.

[27] P. G. Lemarie-Rieusset and E. Zahrouni, *More regular wavelets*, Appl. Comput. Harmon. Anal. 5 (1998), pp. 92–105.

[28] J.-M. Lina and M. Mayrand, *Parametrizations for Daubechies wavelets*, Phys. Rev. E (3) 48 (1993), pp. R4160–R4163.

[29] S. Mallat, *A wavelet tour of signal processing*, Academic Press Inc., San Diego, CA, 1998.

[30] H. Ojanen, *Orthonormal compactly supported wavelets with optimal Sobolev regularity: Numerical results.*, Appl. Comput. Harmon. Anal. 10 (2001), pp. 93–98.

[31] H. Park, *Symbolic computation and signal processing*, J. Symbolic Comput. 37 (2004), pp. 209–226.

[32] H. Park, T. Kalker, and M. Vetterli, *Gröbner bases and multidimensional FIR multirate systems.*, Multidimensional Syst. Signal Process. 8 (1997), pp. 11–30.

[33] D. Pollen, $\mathrm{SU}_I(2, F[z, 1/z])$ *for F a subfield of* $\mathbb{C}$, J. Amer. Math. Soc. 3 (1990), pp. 611–624.

[34] B. Poonen, *Computing rational points on curves*, Number theory for the millennium, III (Urbana, IL, 2000), A K Peters, Natick, MA, 2002, pp. 149–172.

[35] G. Regensburger, *Parametrizing compactly supported orthonormal wavelets by discrete moments*, Appl. Algebra Engrg. Comm. Comput. (2007), to appear

[36] G. Regensburger and O. Scherzer, *Symbolic computation for moments and filter coefficients of scaling functions*, Ann. Comb. 9 (2005), pp. 223–243.

[37] O. Rioul, *Simple regularity criteria for subdivision schemes*, SIAM J. Math. Anal. 23 (1992), pp. 1544–1576.

[38] J. Schneid and S. Pittner, *On the parametrization of the coefficients of dilation equations for compactly supported wavelets*, Computing 51 (1993), pp. 165–173.

[39] I. W. Selesnick and C. S. Burrus, *Maximally Flat Low-Pass FIR Filters with Reduced Delay*, IEEE Trans. Circuits Syst. II, Analog Digit. Signal Process. 45 (1998), pp. 53–68.

[40] I. W. Selesnick, J. E. Odegard, and C. S. Burrus, *Nearly symmetric orthogonal wavelets with non-integer DC groupdelay*, Digital Signal Processing Workshop Proceedings, IEEE, September 1–4 1996, pp. 431–434.

[41] B. G. Sherlock and D. M. Monro, *On the space of orthonormal wavelets*, IEEE Trans. Signal Process. 46 (1998), pp. 1716–1720.

[42] G. Strang and T. Nguyen, *Wavelets and filter banks*, Wellesley-Cambridge Press, Wellesley, MA, 1996.

[43] M. Unser and T. Blu, *Wavelet theory demystified*, IEEE Trans. Signal Process. 51 (2003), pp. 470–483.

[44] L. F. Villemoes, *Energy moments in time and frequency for two-scale difference equation solutions and wavelets*, SIAM J. Math. Anal. 23 (1992), pp. 1519–1543.

[45] H. Volkmer, *Asymptotic regularity of compactly supported wavelets*, SIAM J. Math. Anal. 26 (1995), pp. 1075–1087.

[46] S. H. Wang, A. H. Tewfik, and H. Zou, *Correction to 'parametrization of compactly supported orthonormal wavelets'*, IEEE Trans. Signal Process. 42 (1994), pp. 208–209.

[47] R. O. Wells, Jr., *Parametrizing smooth compactly supported wavelets*, Trans. Amer. Math. Soc. 338 (1993), pp. 919–931.

[48] Y. Zhao and M. N. S. Swamy, *Design of least asymmetric compactly supported orthogonal wavelets via optimization*, Electrical and Computer Engineering, 1999 IEEE Canadian Conference on, vol. 2, May 1999, pp. 817–820.

[49] H. Zou and A. H. Tewfik, *Parametrization of compactly supported orthonormal wavelets*, IEEE Trans. Signal Process. 41 (1993), pp. 1428–1431.

Author information

Georg Regensburger, Johann Radon Institute for Computational and Applied Mathematics (RICAM), Austrian Academy of Sciences,
Altenbergerstraße 69, A-4040 Linz, Austria.
Email: georg.regensburger@oeaw.ac.at

Radon Series Comp. Appl. Math **3**, 215–233

Applications of Gröbner bases in synthesis of multidimensional control systems

Li Xu

Key words. Multidimensional systems, nD systems, control, stabilization, tracking, regulation, coprimeness, skew primeness, matrix equation.

AMS classification. 13P10, 13F20, 11C20, 11C08, 15A24, 93B25, 93B40, 93B50, 93C35

1 Introduction

Stabilization, *asymptotic tracking* and *disturbance rejection* or *regulation* are basic and important requirements for feedback control system synthesis [3, 4, 13, 15, 16, 17, 18]. The purpose of this work is to show the possibility on applications of Gröbner bases in synthesis of multidimensional (nD) feedback control systems [6, 8, 9, 13, 17, 18]. In particular, the following topics will be explored and discussed:

- Definitions and conditions for left, right and double coprimeness and skew primeness of nD matrices over the ring of structurally stable nD rational functions.
- Formulations and solvability conditions for stabilization problem using 1DOF (degree-of-freedom) or 2DOF controller, asymptotic tracking problem and disturbance rejection problem, which lead to the conclusion that these problems can be essentially reduced to the solvability problems of coprime matrix equation and skew prime matrix equation.
- How to solve these two kinds of matrix equations by utilizing Gröbner basis approach.
- Open problems.

2 Coprimeness and skew primeness of nD matrices

Let $\mathbf{R}[\boldsymbol{z}]$ be the ring of real polynomials in $\boldsymbol{z} = (z_1, \ldots, z_n)$, and $\mathbf{G}$ and $\mathbf{H}$ be respectively the rings of nD causal and structurally stable rational functions, i. e., $\mathbf{G} = \{n/d \mid n, d \in \mathbf{R}[\boldsymbol{z}],\ d(0, \ldots, 0) \neq 0\}$, $\mathbf{H} = \{n/d \in \mathbf{G} \mid d(\boldsymbol{z}) \neq 0, \forall \boldsymbol{z} \in \bar{U}^n\}$ with $\bar{U}^n = \{\boldsymbol{z} \in \mathbf{C}^n \mid |\boldsymbol{z}| \leq 1\}$ being the closed unit polydisc on $\mathbf{C}^n$. In addition, let $\mathbf{I} = \{h \in \mathbf{H} \mid h^{-1} \in \mathbf{G}\}$, $\mathbf{J} = \{h \in \mathbf{H} \mid h^{-1} \in \mathbf{H}\}$, and denote by $\mathbf{M}(*)$ the set of matrices with entries in $*$ (e. g., $\mathbf{G}, \mathbf{H}$). $A \in \mathbf{M}(\mathbf{H})$ is said to be $\mathbf{G}$-unimodular (respectively $\mathbf{H}$-unimodular) iff it is square and $\det A \in \mathbf{I}$ ($\mathbf{J}$). $D^{-1}N$, with $D, N \in \mathbf{M}(\mathbf{H})$ and D $\mathbf{G}$-unimodular, is called a left matrix fractional description (MFD) (on $\{\mathbf{G}, \mathbf{H}, \mathbf{I}, \mathbf{J}\}$). Moreover, D, N are said to be left coprime, and correspondingly $D^{-1}N$ left coprime MFD, iff there exist $U, V \in \mathbf{M}(\mathbf{H})$ such that

$$DU + NV = I.$$

The dual definitions on right can be given analogously.

$P \in \mathbf{M}(\mathbf{G})$ is said to admit a left (resp. right) coprime factorization if there exist $D, N \in M(\mathbf{H})$ ($\tilde{D}, \tilde{N} \in M(\mathbf{H})$) such that $P = D^{-1}N$ ($P = \tilde{N}\tilde{D}^{-1}$) and D, N ($\tilde{D}, \tilde{N}$) are left (right) coprime. Moreover, P is said to admit a doubly coprime factorization if it admits both left and right coprime factorizations, which is equivalent to that there exist $D, N, \tilde{D}, \tilde{N}, X_1, Y_1, X_2, Y_2 \in M(\mathbf{H})$ such that $P = D^{-1}N = \tilde{N}\tilde{D}^{-1}$ and the doubly coprime relation

$$\begin{bmatrix} X_2 & Y_2 \\ -N & D \end{bmatrix} \begin{bmatrix} \tilde{D} & -Y_1 \\ \tilde{N} & X_1 \end{bmatrix} = I$$

or equivalently,

$$\begin{bmatrix} \tilde{D} & -Y_1 \\ \tilde{N} & X_1 \end{bmatrix} \begin{bmatrix} X_2 & Y_2 \\ -N & D \end{bmatrix} = I$$

holds.

Further, two matrices $D, N \in M(\mathbf{H})$ are said to be (externally) skew prime iff there exist $U, V \in M(\mathbf{H})$ such that

$$DU + VN = I. \tag{2.1}$$

For $P \in \mathbf{M}(\mathbf{G})$, it is always possible to find $N, D \in \mathbf{M}(\mathbf{H})$ such that $P = D^{-1}N$, but D, N are not in general left coprime even when they possess no left common factor. The following theorem gives the necessary and sufficient condition for the existence of coprime factorizations.

Theorem 2.1 [13, 14]

1. *$P \in \mathbf{G}^{q\times r}$ admits a left coprime factorization iff there exists $D \in \mathbf{H}^{q\times q}$ such that $\det D \neq 0$ and*

$$[I_q \ \ -P]\mathbf{H}^{q+r} = D^{-1}\mathbf{H}^q,$$

i. e., $[I_q \ -P]\mathbf{H}^{q+r}$ *is a free lattice of* $\mathbf{G}^q$*, or equivalently,* $[I_q \ -P]\mathbf{H}^{q+r}$ *is a free* $\mathbf{H}$*-submodule of* $\mathbf{G}^q$ *of rank* q*. Then,* $P = D^{-1}N$*, where* $N = DP \in \mathbf{H}^{q\times r}$*, is a left coprime factorization of* P*.*

2. $P \in \mathbf{G}^{q\times r}$ *admits a right coprime factorization iff there exists* $\tilde{D} \in \mathbf{H}^{r\times r}$ *such that* $\det \tilde{D} \neq 0$ *and*

$$\mathbf{H}^{1\times(q+r)}\begin{bmatrix} P \\ I_r \end{bmatrix} = \mathbf{H}^{1\times r}\tilde{D}^{-1},$$

i. e., $H^{1\times(q+r)}[P^T \ I_r]^T$ *is a free lattice of* $\mathbf{G}^{1\times r}$*, or equivalently,* $\mathbf{H}^{1\times r}\tilde{D}^{-1}$ *is a free* $\mathbf{H}$*-submodule of* $\mathbf{G}^{1\times r}$ *of rank* r*. Then,* $P = \tilde{N}\tilde{D}^{-1}$*, where* $\tilde{N} = P\tilde{D} \in \mathbf{H}^{q\times r}$*, is a right coprime factorization of* P*.*

Theorem 2.2 *If* $P \in \mathbf{G}^{q\times r}$ *admits a left or right coprime factorization, then it admits a doubly coprime factorization.*

Proof. Clear from Corollary 3, Theorem 3 and Corollary 5 of [13]. □

Theorem 2.3 $D, N \in M(\mathbf{H})$ *with* $\det D \neq 0$ *are (externally) skew prime iff there exist* $\bar{N}, \bar{D} \in M(\mathbf{H})$ *such that*

$$ND = \bar{D}\bar{N} \tag{2.2}$$

with D *and* $\bar{N}$ *right coprime and* N *and* $\bar{D}$ *left coprime.*

Proof. Sufficiency: If D, $\bar{N}$ are right coprime and N, $\bar{D}$ are left coprime, and (2.2) holds, then there exist $X_1, X_2, X_3, X_4 \in \mathbf{M}(\mathbf{H})$ such that the doubly coprime relation

$$\begin{bmatrix} X_1 & X_2 \\ -\bar{D} & N \end{bmatrix}\begin{bmatrix} \bar{N} & -X_3 \\ D & X_4 \end{bmatrix} = \begin{bmatrix} \bar{N} & -X_3 \\ D & X_4 \end{bmatrix}\begin{bmatrix} X_1 & X_2 \\ -\bar{D} & N \end{bmatrix} = I$$

holds. It then follows that

$$DX_2 + X_4N = I,$$

i. e., D and N are skew prime.

Necessity: If D and N are skew prime, there exist $U, V \in \mathbf{M}(\mathbf{H})$ such that (2.1) holds, which means that D and V are left coprime (and U and N are right coprime). It follows from Theorem 2.2 that, for the left coprime MFD $D^{-1}V$, there exist $\bar{V}$, $\bar{D} \in \mathbf{M}(\mathbf{H})$ such that $\det \bar{D} \neq 0$ and

$$D^{-1}V = \bar{V}\bar{D}^{-1}$$

with $\bar{D}$ and $\bar{V}$ right coprime. Then, there are $X, Y \in \mathbf{M}(\mathbf{H})$ such that

$$X\bar{V} + Y\bar{D} = I,$$

and thus we have

$$\begin{bmatrix} D & V \\ -X & Y \end{bmatrix}\begin{bmatrix} U & -\bar{V} \\ N & \bar{D} \end{bmatrix} = \begin{bmatrix} I & 0 \\ -XU + YN & I \end{bmatrix}. \tag{2.3}$$

Premultiplying (2.3) by

$$\begin{bmatrix} I & 0 \\ XU - YN & I \end{bmatrix}$$

yields

$$\begin{bmatrix} D & V \\ -\bar{N} & \bar{U} \end{bmatrix} \begin{bmatrix} U & -\bar{V} \\ N & \bar{D} \end{bmatrix} = I \tag{2.4}$$

where $\bar{N} = X - (XU - YN)D, \bar{U} = Y + (XU - YN)V$.

(2.4) is obviously equivalent to

$$\begin{bmatrix} U & -\bar{V} \\ N & \bar{D} \end{bmatrix} \begin{bmatrix} D & V \\ -\bar{N} & \bar{U} \end{bmatrix} = I.$$

Therefore, we have

$$ND = \bar{D}\bar{N}$$

with D and $\bar{N}$ right coprime and N and $\bar{D}$ left coprime. □

The following lemma will be used later.

Lemma 2.4 *Suppose* $V, T, F \in \mathbf{M}(\mathbf{H})$, $\det T \neq 0$, *and* T, F *are left coprime. Then* $VT^{-1}F \in \mathbf{M}(\mathbf{H})$ *iff* $VT^{-1} \in \mathbf{M}(\mathbf{H})$.

Proof. The sufficiency is obvious; the necessity can be shown as the 1D case [16]. □

3 nD stabilization problem

Consider the nD feedback system shown in Figure 3.1 where $P \in \mathbf{M}(\mathbf{G})$ is a linear nD plant and $C \in \mathbf{M}(\mathbf{G})$ is a 1DOF controller. It is easy to see [16] that

$$e = T_{eu}u, \qquad y = T_{yu}u$$

where $e = [e_1^T \; e_2^T]^T$, $u = [u_1^T \; u_2^T]^T$, $y = [y_1^T \; y_2^T]^T$ and

$$T_{eu} = \begin{bmatrix} (I+PC)^{-1} & -P(I+CP)^{-1} \\ C(I+PC)^{-1} & (I+CP)^{-1} \end{bmatrix}$$

$$= \begin{bmatrix} (I+PC)^{-1} & -(I+PC)^{-1}P \\ C(I+PC)^{-1} & I - C(I+PC)^{-1}P \end{bmatrix},$$

$$T_{yu} = \begin{bmatrix} C(I+PC)^{-1} & -C(I+PC)^{-1}P \\ (I+PC)^{-1}PC & (I+PC)^{-1}P \end{bmatrix}.$$

Note that

$$T_{eu} = I - FT_{yu}, \quad T_{yu} = F(T_{eu} - I)$$

where

$$F = \begin{bmatrix} 0 & I \\ -I & 0 \end{bmatrix}.$$

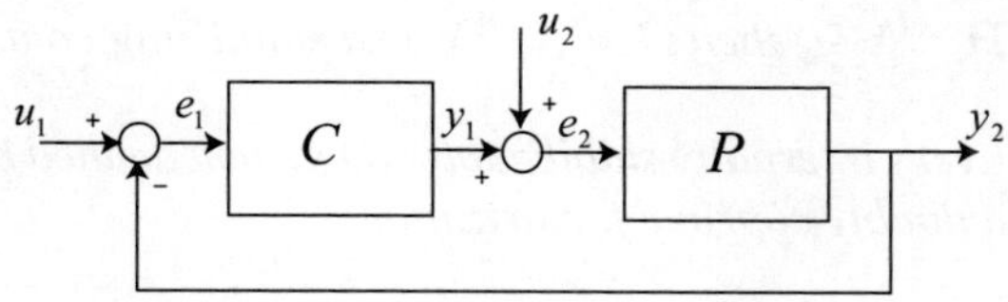

Figure 3.1 A feedback system

If there exists a controller C such that the closed-loop system is structurally stable, i. e., $T_{yu} \in \mathbf{M}(\mathbf{H})$, or equivalently, $T_{eu} \in \mathbf{M}(\mathbf{H})$, then P is said to be internally stabilizable and C is called a stabilizing controller of P.

Theorem 3.1 [13] *A given plant* $P \in \mathbf{G}^{q\times r}$ *is internally stabilizable iff either of the following equivalent statements is true.*

1. *There exists* $S = [U^T\ V^T]^T \in \mathbf{H}^{(q+r)\times q}$ *such that* $\det U \neq 0$ *and*

   (a) $SP = \begin{bmatrix} UP \\ VP \end{bmatrix} \in \mathbf{H}^{(q+r)\times r}$,

   (b) $[I_q\ \ -P]S = U - PV = I_q$.

   Then, $C = VU^{-1}$ *internally stabilizes P, and*

$$U = (I_q - PC)^{-1}, \quad V = C(I_q - PC)^{-1}.$$

2. *There exists* $T = [-X\ Y] \in \mathbf{H}^{r\times(q+r)}$ *such that* $\det Y \neq 0$ *and*

   (a) $PT = \begin{bmatrix} -PX & PY \end{bmatrix} \in \mathbf{H}^{q\times(q+r)}$,

   (b) $T \begin{bmatrix} P \\ I_r \end{bmatrix} = -XP + Y = I_r$.

   Then, $C = Y^{-1}X$ *internally stabilizes P and*

$$Y = (I_r - CP)^{-1}, \quad X = (I_r - CP)^{-1}C.$$

Corollary 3.2 [13]

1. *If $P \in \mathbf{G}^{q \times r}$ admits the left coprime factorization*

$$P = D^{-1}N, \; DX - NY = I_q, \; \det X \neq 0$$

with $[X^T \; Y^T]^T \in \mathbf{H}^{(q+r)\times q}$, then $C = YX^{-1}$ is a stabilizing controller of P.

2. *If $P \in \mathbf{G}^{q \times r}$ admits the right coprime factorization*

$$P = \tilde{N}\tilde{D}^{-1}, \; \tilde{X}\tilde{D} - \tilde{Y}\tilde{N} = I_r, \; \det \tilde{X} \neq 0$$

with $[\tilde{Y} \; \tilde{X}] \in \mathbf{H}^{r \times (q+r)}$, then $C = \tilde{X}^{-1}\tilde{Y}$ is a stabilizing controller of P.

Theorem 3.3 [13] *Every internally stabilizable nD system defined by a transfer matrix $P \in \mathbf{M}(\mathbf{G})$ admits a doubly coprime factorization.*

Theorem 3.4 *$P \in \mathbf{M}(\mathbf{G})$ is internally stabilizable iff it admits a doubly coprime factorization.*

Proof. It is clear from Theorem 3.3 and Corollary 3.2. □

Corollary 3.5 [16] *If $P \in \mathbf{M}(\mathbf{G})$ is internally stabilizable, i. e., there exists a doubly coprime factorization $P = ND^{-1} = \tilde{D}^{-1}\tilde{N}$,*

$$\begin{bmatrix} X_2 & Y_2 \\ -N & D \end{bmatrix} \begin{bmatrix} \tilde{D} & -Y_1 \\ \tilde{N} & X_1 \end{bmatrix} = \begin{bmatrix} \tilde{D} & -Y_1 \\ \tilde{N} & X_1 \end{bmatrix} \begin{bmatrix} X_2 & Y_2 \\ -N & D \end{bmatrix} = I,$$

then the class of all stabilizing controllers of P is given by

$$C = (Y_1 + \tilde{D}Q)(X_1 - \tilde{N}Q)^{-1} = (X_2 - RN)^{-1}(Y_2 + RD)$$

where $Q, R \in \mathbf{M}(\mathbf{H})$ are arbitrary but $\det(X_1 - \tilde{N}Q) \neq 0$, $\det(X_2 - RN) \neq 0$.

4 General nD stabilization problem

Consider the nD general feedback system shown in Figure 4.1 where $P \in \mathbf{M}(\mathbf{G})$ is a linear nD *generalized plant* given by

$$\begin{bmatrix} z_p \\ y_p \end{bmatrix} = P \begin{bmatrix} w_p \\ u_p \end{bmatrix}, \quad P = \begin{bmatrix} P_{11} & P_{12} \\ P_{21} & P_{22} \end{bmatrix},$$

and $C \in \mathbf{M}(\mathbf{G})$ is a linear nD controller described by

$$y_c = [C_1 \; -C_2] \begin{bmatrix} w_c \\ u_c \end{bmatrix},$$

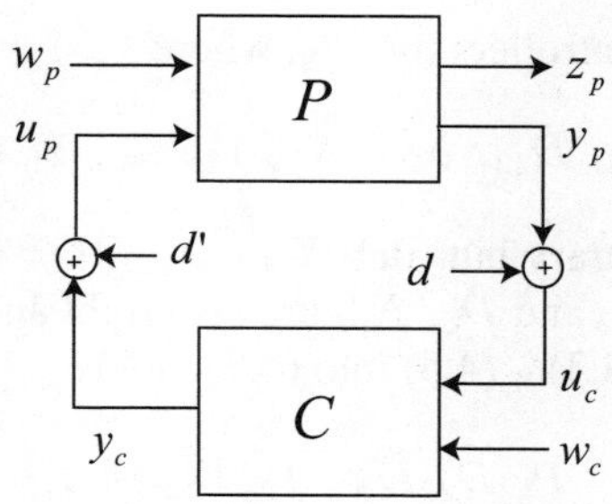

Figure 4.1 A general feedback system

z_p is a vector containing the plant variables to be controlled, y_p and y_c are respectively the utilized output vectors of the plant and controller, w_p and w_c are the vectors of all the exogenous inputs (such as disturbances, initial conditions, reference signals, etc.), u_p and u_c are respectively the utilized inputs to the plant and controller and d, d' are the exogenous unmodeled signals (such as noise, interference, etc.).

The problem considered here is to find a controller C such that the closed-loop feedback system (4.1) is structurally stable, i. e., $\Phi \in \mathbf{M}(\mathbf{H})$:

$$\begin{bmatrix} z_p \\ y_p \\ y_c \end{bmatrix} = \Phi \begin{bmatrix} w_p \\ d' \\ w_c \\ d \end{bmatrix}, \qquad \begin{bmatrix} u_p \\ u_c \end{bmatrix} = \Psi \begin{bmatrix} w_p \\ d' \\ w_c \\ d \end{bmatrix} \tag{4.1}$$

where

$$\Phi = \begin{bmatrix} \Phi_{11} & P_{12}\tilde{\Delta} & P_{12}\tilde{\Delta}C_1 & -P_{12}C_2\Delta \\ \Delta P_{21} & P_{22}\tilde{\Delta} & P_{22}\tilde{\Delta}C_1 & -P_{22}C_2\Delta \\ -C_2\Delta P_{21} & -C_2P_{22}\tilde{\Delta} & \tilde{\Delta}C_1 & -C_2\Delta \end{bmatrix} \tag{4.2}$$

and

$$\Psi = \begin{bmatrix} 0 & 0 & I \\ 0 & I & 0 \end{bmatrix} \Phi + \begin{bmatrix} 0 & I & 0 & 0 \\ 0 & 0 & 0 & I \end{bmatrix}$$

with $\Phi_{11} = P_{11} - P_{12}C_2\Delta P_{21}$, $\Delta = (I + P_{22}C_2)^{-1}$ and $\tilde{\Delta} = (I + C_2P_{22})^{-1}$.

Let $P_{22} \in \mathbf{M}(\mathbf{G})$ be internally stabilizable. Then, by Corollary 3.5, there exists a doubly coprime factorization $P_{22} = N_{22}D_{22}^{-1} = \tilde{D}_{22}^{-1}\tilde{N}_{22}$ satisfying the relation

$$\begin{bmatrix} X_1 & Y_1 \\ -\tilde{N}_{22} & \tilde{D}_{22} \end{bmatrix} \begin{bmatrix} D_{22} & -Y_2 \\ N_{22} & X_2 \end{bmatrix} = \begin{bmatrix} D_{22} & -Y_2 \\ N_{22} & X_2 \end{bmatrix} \begin{bmatrix} X_1 & Y_1 \\ -\tilde{N}_{22} & \tilde{D}_{22} \end{bmatrix} = I \tag{4.3}$$

such that

$$C_2 = (Y_2 + D_{22}Q)(X_2 - N_{22}Q)^{-1} \triangleq N_{c2}D_c^{-1} \tag{4.4}$$

$$= (X_1 - R\tilde{N}_{22})^{-1}(Y_1 + R\tilde{D}_{22}) \triangleq \tilde{D}_c^{-1}\tilde{N}_{c2} \tag{4.5}$$

is the class of stabilizing controllers for P_{22}, where

$$D_{22},\ N_{22},\ \tilde{D}_{22},\ \tilde{N}_{22},\ X_1,\ Y_1,\ X_2,\ Y_2 \in \mathbf{M}(\mathbf{H}),$$

and $Q, R \in \mathbf{M}(\mathbf{H})$ are arbitrary but $\det(X_2 - N_{22}Q) \neq 0$, $\det(X_1 - R\tilde{N}_{22}) \neq 0$. It should be clear that $D_c,\ N_{c2}$ and $\tilde{D}_c,\ \tilde{N}_{c2}$ are also right and left coprime, respectively. Substituting the results of (4.3) – (4.5) into (4.2) yields

$$\Phi = \begin{bmatrix} \Phi_{11} & P_{12}D_{22}\tilde{D}_c & P_{12}D_{22}\tilde{D}_cC_1 & -P_{12}D_{22}\tilde{N}_{c2} \\ D_c\tilde{D}_{22}P_{21} & D_c\tilde{N}_{22} & N_{22}\tilde{D}_cC_1 & -N_{22}\tilde{N}_{c2} \\ -N_{c2}\tilde{D}_{22}P_{21} & -N_{c2}\tilde{N}_{22} & D_{22}\tilde{D}_cC_1 & -N_{c2}\tilde{D}_{22} \end{bmatrix} \tag{4.6}$$

with $\Phi_{11} = P_{11} - P_{12}N_{c2}\tilde{D}_{22}P_{21}$.

In view of the coprimeness of the pairs of $(\tilde{D}_c,\ \tilde{N}_{c2})$, $(D_c,\ N_{c2})$, (D_{22}, N_{22}), it is easy to see that $\Phi \in \mathbf{M}(\mathbf{H})$ iff the conditions of (4.7) – (4.10) hold:

$$P_{11} - P_{12}N_{c2}\tilde{D}_{22}P_{21} \in \mathbf{M}(\mathbf{H}) \tag{4.7}$$

and

$$P_{12}D_{22}[\tilde{D}_c\ \tilde{N}_{c2}] \in \mathbf{M}(\mathbf{H}) \quad \Leftrightarrow \quad P_{12}D_{22} \in \mathbf{M}(\mathbf{H}), \tag{4.8}$$

$$\begin{bmatrix} D_c \\ N_{c2} \end{bmatrix} \tilde{D}_{22}P_{21} \in \mathbf{M}(\mathbf{H}) \quad \Leftrightarrow \quad \tilde{D}_{22}P_{21} \in \mathbf{M}(\mathbf{H}), \tag{4.9}$$

$$\begin{bmatrix} D_{22} \\ N_{22} \end{bmatrix} \tilde{D}_cC_1 \in \mathbf{M}(\mathbf{H}) \quad \Leftrightarrow \quad \tilde{D}_cC_1 \in \mathbf{M}(\mathbf{H}). \tag{4.10}$$

(4.7) – (4.9) reveal the conditions for the generalized plant P to admit a stabilizing controller when P_{22} is internally stabilizable, while (4.10) gives the admissible condition for the 2DOF controller $C = [C_1\ \ -C_2]$. In the sequel, therefore, assume that $[C_1\ C_2] = \tilde{D}_c^{-1}[\tilde{N}_{c1}\ \tilde{N}_{c2}]$, $\tilde{D}_c, \tilde{N}_{c1}, \tilde{N}_{c2} \in \mathbf{M}(\mathbf{H})$, without loss of generality. It is clear that, for the special 1DOF case, i. e., $C_1 = C_2 = \tilde{D}_c^{-1}\tilde{N}_{2c}$, the condition (4.10) is always satisfied.

5 nD regulation and tracking problems

The configuration for nD *generalized* regulation and tracking problems shown in Figure 5.1 will be used where $T_r, T_w \in \mathbf{M}(\mathbf{G})$ are generators of the reference signal r and the disturbance signal w_p, and w_0, r_0 correspond to the initial or boundary conditions of the generators, respectively. P, C and the other variables are defined as in Figure 4.1.

The problems to be considered can be stated as follows: given P, T_w and T_r, to search a controller C for both the 1DOF ($C_1 = C_2$) and 2DOF ($C_1 \neq C_2$) cases such that C stabilizes P and further

Regulation Problem (RP): $T_{z_p w_0} \in \mathbf{M}(\mathbf{H})$, i. e., the transfer matrix from w_0 to z_p is structurally stable.

Tracking Problem (TP): $T_{er_0} \in \mathbf{M}(\mathbf{H})$, i. e., the transfer matrix from r_0 to $e = r - z_p$ is structurally stable.

Regulation and Tracking Problem (RTP): the requirements of RP and TP are satisfied simultaneously.

Assume in the following that P_{22} and P are internally stabilizable, and $T_w = \tilde{D}_w^{-1}\tilde{N}_w$, $T_r = \tilde{D}_r^{-1}\tilde{N}_r$ are both left coprime MFDs. Also assume that a stabilizing controller $C = [C_1\ C_2]$ has been given as in the previous section. Define $\bar{P}_1 = P_{12}D_{22}$, $\bar{P}_2 = \tilde{D}_{22}P_{21}$, $\bar{P}_3 = P_{11} - P_{12}Y_2\tilde{D}_{22}P_{21}$ and note that $\bar{P}_1, \bar{P}_2, \bar{P}_3 \in \mathbf{M}(\mathbf{H})$ by (4.7) – (4.9). Then we have:

Theorem 5.1 *For either the 1DOF or 2DOF C, RP is solvable iff there exist* $V, Q \in \mathbf{M}(\mathbf{H})$ *such that, when* $z_p \neq y_p$

$$\bar{P}_1 Q \bar{P}_2 + V\tilde{D}_w = \bar{P}_3, \tag{5.1}$$

and when $z_p = y_p$

$$N_{22} Q \bar{P}_2 + V\tilde{D}_w = X_2 \bar{P}_2. \tag{5.2}$$

Proof. When $z_p \neq y_p$, it follows from (4.1), (4.4) and (4.6) that

$$\begin{aligned} T_{z_p w_p} &= P_{11} - P_{12}Y_2\tilde{D}_{22}P_{21} - P_{12}D_{22}Q\tilde{D}_{22}P_{21} \\ &= \bar{P}_3 - \bar{P}_1 Q \bar{P}_2. \end{aligned}$$

By Lemma 2.4, then, $T_{z_p w_0} = (\bar{P}_3 - \bar{P}_1 Q \bar{P}_2)\tilde{D}_w^{-1}\tilde{N}_w \in \mathbf{M}(\mathbf{H})$ holds iff

$$(\bar{P}_3 - \bar{P}_1 Q \bar{P}_2)\tilde{D}_w^{-1} \triangleq V \in \mathbf{M}(\mathbf{H})$$

which directly gives (5.1).

When $z_p = y_p$, we have $P_{11} = P_{21}$, $P_{12} = P_{22}$, so $\bar{P}_1 = P_{22}D_{22} = N_{22}$, and further by (4.3)

$$\bar{P}_3 = (I - P_{12}D_{22}Y_1)P_{21} = (I - N_{22}Y_1)P_{21} = X_2\tilde{D}_{22}P_{21} = X_2\bar{P}_2.$$

Substituting these results into (5.1) yields (5.2). □

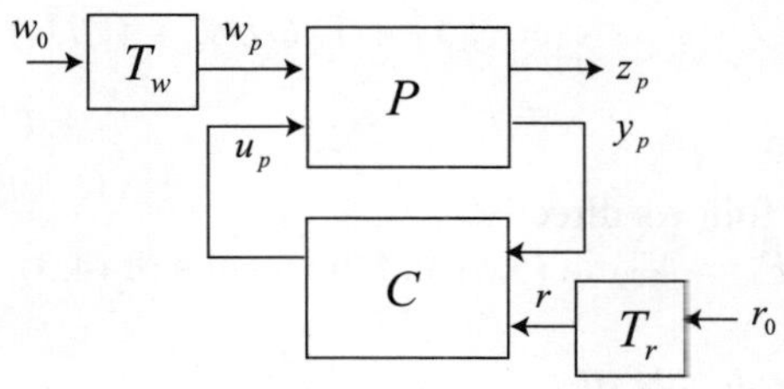

Figure 5.1 The regulation and tracking configuration

Theorem 5.2 *TP is solvable iff*

(a) *for the 2DOF C, there exist* $W, \tilde{N}_{c1} \in \mathbf{M}(\mathbf{H})$ *such that, when* $z_p \neq y_p$,

$$\bar{P}_1 \tilde{N}_{c1} + W\tilde{D}_r = I, \tag{5.3}$$

i. e., $\bar{P}_1$ *and* $\tilde{D}_r$ *are skew prime, and when* $z_p = y_p$,

$$N_{22}\tilde{N}_{c1} + W\tilde{D}_r = I, \tag{5.4}$$

i. e., N_{22} *and* $\tilde{D}_r$ *are skew prime.*

(b) *for the 1DOF C, there exist* $Q, W \in \mathbf{M}(\mathbf{H})$ *such that, when* $z_p \neq y_p$,

$$\bar{P}_1 Q\tilde{D}_{22} + W\tilde{D}_r = I - \bar{P}_1 Y_1, \tag{5.5}$$

and when $z_p = y_p$,

$$N_{22}Q\tilde{D}_{22} + W\tilde{D}_r = X_2\tilde{D}_{22}. \tag{5.6}$$

Proof. (a): Let $C_1 \neq C_2$ and note that the reference signal r in Figure 5.1 is just w_c in Figure 4.1. For $z_p \neq y_p$, we obtain from (4.1) and (4.6) that

$$T_{z_p r} = P_{12}D_{22}\tilde{D}_c C_1 = \bar{P}_1\tilde{N}_{c1}.$$

Then, by $T_{er} = I - T_{z_p r}$ and Lemma 2.4, $T_{er_0} = (I - \bar{P}_1\tilde{N}_{c1})\tilde{D}_r^{-1}\tilde{N}_r \in \mathbf{M}(\mathbf{H})$ holds iff

$$(I - \bar{P}_1\tilde{N}_{c1})\tilde{D}_r^{-1} \triangleq W \in \mathbf{M}(\mathbf{H})$$

which directly gives (5.3). For $z_p = y_p$, noting again $\bar{P}_1 = N_{22}$ in (5.3) we see the result of (5.4).

(b): Let $C_1 = C_2$. So we have $\tilde{N}_{c1} = \tilde{N}_{c2}$. For $z_p \neq y_p$, using the results of (4.3) – (4.5), (5.3) can be rewritten as

$$\begin{aligned}\bar{P}_1\tilde{N}_{c2} + W\tilde{D}_r &= P_{12}N_{c2}\tilde{D}_{22} + W\tilde{D}_r\\ &= P_{12}(Y_2 + D_{22}Q)\tilde{D}_{22} + W\tilde{D}_r\\ &= P_{12}Y_2\tilde{D}_{22} + \bar{P}_1 Q\tilde{D}_{22} + W\tilde{D}_r\\ &= \bar{P}_1 Y_1 + \bar{P}_1 Q\tilde{D}_{22} + W\tilde{D}_r\\ &= I,\end{aligned}$$

thus the result of (5.5) follows directly.

For $z_p = y_p$, using $\bar{P}_1 = N_{22}$ in (5.5) and the results of (4.3), we obtain (5.6). □

Theorem 5.3 *RTP is solvable iff*

(a) *for the 2DOF C, RP and TP are independently solvable, i. e., when* $z_p \neq y_p$, *(5.1) and (5.3), when* $z_p = y_p$, *(5.2) and (5.4), are separately solvable.*

(b) *for the 1DOF C, when $z_p \neq y_p$, (5.1), (5.5) and (5.7), when $z_p = y_p$, (5.2), (5.6) and (5.8), are simultaneously satisfied, respectively, where*

$$V\tilde{D}_w - W\tilde{D}_r P_{21} = P_{11} - P_{21} \tag{5.7}$$

and

$$V\tilde{D}_w = W\tilde{D}_r P_{21}. \tag{5.8}$$

Proof. (a): Obvious. (b): Let $C_1 = C_2$. When $z_p \neq y_p$, (5.1) can be re-expressed as

$$\bar{P}_1 Q\tilde{D}_{22} P_{21} + V\tilde{D}_w = P_{11} - \bar{P}_1 Y_1 P_{21}. \tag{5.9}$$

Then, postmultiplying (5.5) by P_{21} and subtracting the result from (5.9) directly yields (5.7). When $z_p = y_p$, (5.8) can be seen by just using the fact $P_{11} = P_{21}$ to (5.7). □

Now, the question is under what conditions and how we can solve the equations derived above for various cases of RP, TP and RTP. Since the equations of (5.1), (5.5) and the equations of (5.2), (5.6) have respectively the same forms, we only need to consider (5.1) and (5.2). The idea of [3] can be used to give the following results similarly as [4].

Theorem 5.4 *Suppose that $\bar{P}_2\tilde{D}_w^{-1} = N_{2w}D_{2w}^{-1} = \tilde{D}_{2w}^{-1}\tilde{N}_{2w}$, $\bar{P}_3\tilde{D}_w^{-1} = N_{3w}D_{3w}^{-1}$, and $N_{2w}D_{2w}^{-1}$, $\tilde{D}_{2w}^{-1}\tilde{N}_{2w}$, $N_{3w}D_{3w}^{-1}$ are coprime MFDs. Then (5.1) is solvable iff $D_{1w} = D_{3w}^{-1}D_{2w} \in \mathbf{M}(\mathbf{H})$ and there exist $Q, \bar{V} \in \mathbf{M}(\mathbf{H})$ such that*

$$\bar{P}_1 Q + \bar{V}\tilde{D}_{2w} = \bar{N} \tag{5.10}$$

where $\bar{N} \triangleq N_{3w}D_{1w}Y_{2w} \in \mathbf{M}(\mathbf{H})$, and Y_{2w} is determined by the equation $X_{2w}D_{2w} + Y_{2w}N_{2w} = I$.

Proof. Necessity: It follows from (5.1) that

$$\bar{P}_1 Q\bar{P}_2\tilde{D}_w^{-1} + V = \bar{P}_3\tilde{D}_w^{-1}$$

and

$$\bar{P}_1 QN_{2w} + VD_{2w} = N_{3w}D_{3w}^{-1}D_{2w}. \tag{5.11}$$

In view of that D_{3w}, N_{3w} are right coprime and Lemma 2.4, $D_{1w} \triangleq D_{3w}^{-1}D_{2w} \in \mathbf{M}(\mathbf{H})$ must hold true. By Theorem 2.2, we can have the doubly coprime relation

$$\begin{aligned} I &= \begin{bmatrix} X_{2w} & Y_{2w} \\ -\tilde{N}_{2w} & \tilde{D}_{2w} \end{bmatrix} \begin{bmatrix} D_{2w} & -\tilde{Y}_{2w} \\ N_{2w} & \tilde{X}_{2w} \end{bmatrix} \\ &= \begin{bmatrix} D_{2w} & -\tilde{Y}_{2w} \\ N_{2w} & \tilde{X}_{2w} \end{bmatrix} \begin{bmatrix} X_{2w} & Y_{2w} \\ -\tilde{N}_{2w} & \tilde{D}_{2w} \end{bmatrix}. \end{aligned} \tag{5.12}$$

Postmultiplying (5.11) by Y_{2w} and using (5.12) yield

$$\bar{P}_1 Q - (\bar{P}_1 Q\tilde{X}_{2w} - V\tilde{Y}_{2w})\tilde{D}_{2w} = N_{3w}D_{1w}Y_{2w}$$

which is equivalent to (5.10) with $\bar{V} = \bar{P}_1 Q\tilde{X}_{2w} - V\tilde{Y}_{2w}$.

Sufficiency: Suppose that (5.10) is satisfied. Then, we have

$$\bar{P}_1 Q - \bar{V}\tilde{D}_{2w} = \bar{P}_3\tilde{D}_w^{-1} D_{2w} Y_{2w} = \bar{P}_3\tilde{D}_w^{-1}\tilde{Y}_{2w}\tilde{D}_{2w}$$

which is in turn equivalent to

$$\bar{P}_1 Q\tilde{D}_{2w}^{-1} - \bar{V} = \bar{P}_3\tilde{D}_w^{-1}\tilde{Y}_{2w}.$$

Postmultiplying this equation by $\tilde{N}_{2w}$ gives

$$\bar{P}_1 Q\bar{P}_2\tilde{D}_w^{-1} - \bar{V}\tilde{N}_{2w} = \bar{P}_3\tilde{D}_w^{-1}\tilde{Y}_{2w}\tilde{N}_{2w}.$$

In view of that $\tilde{Y}_{2w}\tilde{N}_{2w} = I - D_{2w}X_{2w}$ and $\bar{P}_3\tilde{D}_w^{-1}D_{2w} = N_{3w}D_{1w} \in \mathbf{M}(\mathbf{H})$, we obtain

$$\bar{P}_1 Q\bar{P}_2 - (\bar{V}\tilde{N}_{2w} - N_{3w}D_{1w}X_{2w})\tilde{D}_w = \bar{P}_3.$$

In view of that $D_{1w}, \bar{V} \in \mathbf{M}(\mathbf{H})$, we see that Q and $V \triangleq \bar{V}\tilde{N}_{2w} - N_{3w}D_{1w}X_{2w} \in \mathbf{M}(\mathbf{H})$ is a solution to (5.1). □

Theorem 5.5 *Let $\tilde{D}_{2w}, \tilde{N}_{2w}, D_{2w}, N_{2w}$ as defined in Theorem 5.4. Then, (5.2) is solvable iff $\tilde{N}_{22}$ and $\tilde{D}_{2w}$ are skew prime.*

Proof. It follows from (5.2) that

$$(X_2 - N_{22}Q)\bar{P}_2\tilde{D}_w^{-1} = V.$$

By Lemma 2.4, we see that

$$(X_2 - N_{22}Q)\tilde{D}_{2w}^{-1}\tilde{N}_{2w} = V$$

is satisfied iff

$$(X_2 - N_{22}Q)\tilde{D}_{2w}^{-1} = V'$$

i. e.,

$$N_{22}Q + V'\tilde{D}_{2w} = X_2 \tag{5.13}$$

for some $V' \in \mathbf{M}(\mathbf{H})$ (such that $V = V'\tilde{N}_{2w}$).

Note that the relation of (5.13) can also be derived directly from (5.10) by using the fact that, when $z_p = y_p$, $\bar{P}_1 = N_{22}$, $\bar{P}_3 = X_2\bar{P}_2$ and

$$\begin{aligned}\bar{N} &= N_{3w}D_{1w}Y_{2w} = N_{3w}D_{3w}^{-1}D_{2w}Y_{2w}\\ &= \bar{P}_3\tilde{D}_w^{-1}D_{2w}Y_{2w} = X_2\bar{P}_2\tilde{D}_w^{-1}D_{2w}Y_{2w}\\ &= X_2N_{2w}Y_{2w} = X_2(I - \tilde{X}_{2w}\tilde{D}_{2w}),\end{aligned}$$

and further defining $V' = \bar{V} + X_2\tilde{X}_{2w}$.

Now, what we need to do is to show that the necessary and sufficient condition for (5.13) to be solvable is just the one claimed in the theorem. Necessity: Suppose (5.13) holds true. Premultiplying (5.13) by $\tilde{D}_{22}$ yields

$$\tilde{D}_{22}N_{22}Q + \tilde{D}_{22}V'\tilde{D}_{2w} = \tilde{D}_{22}X_2.$$

By the results of (4.3), we have

$$\tilde{N}_{22}(D_{22}Q + Y_2) + (\tilde{D}_{22}V')\tilde{D}_{2w} = I,$$

i. e., $\tilde{N}_{22}$ and $\tilde{D}_{2w}$ are skew prime.

Sufficiency: Suppose that there exist $X, Y \in \mathbf{M}(\mathbf{H})$ such that

$$\tilde{N}_{22}X + Y\tilde{D}_{2w} = I.$$

Premultiplying this equation by X_2 and using the results of (4.3), we obtain

$$N_{22}(X_1X) + (X_2Y)\tilde{D}_{2w} = X_2.$$

□

Further, the following theorem gives a sufficient condition for the solvability of (5.10).

Theorem 5.6 *If $\bar{P}_1$ is square and nonsingular, and $\bar{P}_1$ and $\bar{N}$ are left coprime, then* (5.10) *is solvable if $\tilde{P}_1$ and $\tilde{D}_{2w}$ are skew prime where $\tilde{P}_1$ is given by*

$$\bar{P}_1^{-1}\bar{N} = \tilde{N}\tilde{P}_1^{-1} \tag{5.14}$$

with $\tilde{N}$, $\tilde{P}_1$ right coprime.

Proof. If $\tilde{P}_1$ and $\tilde{D}_{2w}$ are skew prime, then there exist X and $Y \in \mathbf{M}(\mathbf{H})$ such that

$$\tilde{P}_1X + Y\tilde{D}_{2w} = I. \tag{5.15}$$

Premultiplying (5.15) by $\bar{N}$ and employing the relation of (5.14), we see that

$$\bar{P}_1(\tilde{N}X) + (\bar{N}Y)\tilde{D}_{2w} = \bar{N}$$

which gives the solution $Q \triangleq \tilde{N}X, \bar{V} \triangleq \bar{N}Y$ to (5.10). □

For the 1D and 2D cases, the skew primeness of $\tilde{P}_1$ and $\tilde{D}_{2w}$ stated in Theorem 5.6 is also a necessary condition for the solvability of (5.10) [17]. However, at this stage, we have not yet found a proof of the necessity for the nD ($n \geq 3$) case.

6 Solution of coprime matrix equation by Gröbner bases

It has been clarified in the previous sections that, for a given plant $P \in \mathbf{M}(\mathbf{G})$, the synthesis of stabilizing controllers can be reduced to the problems of finding left or/and right coprime MFDs of P, and solving the corresponding coprime matrix equations, while the solution of RP and TP with respect to P can be reduced to the solution of certain skew prime matrix equations. In view of Theorem 2.3, the solution of a skew prime matrix equation can be essentially reduced to the solution problems of left and right coprime matrix equations. Therefore, in the following, we focus ourselves on the problems:

- how to obtain a coprime MFD for a given P, say a left coprime MFD $P = D^{-1}N$, and
- how to solve the left coprime matrix equation

$$DX + NY = I. \tag{6.1}$$

It is noted that $\mathbf{R}[\boldsymbol{z}] \subset \mathbf{H}$, and thus, without loss of generality, we can only consider the left coprime MFD $P = D^{-1}N$ with $D, N \in \mathbf{M}(\mathbf{R}[\boldsymbol{z}])$ and the equation of (6.1) can be equivalently transformed to the following polynomial matrix equation:

$$D\bar{X} + N\bar{Y} = V \tag{6.2}$$

where $\bar{X}, \bar{Y}$ and $V \in \mathbf{M}(\mathbf{R}[\boldsymbol{z}])$ and $\det V(\boldsymbol{z}) \neq 0$ on $\bar{U}^n$. Further, applying Cauchy-Binet theorem to (6.2) yields that

$$\sum_{i=1}^{\beta} a_i(\boldsymbol{z}) x_i(\boldsymbol{z}) = \det V(\boldsymbol{z})$$

where $a_i(\boldsymbol{z})$ are the maximal order minors of the matrix $F = [D\ N]$ and $x_i(\boldsymbol{z})$ are the maximal order minors of $[\bar{X}^T\ \bar{Y}^T]^T$. Let $\mathcal{I}$ be the ideal generated by $a_i(\boldsymbol{z})$ ($i = 1, \ldots, \beta$) and $\mathcal{V}(\mathcal{I})$ the variety of $\mathcal{I}$. It is obvious that a necessary condition for D and N to be left coprime is that $\mathcal{V}(\mathcal{I}) \cap \bar{U}^n = \emptyset$, i. e., $a_i(\boldsymbol{z})$ ($i = 1, \ldots, \beta$) possess no common zeros on $\bar{U}^n$.

For 2D case, it is well known that factor coprimeness of two 2D polynomial matrices is equivalent to their minor coprimeness [21], and a left factor/minor coprime MFD over $\mathbf{R}([z_1, z_2])$ for a given $P \in \mathbf{M}(\mathbf{G})$ (with $n = 2$) can always be obtained [5]. Based on this fact, the following theorem has been shown [1, 6].

Theorem 6.1 [1, 6] $P \in \mathbf{M}(\mathbf{G})$ ($n = 2$) *admits a left coprime factorization iff, for any left MFD* $P = D^{-1}N$ *with* $D, N \in \mathbf{M}(\mathbf{R}[z_1, z_2])$ *being left factor/minor coprime,* $\mathcal{V}(\mathcal{I}) \cap \bar{U}^2 = \emptyset$.

Theorem 6.1 implies that a left coprime MFD for a 2D causal plant P can be obtained, if it exists, by using any left factor coprime 2D polynomial MFD of P.

As $\mathcal{V}(\mathcal{I})$ is of zero dimension when $D(z_1, z_2)$ and $N(z_1, z_2)$ are left minor coprime, it is possible in principle to verify the existence condition given in Theorem 6.1 by explicit computation of $\mathcal{V}(\mathcal{I})$. However, the direct computation of $\mathcal{V}(\mathcal{I})$ may be quite troublesome, and more efficient methods have been pursued [2, 7, 18]. We show, in the following, a test by Gröbner basis approach [18].

Consider $P(z_1, z_2) = D^{-1}(z_1, z_2)N(z_1, z_2)$ where $D(z_1, z_2) \in \mathbf{R}[z_1, z_2]^{m\times m}$ and $N(z_1, z_2) \in \mathbf{R}[z_1, z_2]^{m\times l}$ are left factor coprime. By the results of [12], then, we always have $X_1(z_1, z_2)$, $Y_1(z_1, z_2)$, $X_2(z_1, z_2)$ and $Y_2(z_1, z_2) \in \mathbf{M}(\mathbf{R}[z_1, z_2])$ such that

$$D(z_1, z_2)X_1(z_1, z_2) + N(z_1, z_2)Y_1(z_1, z_2) = V_1(z_1),$$
$$D(z_1, z_2)X_2(z_1, z_2) + N(z_1, z_2)Y_2(z_1, z_2) = V_2(z_2)$$

where $V_1(z_1)$, $V_2(z_2)$ are diagonal 1D polynomial matrices with non-zero determinants.

Decompose $V_1(z_1)$ and $V_2(z_2)$ as

$$V_1(z_1) = V_{1u}(z_1)V_{1s}(z_1),$$
$$V_2(z_2) = V_{2u}(z_2)V_{2s}(z_2)$$

such that all the entries of $V_{1s}(\xi)$ and $V_{1s}(\xi)$ are 1D stable polynomials, while all the entries of $V_{1u}(\xi)$ and $V_{2u}(\xi)$ are 1D completely unstable polynomials, i. e., have only unstable zeros in $\bar{U}$ [18].

Define

$$\mathcal{V}(\det V_{1u}(z_1), \det V_{2u}(z_2)) = \left\{(z_1, z_2) \in \mathbf{C}^2 \mid \det V_{1u}(z_1) = 0,\ \det V_{2u}(z_2) = 0\right\}$$

and let

$$F(z_1, z_2) = \begin{bmatrix} D(z_1, z_2) & N(z_1, z_2) \end{bmatrix} = \begin{bmatrix} \vec{f}_1 & \dots & \vec{f}_k \end{bmatrix}$$

where $k = m + l$ and $\vec{f}_i$ $(i = 1, \dots, k)$ are $m \times 1$ 2D polynomial vectors.

Theorem 6.2 [18] *The following statements are equivalent:*

(i) $\mathcal{V}(\mathcal{I}) \cap \bar{U}^2 = \emptyset$, *or equivalently,* (6.2) *is solvable when* $r = 2$.

(ii) *For any* $(z_{10}, z_{20}) \in \mathcal{V}(\det V_{1u}(z_1), \det V_{2u}(z_2))$, $F(z_{10}, z_{20})$ *is of full rank.*

(iii) *A non-zero constant is an element in the Gröbner basis of the ideal generated by* $\det V_{1u}(z_1)$, $\det V_{2u}(z_2)$ *and* $a_i(z_1, z_2)$, $i = 1, \dots, \beta$.

(iv) *For* $i = 1, \dots, m$, $\vec{e}_i$ *is an element of the Gröbner basis of the module generated by*

$$\left\{ \vec{f}_1, \dots, \vec{f}_k, \begin{bmatrix} 0 \\ \vdots \\ 0 \\ \det V_{1u}(z_1) \\ 0 \\ \vdots \\ 0 \end{bmatrix} *, \begin{bmatrix} 0 \\ \vdots \\ 0 \\ \det V_{2u}(z_2) \\ 0 \\ \vdots \\ 0 \end{bmatrix} * \right\}$$

where $\vec{e}_i$ *denotes an* $m \times 1$ *vector having* 1 *at the* i*th position and* 0 *at the other positions, and* $*$ *denotes the* i*th position of the associated vectors.*

Next, we show a procedure for constructing a solution to (6.2) by using Gröbner basis approach [18].

Theorem 6.3 [18] *Suppose that* $\mathcal{V}(\mathcal{I}) \cap \bar{U}^2 = \emptyset$. *Then the polynomial* $s(z_1, z_2)$ *defined as*

$$s(z_1, z_2) = \det V_{1s}(z_1) \det V_{2s}(z_2)$$

vanishes on $\mathcal{V}(\mathcal{I})$ *and is stable, namely, devoid of zeros in* $\bar{U}^2$.

2D solution procedure

Input: $F(z_1, z_2) = [D(z_1, z_2)\ N(z_1, z_2)] = [\vec{f_1}, \ldots, \vec{f_k}]$, and a stable polynomial $s(z_1, z_2) \in \mathbf{R}[z_1, z_2]$ vanishing over $\mathcal{V}(\mathcal{I})$.

Output: $\bar{X}(z_1, z_2)$, $\bar{Y}(z_1, z_2)$ and $V(z_1, z_2)$ for (6.2).

Step 1. Solve the following equation for $i = 1, \ldots, m$ using the Gröbner basis approach:

$$\bar{x}_{i,1}(z_1, z_2, t)\vec{f_1}(z_1, z_2) + \ldots + \bar{x}_{i,k}(z_1, z_2, t)\vec{f_k}(z_1, z_2) + \tilde{x}_i(z_1, z_2, t)\begin{bmatrix} 0 \\ \vdots \\ 1 - ts(z_1, z_2) \\ \vdots \\ 0 \end{bmatrix} * = \begin{bmatrix} 0 \\ \vdots \\ 1 \\ \vdots \\ 0 \end{bmatrix} *$$

where t is a new indeterminate.

Step 2. Substituting $t = 1/s$ into the above equations and clearing out the denominators yields

$$F(z_1, z_2)\vec{f_i}'(z_1, z_2) = [0, \ldots, s(z_1, z_2)^{r_i}, \ldots, 0]^T$$

or equivalently,

$$F(z_1, z_2)[\vec{f_1}', \ldots, \vec{f_m}'] = \begin{bmatrix} s^{r_1} & & \\ & \ddots & \\ & & s^{r_m} \end{bmatrix} \triangleq V(z_1, z_2)$$

where $\vec{f_i}'(z_1, z_2)$ $(i = 1, \ldots, m)$ are $k \times 1$ 2D polynomial vectors and r_i $(i = 1, \ldots, m)$ are non-negative integers.

Step 3. Partitioning $[\vec{f_1}'(z_1, z_2), \ldots, \vec{f_m}'(z_1, z_2)]$ as $[\bar{X}(z_1, z_2)^T\ \bar{Y}(z_1, z_2)^T]^T$, we have that

$$D(z_1, z_2)\bar{X}(z_1, z_2) + N(z_1, z_2)\bar{Y}(z_1, z_2) = V(z_1, z_2)$$

where $\det V(z_1, z_2) = s(z_1, z_2)^{r_1 + \ldots + r_m}$ is obviously a stable 2D polynomial.

For a general nD ($n \geq 3$) case, we encounter the difficulties that factor coprimeness of two nD polynomial matrices does not imply the minor coprimeness of them, and we do not know how to construct a minor coprime MFD for a given $P \in \mathbf{M}(\mathbf{G})$. It is still an open problem to construct a coprime MFD for a given $P \in \mathbf{M}(\mathbf{G})$, though its existence condition has been shown recently by the result of Theorem 3.3 [13].

By introducing the concept of reduced minors of an nD polynomial matrix, it has been shown that the nD stabilization problem can be characterized by using polynomial MFDs of the given plant $P \in \mathbf{M}(\mathbf{G})$ which are not necessarily minor coprime [8, 9].

Let $a_1(z), \ldots, a_\beta(z)$ denote the $m \times m$ minors of the nD polynomial matrix $F(z) = [D(z)\ N(z)]$. Extracting a greatest common divisor $d(z)$ of $a_i(z)$ $(i = 1, \ldots, \beta)$ yields

$$a_i(z) = d(z) b_i(z), \quad i = 1, \ldots, \beta.$$

Then, $b_i(z), \ldots, b_\beta(z)$ are called the reduced minors of $F(z)$ [9].

It has been shown that $P(z) = D(z)^{-1} N(z)$ (not necessarily left minor coprime) is stabilizable iff $b_1(z), \ldots, b_\beta(z)$ have no common zeros in $\bar{U}^n$, or equivalently, there exist $x_1(z), \ldots, x_\beta(z)$ such that

$$\sum_{i=1}^{\beta} b_i(z) x_i(z) = s(z) \tag{6.3}$$

where $s(z)$ is stable, i.e., $s(z) \neq 0$ for any $z \in \bar{U}^n$. Further, it has been shown that if the solution $x_i(z)$ $(i = 1, \ldots, \beta)$ can be found, then a stabilizing controller $C(z) = Y(z) X(z)^{-1}$ can be constructively obtained [8, 9].

7 Open problems

Let $\mathcal{I}$ be the ideal generated by $b_1(z), \ldots, b_\beta(z)$, and $\mathcal{V}(\mathcal{I})$ the variety of $\mathcal{I}$.

Problem 1. [9] Find an efficient method to determine whether or not $\mathcal{V}(\mathcal{I}) \cap \bar{U}^n = \emptyset$.

Problem 2. [19] Suppose that $\mathcal{V}(\mathcal{I}) \cap \bar{U}^n = \emptyset$. Find a constructive method to obtain a stable polynomial $\tilde{s}(z)$ such that $\tilde{s}(z)$ vanishes on $\mathcal{V}(\mathcal{I})$.

For the case when $\mathcal{I}$ is of zero dimension, i.e., $\mathcal{V}(\mathcal{I})$ consists of only a finite number of points, such a stable polynomial $\tilde{s}(z)$ can be constructed by utilizing Gröbner bases [11, 20]. Moreover, if $\mathcal{V}(\mathcal{I})$ is finite w.r.t. the variables $z_1, \ldots, z_{n-2}$, the problem can also be solved by using Gröbner bases [10, 11]. It should be noted that, once this problem is solved, it will be ready to construct a solution to (6.3) by Gröbner basis approach [18, 20].

Problem 3. Find a constructive method to obtain a left or/and right coprime MFDs for a given stabilizable $P \in \mathbf{M}(\mathbf{G})$.

Problem 4. Find a constructive method to construct the matrices $\bar{D}$ and $\bar{N}$ defined in Theorem 2.3 when $n \geq 3$. (Solution for the 2D case can be found in [17].)

8 Conclusions

Necessary and sufficient solvability conditions have been given for nD stabilization, regulation and tracking problems under various configurations and/or cases, which show that these problems can be essentially reduced to the solvability problems of coprime and skew prime matrix equations. It has been shown that for the 2D case these equations, and thus the synthesis problems of various kinds of 2D controllers, can be constructively solved by utilizing Gröbner basis approach. However, for the general nD ($n \geq 3$) cases some substantial challenges still remain, which are summarized as several open problems.

Bibliography

[1] M. Bisiacco, E. Fornasini, and G. Marchesini, Controller Design for 2D Systems, in: *Frequency Domain and State Space Methods for Linear Systems* (ed. C. I. Byrnes and A. Lindquist), North Holland: Elservier 1986, pp. 99–113.

[2] M. Bisiacco, E. Fornasini, and G. Marchesini, Linear Algorithms for Computing Closed Loop Polynomials of 2D Systems, in: *Proc. IEEE Symp. on Circ. and Syst.*, Helsinki, Finland, 1988, pp. 345–348.

[3] L. F. Cheng and J. B. Pearson, Synthesis of linear multivariable regulators, *IEEE Trans. Automat. Control* 26(1) (1981), pp. 194–202.

[4] O. R. Gonzalez and P. J. Antsaklis, Internal models in regulation, stabilization and tracking, *Int. J. Control* 53 (1991), pp. 411–430.

[5] J. P. Guiver and N. K. Bose, Polynomial Matrix Primitive Factorization over Arbitrary Coefficient Field and Related Results, *IEEE Trans. Circ. and Syst.* 29 (1982), pp. 649–657.

[6] J. P. Guiver and N. K. Bose, Causal and weakly causal 2-D filters with applications in stabilizations, in: *Multidimensional System Theory* (ed. N. K. Bose), Dordrecht: Reidel 1985, p. 52.

[7] E. Fornasini, A note on output feedback stabilizability of multivariable 2D systems, *Systems & Control Letters* 10 (1988), pp. 45–50.

[8] Z. Lin, Feedback stabilizability of MIMO nD linear systems, *Multidimensional Syst. Signal Process.* 9 (1998), pp. 149–172.

[9] Z. Lin, Output feedback stabilizability and stabilization of linear n-D systems, in: *Multidimensional Signals, Circuits and Systems* (ed. K. Galkowski, J. Wood), Taylor and Francis, 2001, pp. 59–76.

[10] Z. Lin, J. Lam, K. Galkowski, and S. Xu, A constructive approach to stabilizability and stabilization of a class of nD systems, *Multidimensional Syst. Signal Process.* 12 (2001), pp. 329–343.

[11] Z. Lin, L. Xu, and Q. Wu, Applications of Gröbner bases to signal and image processing: a survey, *Linear Algebra and Its Applications* 391 (2004), pp. 169–202.

[12] M. Morf, B. Levy, and S. Y. Kung, New result in 2-D systems theory, Part I: 2-D polynomial matrices, factorization and coprimeness, *Proc. IEEE* 65 (1977), pp. 861–872.

[13] A. Quadrat, Every internally stabilizable multidimensional system admits a doubly coprime factorization, *Proc. MTNS04*, Leuven, Belgium, 2004.

[14] A. Quadrat, A lattice approach to analysis and synthesis problems, *Math. Control, Signals, Systems* 18(2) 2006, pp. 147–186.

[15] R. Saeks and J. Murray, Feedback System Design: the tracking and disturbance rejection problems, *IEEE Trans. Automat. Control* 26 (1981), pp. 203–217.

[16] M. Vidyasagar, *Control System Synthesis: A Factorization Approach*, MIT Press, 1985.

[17] L. Xu, O. Saito, and K. Abe, Bilateral Polynomial Matrix Equations in Two Indeterminates, *Multidimensional Syst. Signal Process.* 1(4) (1990), pp. 363–379.

[18] L. Xu, O. Saito, and K. Abe, Output Feedback Stabilizability and Stabilization Algorithms for 2D Systems, *Multidimensional Syst. Signal Process.* 5(1) (1994), pp. 41–60.

[19] L. Xu, J. Ying, Z. Lin, O. Saito, and Y. Anazawa, Open Problems in Control of Linear Discrete Multidimensional Systems, in: *Unsolved Problems in Mathematical Systems and Control Theory*, (ed. V. Blondel and A. Megretski), Princeton University Press, 2004, pp. 221–228.

[20] L. Xu, J. Q. Ying, and O. Saito, Feedback Stabilization for a class of MIMO n-D systems by Gröbner basis approach, in *Abstract of 1st International Workshop on Multidimensional systems*, Poland, 1998, pp. 88–90.

[21] D. Youla and G. Gnavi, Notes on n-Dimensional System Theory, *IEEE Trans. Circuits and Systems* 26 (1979), pp. 105–111.

Author information

Li Xu, Department of Electronics and Information Systems,
Akita Prefectural University,
Akita 015-0055, Japan.
Email: xuli@akita-pu.ac.jp

Radon Series Comp. Appl. Math **3**, 235–251

State representations of time-varying linear systems

Eva Zerz

Key words. Linear systems, time-varying systems, behavioral approach, realization theory, state representations, skew polynomial rings.

AMS classification. 93C05, 93C15, 93B25, 16S32, 16S36

We construct a special type of first order representation of a linear time-varying system given by linear ordinary differential equations with rational or meromorphic coefficients. Under certain conditions, such a representation yields a classical state space model. We show that every system admits a partition of its variables into inputs and outputs such that a state representation can be obtained. All the problems addressed in this paper can be solved algorithmically using non-commutative computer algebra over the polynomial ring $K[s]$, where K is the field of rational or meromorphic functions, and we have the commutator rule $sk - ks = k'$ for $k \in K$, reflecting the product rule of differentiation.

1 Introduction

The focus of behavioral systems theory was on time-invariant linear systems for a long time, since this is an important special case with a rich structure. The theory can be called well developed in most respects by now. Recently, there have been several advances, from different angles, to tackle the time-varying counterpart [3, 4, 10, 12]. Algebraically speaking, one has to deal with non-commutative polynomial rings (for an introduction, see classical textbooks such as [6, 11], and for applications in the systems theoretic context, see [8, 9, 16]). This is more or less a technicality that can be overcome, and today, there are efficient computer algebra systems for the manipulation of non-commutative polynomial matrices, see e. g. [1, 5, 14, 15, 19]. However, more importantly, the underlying analytical problem is to find a suitable signal space that will make the algebra work. A possible solution was proposed in [18, 21]. The present paper continues this line of research and addresses the problems of row-proper representations, input-output structures with proper transfer matrices, and state representations. We stress the fact that the algebraic results presented in section 2 are known [3, 8, 16]. However, concrete interpretations in terms of signals are given in the subsequent sections. The main tool for obtaining a state representation is the precursory construction of an "output-nulling" representation (see Theorem 4.1), and in this aspect, our construction differs from the one given in [3, 8, 9]. A preliminary version of the present paper can be found in [20]. However, the material of section 5 is new.

2 Algebraic preliminaries

Let K denote either the field of real-rational functions, or the field of real-meromorphic functions, and let $\mathcal{D}$ denote the ring of linear ordinary differential operators with coefficients in K.

Each $0 \neq d \in \mathcal{D} = K[\frac{d}{dt}]$ possesses a unique representation

$$d = a_n \frac{d^n}{dt^n} + a_{n-1} \frac{d^{n-1}}{dt^{n-1}} + \ldots + a_1 \frac{d}{dt} + a_0, \tag{2.1}$$

where $a_i \in K$ and $a_n \neq 0$. One calls the non-negative integer n the degree of d, and one sets $\deg(0) := -\infty$.

The ring $\mathcal{D}$ is not commutative, because $\frac{d}{dt} ta = a + t\frac{d}{dt} a$ holds for all differentiable functions a, and thus one obtains the commutator rule $\frac{d}{dt} t - t\frac{d}{dt} = 1$. More generally, for $k \in K$, one has

$$\frac{d}{dt} k - k \frac{d}{dt} = k',$$

where $k' \in K$ is the derivative of k. A simple inductive argument shows that

$$\frac{d^n}{dt^n} k - k \frac{d^n}{dt^n} = \sum_{i=0}^{n-1} \binom{n}{i} k^{(n-i)} \frac{d^i}{dt^i}. \tag{2.2}$$

Thus, d from (2.1) can also be written in the form

$$d = \frac{d^n}{dt^n} a_n + \frac{d^{n-1}}{dt^{n-1}} \tilde{a}_{n-1} + \ldots + \frac{d}{dt} \tilde{a}_1 + \tilde{a}_0$$

for some suitably defined coefficients $\tilde{a}_i \in K$ (note that the degree n and the leading coefficient a_n remain unchanged).

Another consequence of (2.2) is the fact that

$$\deg(d_1 d_2) = \deg(d_1) + \deg(d_2)$$

for all $d_1, d_2 \in \mathcal{D}$. This is obviously true if one of the two factors d_i happens to be zero, according to the convention $\deg(0) = -\infty$. If the factors d_i are both non-zero, let $d_1 = a_n \frac{d^n}{dt^n} + \text{l.o.t}$ and $d_2 = b_m \frac{d^m}{dt^m} + \text{l.o.t}$, where l.o.t denotes the lower order terms. Then $d_1 d_2 = a_n \frac{d^n}{dt^n} b_m \frac{d^m}{dt^m} + \text{l.o.t} = a_n b_m \frac{d^{n+m}}{dt^{n+m}} + \text{l.o.t}$. Since the coefficients a_n, b_m are elements of the field K, we have $a_n b_m \neq 0$ provided that $a_n \neq 0$ and $b_m \neq 0$. This also implies that $\mathcal{D}$ is a domain, that is, it contains no zero-divisors.

Theorem 2.1 [11] *The ring $\mathcal{D}$ is simple, that is, the only ideals that are both right and left ideals are the trivial ones, i. e., 0 and $\mathcal{D}$ itself. Moreover, $\mathcal{D}$ is a right and left principal ideal domain, that is, every left ideal and every right ideal can be generated by one single element.*

In fact, the ring $\mathcal{D}$ is even a right and left Euclidean domain [6], which means that there exists a right and left "division with remainder," where the Euclidean function is given by the degree. Anyhow, Theorem 2.1 implies that $\mathcal{D}$ is left and right Noetherian, and thus, it has the left and right Ore property. The left Ore property means that any two elements of $\mathcal{D}$ possess a left common multiple, that is,

$$\forall (a, b) \in \mathcal{D}^2 \; \exists (c, d) \neq (0, 0) \in \mathcal{D}^2 : ca = db.$$

The left Ore property is necessary and sufficient for $\mathcal{D}$ to admit a skew field $\mathcal{K}$ of left fractions [2]

$$\mathcal{K} = \{d^{-1} n \mid d, n \in \mathcal{D}, d \neq 0\}.$$

Since $\mathcal{D}$ enjoys also the right Ore property, we have

$$\mathcal{K} = \{\tilde{n} \tilde{d}^{-1} \mid \tilde{d}, \tilde{n} \in \mathcal{D}, \tilde{d} \neq 0\}.$$

Indeed, given a right fraction $k = \tilde{n} \tilde{d}^{-1}$, we can find $n, d \in \mathcal{D}$ such that $d \tilde{n} = n \tilde{d}$ due to the left Ore property (note that $\tilde{d} \neq 0$ implies $d \neq 0$) and thus $k = d^{-1} n$, a left fraction. Conversely, the right Ore property guarantees that left fractions can be rewritten as right fractions. Therefore, $\mathcal{K}$ contains both all expressions of the form $d^{-1} n$ and all expressions of the form $\tilde{n} \tilde{d}^{-1}$, with the obvious rules of calculating with non-commutative fractions (which is well known in systems theory from the study of matrix fraction descriptions).

With this construction of the quotient field $\mathcal{K}$, the rank of a matrix $R \in \mathcal{D}^{g \times q}$ is well-defined via [13]

$$\operatorname{rank}(R) = \dim_{\mathcal{K}} \mathcal{K}^{1 \times g} R = \dim_{\mathcal{K}} R \mathcal{K}^q.$$

Thus, the equality of row rank and column rank, which is a standard fact in classical linear algebra, carries over to the non-commutative setting. This should not be confused, however, with the statement that the rank of a matrix coincides with the rank of its transpose.

Example 2.2 Let $a, b \in \mathcal{D}$ be such that $ab \neq ba$. Then the matrix

$$R = \begin{bmatrix} 1 & a \\ b & ba \end{bmatrix}$$

has rank 1 (as would be expected from the commutative case), but the rank of its transpose

$$R^T = \begin{bmatrix} 1 & b \\ a & ba \end{bmatrix}$$

equals 2. This is due to the fact that for determining the rank, we are only allowed to take left linear combinations of the rows, and right linear combinations of the columns.

This simple example also shows that the concept of the determinant of a square matrix does not generalize (in a straightforward way) to the non-commutative situation: For instance, it is unclear whether the determinant of R should be defined as $ba-ba = 0$ or as $ba - ab \neq 0$.

The lack of a well-defined determinant leads to the non-applicability of concepts building upon it, such as Cramer's rule. Therefore, new proofs have to be given for several of the results below, which are well known in the constant coefficient case.

2.1 Proper rational matrices

An element $h \in \mathcal{K}$ is called proper if it can be written in the form $h = d^{-1}n$ for some $0 \neq d \in \mathcal{D}$ and $n \in \mathcal{D}$ with $\deg(d) \geq \deg(n)$. Equivalently, we have $h = \tilde{n}\tilde{d}^{-1}$ with $\deg(\tilde{d}) \geq \deg(\tilde{n})$. This follows from $n\tilde{d} = d\tilde{n}$ and thus $\deg(n) + \deg(\tilde{d}) = \deg(d)+\deg(\tilde{n})$, which implies that $\deg(d) \geq \deg(n)$ is equivalent to $\deg(\tilde{d}) \geq \deg(\tilde{n})$. One can check easily that the sum and the product of two proper functions is again proper. Thus the proper functions form a subring of the field $\mathcal{K}$.

We say that $H \in \mathcal{K}^{p\times m}$ is proper if all its entries are proper. Equivalently, a proper matrix is one that can be written in the form $H = d^{-1}N$ for some $0 \neq d \in \mathcal{D}$ and $N \in \mathcal{D}^{p\times m}$ such that $\deg(d) \geq \deg(n_{ij})$ for all entries n_{ij} of N. This can be achieved by taking a left common multiple of all denominators that appear in $h_{ij} = d_{ij}^{-1}\nu_{ij}$, say $d = c_{ij}d_{ij}$. Then $h_{ij} = d^{-1}c_{ij}\nu_{ij}$, and we set $n_{ij} = c_{ij}\nu_{ij}$. Again, it is equivalent to require the analogous property for a right fractional representation $H = \tilde{N}\tilde{d}^{-1}$.

The notion of strict properness is defined analogously by requiring that the degree of the denominator is strictly greater than the degree of the numerator. Next, we address the question under which conditions on P, Q we have properness of the matrix fraction $H = P^{-1}Q$.

Lemma 2.3 *Let $P \in \mathcal{D}^{p\times p}$ be a matrix of full rank, and let $Q \in \mathcal{D}^{p\times m}$ be arbitrary. If $H = P^{-1}Q$ is proper, then we have $\deg(P_{i-}) \geq \deg(Q_{i-})$ for $1 \leq i \leq p$, where M_{i-} denotes the i-th row of a matrix M, and the degree of a matrix is defined to be maximum of the degrees of its entries.*

Proof. We have $Q = PH$, and $H = \tilde{N}\tilde{d}^{-1}$ with $\deg(\tilde{d}) \geq \deg(\tilde{n}_{ij})$ for all i, j. Hence

$$Q_{i-} = P_{i-}H = P_{i-}\tilde{N}\tilde{d}^{-1}$$

and thus $Q_{i-}\tilde{d} = P_{i-}\tilde{N}$. This implies

$$\deg(Q_{i-}) + \deg(\tilde{d}) \leq \deg(P_{i-}) + \deg(\tilde{N})$$

and hence

$$\deg(Q_{i-}) - \deg(P_{i-}) \leq \deg(\tilde{N}) - \deg(\tilde{d}) \leq 0.$$

□

2.2 Row-proper polynomial matrices

As may be expected from the constant coefficient case, the converse of the statement of Lemma 2.3 is not true in general. We need the additional property of row-properness (or row-reducedness). For this, let $\kappa_i(R)$ denote the i-th row degree of $R \in \mathcal{D}^{g\times q}$, that is, $\kappa_i(R) = \deg(R_{i-})$ for $1 \leq i \leq g$. Note that every $R \in \mathcal{D}^{g\times q}$ can be written as

$$R = \Delta R^{hr} + L \tag{2.3}$$

where $\Delta = \operatorname{diag}(s^{\kappa_1(R)}, \ldots, s^{\kappa_g(R)})$, $s = \frac{d}{dt}$, and $R^{hr} \in K^{g\times q}$ ("highest row coefficient matrix"). The matrix L ("lower order terms") is such that $\kappa_i(L) < \kappa_i(R)$ for all i. We say that a matrix R is row-proper if the matrix R^{hr} has full row rank. This implies that R itself has full row rank. If R is square, row-properness signifies that R^{hr} is non-singular in the sense that $\det(R^{hr}) \neq 0$ (since the coefficient field K is commutative, we can work with determinants as usually). The following theorem can be found in [8], but we give a new proof for the sake of completeness.

Theorem 2.4 [8] *Let $P \in \mathcal{D}^{p\times p}$ be row-proper, and let $Q \in \mathcal{D}^{p\times m}$ be arbitrary. Then $H = P^{-1}Q$ is proper if and only if $\kappa_i(P) \geq \kappa_i(Q)$ for all i.*

Proof. In view of the lemma, it suffices to show the "if" part. Let $P = \Delta P^{hr} + L$ be decomposed according to (2.3). The matrix P^{hr} is invertible. Then

$$\begin{aligned} H = P^{-1}Q &= (\Delta P^{hr}(I + {P^{hr}}^{-1}\Delta^{-1}L))^{-1}Q \\ &= (I + {P^{hr}}^{-1}\Delta^{-1}L)^{-1}{P^{hr}}^{-1}\Delta^{-1}Q. \end{aligned}$$

By construction, the matrix $X := {P^{hr}}^{-1}\Delta^{-1}L$ is strictly proper, and $Z := {P^{hr}}^{-1}\Delta^{-1}Q$ is proper. It suffices to show that the strict properness of X implies that $Y := (I+X)^{-1}$ is proper (then $H = YZ$ is also proper). Let $Y = d^{-1}N$ and $X = \tilde{N}\tilde{d}^{-1}$. Since $Y(I + X) = I$, we obtain

$$N\tilde{d} + N\tilde{N} = d\tilde{d}I.$$

Thus $\deg(N\tilde{d}+N\tilde{N}) = \deg(d)+\deg(\tilde{d})$. Since $\deg(\tilde{N}) < \deg(\tilde{d})$, we obtain $\deg(N) = \deg(d)$. □

3 Behaviors

We deal exclusively with linear K-differential behaviors $\mathcal{B}$, that is, solution sets of systems of linear differential equations with coefficients in K. Such a system takes the form $Rw = 0$ for some $R \in \mathcal{D}^{g \times q}$. Of course, the question arises in which function space we are going to look for solutions w, that is, we have to specify $\mathcal{A}$ such that

$$\mathcal{B} = \ker_{\mathcal{A}}(R) = \{w \in \mathcal{A}^q \mid Rw = 0\}.$$

Our choice is the set $\mathcal{A} = \mathcal{C}^\infty_{ae}(\mathbb{R}, \mathbb{R})$ which consists of all functions a for which there exists a finite (discrete) set $\mathbb{E}(a) \subset \mathbb{R}$ such that $a \in \mathcal{C}^\infty(\mathbb{R} \setminus \mathbb{E}(a), \mathbb{R})$. That is, a is smooth almost everywhere, which explains the notation. By construction, the set $\mathcal{A} = \mathcal{C}^\infty_{ae}(\mathbb{R}, \mathbb{R})$ is a real vector space, and a $\mathcal{D}$-module, where K denotes the rational (meromorphic) functions. Moreover, it was shown in [18] that $\mathcal{A}$ is an injective cogenerator (see also [3, 10] for detailed explanations of this concept). This means that

$$\ker_{\mathcal{A}}(R) = \operatorname{im}_{\mathcal{A}}(M) \tag{3.1}$$

is equivalent to

$$\operatorname{im}(\cdot R) = \ker(\cdot M) \tag{3.2}$$

where two matrices $R \in \mathcal{D}^{g \times q}$ and $M \in \mathcal{D}^{q \times l}$ are interpreted as mappings

$$R : \mathcal{A}^q \to \mathcal{A}^g \quad \text{and} \quad M : \mathcal{A}^l \to \mathcal{A}^q$$

in (3.1), and as mappings

$$\cdot R : \mathcal{D}^{1 \times g} \to \mathcal{D}^{1 \times q} \quad \text{and} \quad \cdot M : \mathcal{D}^{1 \times q} \to \mathcal{D}^{1 \times l}$$

in (3.2), respectively. The injective cogenerator property is an extremely powerful tool in behavioral systems theory, because it makes it possible to translate any analytic statement that can be formulated in terms of kernels and images w. r. t. the signal space $\mathcal{A}$, into an algebraic statement in terms of kernels and images w. r. t. the operator ring $\mathcal{D}$, and vice versa.

Thus, let $\mathcal{A} = \mathcal{C}^\infty_{ae}(\mathbb{R}, \mathbb{R})$ and

$$\mathcal{B} = \ker_{\mathcal{A}}(R) = \{w \in \mathcal{A}^q \mid Rw = 0\}$$

for some $R \in \mathcal{D}^{g \times q}$. One calls R a (kernel) representation of $\mathcal{B}$.

The most important consequences of the injective cogenerator property are summarized in the following theorem.

Theorem 3.1 [3, 10, 18] *Let $M \in \mathcal{D}^{g \times l}$ and $N \in \mathcal{D}^{g \times q}$ be given, and let R be a $\mathcal{D}$-matrix whose rows generate the left kernel of M, that is, $\operatorname{im}(\cdot R) = \ker(\cdot M)$. Then we have*

$$\exists x \in \mathcal{A}^l : Mx = Nw \quad \Leftrightarrow \quad RNw = 0$$

for all $w \in \mathcal{A}^q$ ("elimination of latent variables"). If R_1 and R_2 represent the same behavior, then we have $R_1 = XR_2$ and $R_2 = YR_1$ for some $\mathcal{D}$-matrices X, Y. In particular, any two representations of a behavior have the same rank.

3.1 Existence of full row rank representations

A matrix $U \in \mathcal{D}^{g\times g}$ is called unimodular if there exists a matrix $U^{-1} \in \mathcal{D}^{g\times g}$ with $UU^{-1} = U^{-1}U = I_g$.

Theorem 3.2 (Jacobson form) [6] *Let $R \in \mathcal{D}^{g\times q}$. Then there exist unimodular matrices $U \in \mathcal{D}^{g\times g}$ and $V \in \mathcal{D}^{q\times q}$ such that*

$$URV = \begin{bmatrix} D & 0 \\ 0 & 0 \end{bmatrix}$$

where $D = \operatorname{diag}(1,\dots,1,d) \in \mathcal{D}^{p\times p}$ for some $0 \neq d \in \mathcal{D}$, and $p = \operatorname{rank}(R)$.

Since $\mathcal{D}$ is Euclidean, the Jacobson form of R can be obtained by performing elementary row and column operations on R. The Jacobson form is also known as Teichmüller-Nakayama form [12].

Example 3.3 Consider

$$R = \begin{bmatrix} \frac{d}{dt}+t & -1 & -1 \\ 0 & \frac{d}{dt}-\frac{1}{t} & -t \end{bmatrix} \in K[\tfrac{d}{dt}]^{2\times 3} \tag{3.3}$$

where K denotes the rational functions. Then we obtain, with

$$U = \begin{bmatrix} 1 & 0 \\ -t & 1 \end{bmatrix} \quad \text{and} \quad V = \begin{bmatrix} 0 & 0 & 1 \\ 0 & 1 & t \\ -1 & -1 & \frac{d}{dt} \end{bmatrix}$$

the Jacobson form

$$URV = \begin{bmatrix} 1 & 0 & 0 \\ 0 & d & 0 \end{bmatrix},$$

where $d = \frac{d}{dt} + t - \frac{1}{t}$.

Theorem 3.4 [3, 12, 18] *Let $\mathcal{B} = \{w \in \mathcal{A}^q \mid Rw = 0\}$ for some $R \in \mathcal{D}^{g\times q}$. Then $\mathcal{B}$ can be represented by a matrix with full row rank.*

3.2 Input-output structures

Since any two representations of a behavior have the same rank, the integer $p := \operatorname{rank}(R)$ is an invariant of $\mathcal{B}$, sometimes called the output-dimension of $\mathcal{B}$.

Let $R \in \mathcal{D}^{p\times q}$ be a full row rank representation of $\mathcal{B}$. Then there exists a $p \times p$ submatrix P of R of full rank. Without loss of generality, arrange the columns of R such that $R = [-Q, P]$. Let $w = [u^T, y^T]^T$ be arranged accordingly. If $p < q$, then this is called an input-output structure of $\mathcal{B}$, and $H = P^{-1}Q \in \mathcal{K}^{p\times m}$, where $m = q - p$, is called its transfer matrix. One can show that the transfer matrix depends only on $\mathcal{B}$ and the chosen partition of the variables, and not on the specific choice of the representation. An input-output structure is called causal if its transfer matrix is proper. If $p = q$, we call $\mathcal{B}$ autonomous.

The terms "input-output structure" and "autonomous" are motivated by the fact that

$$\forall u \in \mathcal{A}^m \; \exists y \in \mathcal{A}^p : \; Py = Qu,$$

that is, for every choice of the input u, there exists an output y such that $[u^T, y^T]^T \in \mathcal{B}$. This follows from

$$\ker(\cdot P) = 0,$$

which implies, since $\mathcal{A}$ is an injective cogenerator,

$$\mathrm{im}_{\mathcal{A}}(P) = \mathcal{A}^p.$$

This means that $P : \mathcal{A}^p \to \mathcal{A}^p$ is surjective, i. e., for all $v \in \mathcal{A}^p$ there exists $y \in \mathcal{A}^p$ such that $Py = v$. In particular, this is true for $v = Qu$. Then one says that u is a vector of free variables of $\mathcal{B}$, and one can show that a system is autonomous if and only if it has no free variables.

3.3 Existence of row-proper representations

Theorem 3.5 *Let $R \in \mathcal{D}^{p \times q}$ be a matrix with full row rank. There exists a unimodular matrix $U \in \mathcal{D}^{p \times p}$ such that UR is row-proper. Thus any $\mathcal{B}$ possesses a row-proper representation.*

Proof. If R is row-proper, we are finished. Therefore, assume otherwise, that is, let $\mathrm{rank}(R^{hr}) < p$. We show that there exists a unimodular matrix U such that

$$\sum_{i=1}^{p} \kappa_i(UR) < \sum_{i=1}^{p} \kappa_i(R).$$

Iteratively, this yields the result, because a strictly decreasing sequence of non-negative integers must be finite, that is, after finitely many steps we reach a row-proper matrix. Since $\mathrm{rank}(R^{hr}) < p$, there exists $0 \neq \alpha = (\alpha_1, \ldots, \alpha_p) \in K^{1 \times p}$ with

$$\sum_{j=1}^{p} \alpha_j R^{hr}_{j-} = 0.$$

Choose j^* with $\alpha_{j^*} \neq 0$ and

$$\kappa_{j^*}(R) \geq \kappa_j(R) \quad \text{for all } j \text{ with } \alpha_j \neq 0.$$

Without loss of generality, let $\alpha_{j^*} = 1$, that is, we have

$$R^{hr}_{j^*-} + \sum_{j \neq j^*} \alpha_j R^{hr}_{j-} = 0.$$

Now we perform the elementary operation: R_{j^*-} plus $\sum_{j \neq j^*} \alpha_j s^{\kappa_{j^*}(R) - \kappa_j(R)} R_{j-}$. Since $\alpha_j s^{\kappa_{j^*}(R)} = s^{\kappa_{j^*}(R)} \alpha_j + \text{l.o.t}$, all the terms of order κ_{j^*} in the j^*-th row cancel. Thus the new matrix $R' = UR$ satisfies $\kappa_{j^*}(R') < \kappa_{j^*}(R)$, and $\kappa_j(R') = \kappa_j(R)$ for all $j \neq j^*$. This establishes the claim. □

3.4 Existence of causal input-output structures

Let $R \in \mathcal{D}^{p\times q}$ be a row-proper representation of $\mathcal{B}$ with $p < q$. Consider the decomposition

$$R = \Delta R^{hr} + L$$

from (2.3). The matrix R^{hr} has full row rank by assumption. Thus we have, up to a permutation of columns,

$$R^{hr} = [-Q^{hr}, P^{hr}] \tag{3.4}$$

where $P^{hr} \in K^{p\times p}$ is non-singular. Then the input-output structure resulting from partitioning the system variables according to the partition of (3.4) will lead to a causal input-output structure. The following theorem is a direct consequence of this consideration.

Theorem 3.6 *Any non-autonomous behavior $\mathcal{B}$ admits a causal input-output structure.*

Example 3.7 Consider the matrix from (3.3). The highest row coefficient matrix is

$$R^{hr} = \begin{bmatrix} 1 & 0 & 0 \\ 0 & 1 & 0 \end{bmatrix}.$$

Therefore, if one chooses the first two components of w as outputs, and the third component as an input (which is natural, since R actually represents a state space system), we get the transfer matrix (writing s for $\frac{d}{dt}$)

$$H = \begin{bmatrix} s+t & -1 \\ 0 & s-\frac{1}{t} \end{bmatrix}^{-1} \begin{bmatrix} 1 \\ t \end{bmatrix} = s^{-2} \begin{bmatrix} s \\ ts+2 \end{bmatrix},$$

which is proper.

4 Realization theory

4.1 Reduction to first order

Theorem 4.1 *Let $R \in \mathcal{D}^{g\times q}$. There exist an integer n and matrices $F \in K^{n\times n}$, $G \in K^{n\times q}$, $M \in K^{g\times n}$, $N \in K^{g\times q}$ such that for all $w \in \mathcal{A}^q$, we have*

$$Rw = 0 \quad \Leftrightarrow \quad \exists x \in \mathcal{A}^n : \begin{cases} \dot{x} &= Fx + Gw \\ 0 &= Mx + Nw. \end{cases}$$

We may assume, without loss of generality, that F and M are real matrices and that the matrix pair (F, M) is observable. Also, we may choose $n = \sum_{i=1}^{g} \kappa_i(R)$. If additionally, R is row-proper, then we may choose N with full row rank.

Proof. Without loss of generality, let the rows of R be arranged such that $\kappa_i(R) \geq 1$ for $1 \leq i \leq k$ and $\kappa_i(R) = 0$ for $k+1 \leq i \leq g$. Consider one of the first k rows of R, say, $R_{i-} = s^{d_i} R_{i,d_i} + \ldots + sR_{i,1} + R_{i,0}$, that is, $R_{i,k} \in K^{1\times q}$ denotes the right coefficient of s^k in the i-th row of R. By assumption, we have $d_i \geq 1$. Let

$$F_i = \begin{bmatrix} 0 & \cdots & \cdots & 0 \\ 1 & \ddots & & \vdots \\ & \ddots & \ddots & \vdots \\ & & 1 & 0 \end{bmatrix} \in \mathbb{R}^{d_i \times d_i}, \quad G_i = \begin{bmatrix} R_{i,0} \\ \vdots \\ R_{i,d_i-1} \end{bmatrix},$$

$M_i = [0, \ldots, 0, 1] \in \mathbb{R}^{1\times d_i}$ and $N_i = R_{i,d_i}$. The equation

$$\begin{bmatrix} sI - F_i \\ -M_i \end{bmatrix} x^i = \begin{bmatrix} G_i \\ N_i \end{bmatrix} w,$$

where $s = \frac{d}{dt}$, can be pre-multiplied by the unimodular matrix

$$U = \begin{bmatrix} 1 & s & \cdots & s^{d_i} \\ & \ddots & \ddots & \vdots \\ & & \ddots & s \\ & & & 1 \end{bmatrix}$$

which yields the equivalent equation

$$\begin{bmatrix} 0 \\ -I \end{bmatrix} x^i = \begin{bmatrix} R_{i-} \\ * \end{bmatrix} w. \tag{4.1}$$

Given w, such an $x^i \in \mathcal{A}^{d_i}$ exists if and only if $R_{i-}w = 0$.

Thus we realize each of the first k rows of R, and we combine the results in the block matrices

$$F = \begin{bmatrix} F_1 & & \\ & \ddots & \\ & & F_k \end{bmatrix}, \quad G = \begin{bmatrix} G_1 \\ \vdots \\ G_k \end{bmatrix},$$

$$M = \begin{bmatrix} M_1 & & \\ & \ddots & \\ & & M_k \\ & 0 & \end{bmatrix}, \quad N = \begin{bmatrix} N_1 \\ \vdots \\ N_k \\ R^2 \end{bmatrix},$$

where R^2 denotes the matrix consisting of the last $g-k$ rows of R, whose entries are in K, by assumption. Thus we obtain the desired equivalence if we set $x^T = [x^{1T}, \ldots, x^{kT}]$, each $x^i \in \mathcal{A}^{\kappa_i(R)}$ being the vector of latent variables constructed in the realization of the i-th row of R. The statements on F and M can easily be checked. Note that by construction, we have $N = R^{hr}$, thus establishing also the last claim. □

The representation obtained in Theorem 4.1 is called an "output-nulling representation" by some authors. It is a special kind of differential-algebraic equation, in which the differential part $\dot{x} = Fx + Gw$ and the algebraic part $0 = Mx + Nw$ are separated from each other ("semi-explicit form"). One could also use the usual elimination of latent variables from Theorem 3.1 for the proof of Theorem 4.1. However, the argument given above yields a concrete formula for x in terms of w and its derivatives. Indeed, a closer inspection of (4.1) shows that

$$-x_j^i = (R_{i,j} + sR_{i,j+1} + \ldots + s^{d_i-j} R_{i,d_i})w,$$

where x_j^i is the j-th component of x^i, and thus

$$-x_j^i = \sum_{k=0}^{d_i-j} s^k R_{i,j+k} w = \sum_{k=0}^{d_i-j} (R_{i,j+k} w)^{(k)}. \tag{4.2}$$

This relation will be crucial in the final section of the paper.

Example 4.2 For the matrix R from (3.3), we obtain

$$F = \begin{bmatrix} 0 & 0 \\ 0 & 0 \end{bmatrix}, \quad G = \begin{bmatrix} t & -1 & -1 \\ 0 & -\frac{1}{t} & -t \end{bmatrix}$$

$$M = \begin{bmatrix} 1 & 0 \\ 0 & 1 \end{bmatrix}, \quad N = \begin{bmatrix} 1 & 0 & 0 \\ 0 & 1 & 0 \end{bmatrix}.$$

For the Jacobson form computed in that example, we obtain $F = 0$,

$$G = \begin{bmatrix} 0 & t - \frac{1}{t} & 0 \end{bmatrix}, \quad M = \begin{bmatrix} 0 \\ 1 \end{bmatrix}, \quad N = \begin{bmatrix} 1 & 0 & 0 \\ 0 & 1 & 0 \end{bmatrix}.$$

4.2 State representations

Consider a first order latent variable representation

$$\mathcal{B} = \{w \in \mathcal{A}^q \mid \exists x \in \mathcal{A}^n : \dot{x} = Fx + Gw, 0 = Mx + Nw\},$$

where N has full row rank. Such a representation can always be achieved by first reducing any given representation to row-properness, and by then applying the realization technique described above. If $\mathcal{B}$ is autonomous, then N is square of full rank. Putting $C := -N^{-1}M$ and $A := F - GN^{-1}M$, one gets a state space representation of the form

$$\mathcal{B} = \{w \in \mathcal{A}^q \mid \exists x \in \mathcal{A}^n : \dot{x} = Ax, w = Cx\}.$$

If $\mathcal{B}$ is not autonomous, then we have, up to a permutation of the columns of N, and a corresponding permutation of the components of w,

$$N = [-N_1, N_2],$$

where N_2 is square and non-singular. Note that when the first order representation results from the procedure described above, then we may assume that $N = R^{hr}$ and the partition of N coincides with the one constructed in (3.4). In any case, it yields a causal input-output structure: We set $G = [-G_1, G_2]$, $w = [u^T, y^T]^T$ and obtain

$$\mathcal{B} = \left\{ \begin{bmatrix} u \\ y \end{bmatrix} \in \mathcal{A}^{m+p} \,\middle|\, \exists x \in \mathcal{A}^n : \dot{x} = Ax + Bu,\ y = Cx + Du \right\},$$

where $A = F - G_2 N_2^{-1} M$, $B = -G_1 + G_2 N_2^{-1} N_1$, $C = -N_2^{-1} M$, $D = N_2^{-1} N_1$. Thus we have obtained a state-space representation of $\mathcal{B}$. We have $\begin{bmatrix} u \\ y \end{bmatrix} \in \mathcal{B}$ if and only if

$$\exists x \in \mathcal{A}^n : \begin{bmatrix} \frac{d}{dt} I - A \\ C \end{bmatrix} x = \begin{bmatrix} B & 0 \\ -D & I \end{bmatrix} \begin{bmatrix} u \\ y \end{bmatrix}.$$

Using Theorem 3.1, we compute a matrix $[-X, Y]$ whose rows generate the left kernel of

$$\begin{bmatrix} sI - A \\ C \end{bmatrix},$$

where s stands once more for $\frac{d}{dt}$ for notational simplicity. Without loss of generality, Y is square of full rank. Then

$$\mathcal{B} = \left\{ \begin{bmatrix} u \\ y \end{bmatrix} \in \mathcal{A}^{m+p} \,\middle|\, Yy = (XB + YD)u \right\},$$

from which one can see that the transfer matrix of $\mathcal{B}$ w. r. t. the chosen input-output structure is

$$H = Y^{-1}(XB + YD) = Y^{-1}XB + D = C(sI - A)^{-1}B + D,$$

as might be expected. The rightmost matrix is proper. Note that the usual argument from the commutative case, which uses the representation of $(sI - A)^{-1}$ as a fraction of its adjugate matrix and its determinant does not work here. Nevertheless, $(sI - A)^{-1}$ can be shown to be strictly proper also in the non-commutative case, similarly as in the proof of Theorem 2.4.

If the state space system results from the procedure from Theorem 4.1, then it can be assumed to be observable. This is intuitively clear, because the state is not changed in the transition from output-nulling to state representation, and we know that x can be reconstructed from w via (4.2). More formally, one can show that the Kalman matrix

$$O_{(A,C)} = [C^T, (A^T + \tfrac{d}{dt})C^T, \ldots, (A^T + \tfrac{d}{dt})^{n-1} C^T] \in K^{n \times nq}$$

has full row rank. This follows from $A = F - G_2 N_2^{-1} M$, $C = -N_2^{-1} M$, and thus we have

$$O_{(A,C)} = O_{(F,M)} \begin{bmatrix} -(N_2^{-1})^T & & * \\ & \ddots & \\ 0 & & -(N_2^{-1})^T \end{bmatrix}.$$

Since the rightmost matrix is invertible over K, and since (F, M) can be assumed to be observable, the result follows.

Theorem 4.3 *Any behavior admits an observable state representation (in the non-autonomous case, this is achieved by selecting a causal input-output structure).*

5 Property of state

Consider a state representation of $\mathcal{B}$, that is,

$$w = \begin{bmatrix} u \\ y \end{bmatrix} \in \mathcal{B} \quad \Leftrightarrow \quad \exists x \in \mathcal{A}^n : \left\{ \begin{array}{rcl} \dot{x} & = & Ax + Bu \\ y & = & Cx + Du. \end{array} \right.$$

Let $t_0 \in \mathbb{R}$ be such that there exists an open neighborhood I of t_0 in which A, B, C, D are all smooth. Let $u \in \mathcal{A}^m$ be smooth on I. By the unique solvability of the initial value problem $\dot{x}(t) = A(t)x(t) + B(t)u(t)$, $x(t_0) = x_0$ on I, it is clear that $y|_I$ is uniquely determined by $u|_I$ and $x(t_0)$. This is the the usual property of state: provided that the input is known, the state contains all the information about the system's past needed for determining its future (here, "past" and "future" refer to $t \in I$ with $t < t_0$ or $t > t_0$, respectively). The observability of (A, C) reflects the fact that the mapping from $x(t_0)$ to $y|_I$ (setting $u|_I = 0$) is injective, that is, the state represents precisely the required information on the system's past, without any redundancy.

In what follows, we give another interpretation of the state property in a different setting: Let $I \subseteq \mathbb{R}$ be an open interval. Let D denote the set of smooth functions on I. Consider $R \in D[\frac{d}{dt}]^{g \times q}$, and let

$$\mathcal{B} = \{w \in L^1_{\mathrm{loc}}(I, \mathbb{R})^q \mid Rw = 0\},$$

where $Rw = 0$ is understood in the distributional sense. As usual, we identify the space of locally integrable functions $L^1_{\mathrm{loc}}(I, \mathbb{R})$ with a subspace of the space of distributions $\mathfrak{D}'(I, \mathbb{R})$. Clearly, $L^1_{\mathrm{loc}}(I, \mathbb{R})$ is not a $\mathcal{D}$-module, since the derivative of a locally integrable function is not necessarily locally integrable. However, not even $\mathfrak{D}'(I, \mathbb{R})$ is a $\mathcal{D}$-module, since the product of a rational function and a distribution is not necessarily a distribution. Therefore we have to restrict to $R \in D[\frac{d}{dt}]^{g \times q}$ for $Rw = 0$ to make sense as a distributional differential equation (distributions are arbitrarily often differentiable, and the product of a distribution and a smooth function is well-defined).

Consider $w_1, w_2 \in \mathcal{B}$ and let $t_0 \in I$. Define the concatenation of w_1, w_2 at t_0, denoted by $w := w_1 \wedge_{t_0} w_2$, by

$$w(t) = \left\{ \begin{array}{ll} w_1(t) & \text{if } t < t_0 \\ w_2(t) & \text{if } t > t_0. \end{array} \right.$$

The value of w at t_0 can be assigned arbitrarily. The concatenation w is also locally integrable and thus, we may ask whether $w \in \mathcal{B}$.

For this, write the i-th row of R as

$$R_{i-} = s^{d_i} R_{i,d_i} + \ldots + s R_{i,1} + R_{i,0}$$

where $d_i = \kappa_i(R)$. The following lemma was shown in the constant coefficient case in [17].

Lemma 5.1 *If $w \in \mathcal{B}$, then the functions x^i_j defined in (4.2) are all continuous, in particular, they are locally integrable.*

Proof. By construction, we have $\dot{x}^i_{j+1} = x^i_j + R_{i,j}w$ for $1 \le i \le g$ and $1 \le j \le d_i - 1$. Since $R_{i,j}$ is smooth, we have $R_{i,j}w \in L^1_{\text{loc}}(I, \mathbb{R})$. Since the primitive of a locally integrable function is continuous, we obtain

$$x^i_j \text{ locally integrable} \quad \Rightarrow \quad x^i_{j+1} \text{ continuous.}$$

However, $\dot{x}^i_1 = R_{i,0}w$ which is locally integrable. □

Theorem 5.2 *Let $w_1, w_2 \in \mathcal{B}$ and let $t_0 \in I$. Then $w = w_1 \wedge_{t_0} w_2 \in \mathcal{B}$ if and only if*

$$x_1(t_0) = x_2(t_0),$$

where x_k is derived from w_k according to (4.2) for $k = 1, 2$. This is equivalent to

$$\sum_{k=0}^{d_i - j} (R_{i,k+j}(w_2 - w_1))^{(k)}(t_0) = 0 \tag{5.1}$$

for $1 \le i \le g$ and $1 \le j \le d_i$. In particular, this is true if w_i are smooth enough at t_0 and $w_1^{(i)}(t_0) = w_2^{(i)}(t_0)$ for $0 \le i \le \max\{d_i\} - 1$.

Proof. We set $x := x_1 \wedge_{t_0} x_2$. Then $\dot{x} = \dot{x}_1 \wedge_{t_0} \dot{x}_2 + (x_2(t_0) - x_1(t_0))\delta_{t_0}$. Similarly, we have

$$\begin{aligned} R_{i,k}w &= R_{i,k}w_1 \wedge_{t_0} R_{i,k}w_2 \\ &= ((\dot{x}_1)^i_{k+1} - (x_1)^i_k) \wedge_{t_0} ((\dot{x}_2)^i_{k+1} - (x_2)^i_k) \\ &= \dot{x}^i_{k+1} - ((x_2)^i_{k+1} - (x_1)^i_{k+1})(t_0)\delta_{t_0} - x^i_k, \end{aligned}$$

where we define $x^i_0 = x^i_{d_i+1} = 0$. Summing these expressions, we obtain

$$\begin{aligned} \sum_{k=0}^{d_i} (R_{i,k}w)^{(k)} &= \sum_{k=0}^{d_i} ((x^i_{k+1})^{(k+1)} - (x^i_k)^{(k)}) \\ &\quad - \sum_{k=0}^{d_i} ((x_2)^i_{k+1} - (x_1)^i_{k+1})(t_0)\delta^{(k)}_{t_0}. \end{aligned}$$

This yields a telescopic sum, and

$$\sum_{k=0}^{d_i} (R_{i,k}w)^{(k)} = -\sum_{k=0}^{d_i} ((x_2)^i_{k+1} - (x_1)^i_{k+1})(t_0)\delta^{(k)}_{t_0}.$$

Since the derivatives of the Dirac delta distribution are linearly independent, $Rw = 0$ is equivalent to $x_1(t_0) = x_2(t_0)$, that is, (5.1). □

In particular, systems given by first order representations

$$\mathcal{B} = \{w \in L^1_{\mathrm{loc}}(I, \mathbb{R})^q \mid E\dot{w} + Fw = 0\},$$

where $E, F \in D^{g\times q}$, are Markovian [17] in the sense that for every $w_1, w_2 \in \mathcal{B}$ that are continuous at $t_0 \in I$ with $w_1(t_0) = w_2(t_0)$, we have $w_1 \wedge_{t_0} w_2 \in \mathcal{B}$. In the time-invariant case, the existence of a first order representation is actually equivalent to Markovianity. A similar, but somewhat weaker statement holds in the time-varying case, because the proof of the converse direction uses the reduction to row-proper form. However, unimodular transformations over $\mathcal{D} = K[\frac{d}{dt}]$ are not well-defined operations on $R \in D[\frac{d}{dt}]$, even if we restrict D to be the polynomial (analytic) functions.

Finally, we discuss some examples that illustrate the different solution concepts of this paper.

Example 5.3 1. Consider $\dot{w} = 0$. This has only the constant solutions in $\mathfrak{D}'(\mathbb{R}, \mathbb{R})$, and therefore, also in $L^1_{\mathrm{loc}}(\mathbb{R}, \mathbb{R})$. In $\mathcal{A} = \mathcal{C}^\infty_{ae}(\mathbb{R}, \mathbb{R})$, every piecewise constant function is a solution.

2. The equation $t\dot{w} = 0$ has the same solutions as $\dot{w} = 0$ over $\mathcal{A}$, because the two equations differ only by a unit in $\mathcal{D}$. In $L^1_{\mathrm{loc}}(\mathbb{R}, \mathbb{R})$ however, we get some additional solutions of the form cH, where $c \in \mathbb{R}$ and H denotes the Heaviside function. These satisfy $t\dot{w} = 0$ in the distributional sense, because $\dot{H} = \delta$ and $t\delta = 0$. This means that the set of locally integrable solutions is given by the functions of the form $w(t) = c_1$ for $t < 0$ and $w(t) = c_2$ for $t > 0$. The state of this system according to (4.2) equals $x = -tw$. Thus we can concatenate any two solutions at $t_0 = 0$, whereas at $t_0 > 0$, w_1 and w_2 can only be concatenated if $w_1(t_0) = w_2(t_0)$ (and then w_1 and w_2 are anyhow identical on $t > 0$).

3. Consider $t\dot{w} - w = 0$. In $\mathcal{A}$, any piecewise linear function is a solution. In $L^1_{\mathrm{loc}}(\mathbb{R}, \mathbb{R})$, we have the solutions $w(t) = c_1 t + c_2 tH(t)$, where $c_1, c_2 \in \mathbb{R}$. Two solutions w_k can be concatenated at t_0 if and only if $t_0 w_1(t_0) = t_0 w_2(t_0)$. Again, any two solutions can be concatenated at $t_0 = 0$, but in this example, we have $w(0) = 0$ for any solution, and thus the concatenation is always continuous.

4. Consider $t\dot{w} + w = 0$. In $\mathcal{A}$, the solutions are all the functions that are piecewise of the form $\frac{c}{t}$ for some constants $c \in \mathbb{R}$. On the other hand $w(t) = \frac{c}{t}$ is not locally integrable on $\mathbb{R}$ (unless $c = 0$). Therefore

$$\{w \in L^1_{\mathrm{loc}}(\mathbb{R}, \mathbb{R}) \mid t\dot{w} + w = 0\} = \{0\}.$$

On any interval I with $0 \notin I$ however, the locally integrable solutions are those of the form $w(t) = \frac{c}{t}$ for some $c \in \mathbb{R}$, and two of them are concatenable at $t_0 \in I$ if and only if their values at t_0 coincide.

6 Conclusion

We have discussed several aspects of the concept of state for time-varying systems given by linear ordinary differential equations with rational (meromorphic) coefficients. For simplicity, we have limited our discussion to the real case (i. e., real-valued signals and real-rational or real-meromorphic coefficients), but the complex case is completely analogous. Although the relevant ring of differential operators is non-commutative, most of the well-known results from the time-invariant case carry over to this setting, but new proofs have to be given in several places, because their time-invariant counterparts use arguments that are not valid in the non-commutative setting.

In particular, we have seen that every system representation can be reduced to row-properness, and hence it is possible to divide the system variables into inputs and outputs in a way that guarantees that the resulting transfer matrix is proper. Once this selection is made, a classical state representation can be constructed, in fact, we even obtain an observable state space model.

The state has then the usual interpretation as being precisely the additional information needed to associate an output to an input in a unique way. This interpretation carries over to the case where one considers locally integrable instead of smooth solutions on an interval in which the coefficients have no poles.

The constructions of this paper are algorithmic in nature. However, to the author's best knowledge, there currently exist only preliminary implementations of the specific non-commutative computer algebraic methods needed here: for instance, the MAPLE package "Janet" [1] contains a neat functionality to compute the Jacobson form, but this is mainly intended for toy examples and not optimized for serious problem sizes. The report [7] also describes an implementation, but this program is not yet publicly available. Most other program packages dealing with non-commutative polynomials work over the Weyl algebra or more generally, Ore algebras [5] or G-algebras [14, 15], but they usually do not cope with rational rather than polynomial coefficients. Therefore, the improvement and extension of the existing implementation in "Janet" is a project for future work.

Bibliography

[1] Y. A. Blinkov, C. F. Cid, V. P. Gerdt, W. Plesken, and D. Robertz, The MAPLE Package "Janet", Part II: Linear Partial Differential Equations, *Proc. Computer Algebra Scientific Computing (CASC)*, Passau, 2003.

[2] N. Bourbaki, *Algebra*, Hermann, 1974.

[3] H. Bourlès, Structural properties of discrete and continuous linear time-varying systems: a unified approach, in: F. Lamnabhi-Lagarrigue (Ed.) et al., *Advanced Topics in Control Systems Theory*, Springer Lecture Notes in Control and Information Sciences 311 (2005), pp. 225–280.

[4] K. Çamlibel, M. N. Belur, A. J. Sasane, and J. C. Willems, On a Class of Time-varying Behaviors, *Proc. Int. Symp. Math. Theory Networks Systems (MTNS)*, Notre Dame, 2002.

[5] F. Chyzak, A. Quadrat, and D. Robertz, OreModules: A symbolic package for the study of multidimensional linear systems, in: J. Chiasson and J.-J. Loiseau (Eds.), *Applications of Time-Delay Systems*, Springer; see also `wwwb.math.rwth-aachen.de/OreModules`, to appear.

[6] P. M. Cohn, *Free Rings and their Relations*, Academic Press, 1971.

[7] G. Culianez, Formes de Hermite et de Jacobson: Implémentations et Applications, Rapport de stage, INRIA Sophia Antipolis, 2005.

[8] E. Delaleau and J. Rudolph, An intrinsic characterization of properness for linear time-varying systems, *J. Math. Syst. Estim. Control* 5 (1995), pp. 1–18.

[9] M. Fliess, Some basic structural properties of generalized linear systems, *Syst. Control Lett.* 15 (1990), pp. 391–396.

[10] S. Fröhler and U. Oberst, Continuous time-varying linear systems, *Syst. Control Lett.* 35 (1998), pp. 97–110.

[11] K. R. Goodearl and R. B. Warfield, Jr., *An Introduction to Noncommutative Noetherian Rings*, Cambridge University Press, 1989.

[12] A. Ilchmann and V. Mehrmann, A behavioral approach to time-varying linear systems, Part 1: General theory, *SIAM J. Control Optim.* 44 (2005), pp. 1725–1747.

[13] T. Y. Lam, On the equality of row rank and column rank, *Expositiones Math.* 18 (2000), pp. 161–163.

[14] V. Levandovskyy, Non-commutative Computer Algebra for Polynomial Algebras: Gröbner Bases, Applications, and Implementation, Ph.D. Thesis, University of Kaiserslautern, 2005.

[15] V. Levandovskyy and H. Schönemann, Plural – a Computer Algebra System for Noncommutative Polynomial Algebras, *Proc. Int. Symp. Symbolic Algebraic Computation (ISSAC)*, 2003; see also `www.singular.uni-kl.de/plural`.

[16] J.-F. Pommaret and A. Quadrat, Generalized Bézout identity, *Appl. Algebra Eng. Commun. Comput.* 9 (1998), pp. 91–116.

[17] P. Rapisarda and J. C. Willems, State maps for linear systems, *SIAM J. Control Optim.* 35 (1997), pp. 1053–1091.

[18] E. Zerz, An algebraic analysis approach to linear time-varying systems, *IMA J. Math. Control Inf.* 23 (2006), pp. 113–126.

[19] E. Zerz and V. Levandovskyy, Algebraic Systems Theory and Computer Algebraic Methods for Some Classes of Linear Control Systems, *Proc. Int. Symp. Math. Theory Networks Systems (MTNS)*, Kyoto, 2006.

[20] E. Zerz, First Order Representations of Time-Varying Linear Systems, *Proc. IEEE Int. Symp. Computer-Aided Control Systems Design (CACSD)*, Munich, 2006.

[21] E. Zerz, An Algebraic Analysis Approach to Linear Time-varying Behaviors, *Proc. IFAC World Congress*, Prague, 2005.

Author information

Eva Zerz, Lehrstuhl D für Mathematik,
RWTH Aachen,
Templergraben 64, D-52062 Aachen, Germany.
Email: `eva.zerz@math.rwth-aachen.de`

